Timber
Structure, Properties, Conversion and Use

ONE WEEK LOAN

Timber
Structure, Properties, Conversion and Use

H.E. Desch (Deceased)

Revised by

J.M. Dinwoodie

OBE, BSc(For), MTech, PhD, DSc

Building Research Establishment

Seventh Edition

First edition 1938
Second edition 1947
Third edition 1953
Fourth edition 1968
Fifth edition 1973
Sixth edition 1981
Seventh edition 1996

Published 1996 by
MACMILLAN PRESS LTD
Houndmills, Basingstoke, Hampshire RG21 6XS
and London
Companies and representatives throughout the world

ISBN 0–333–60905–0

A catalogue record for this book is available from the
British Library

10   9   8   7   6   5   4   3   2   1
05   04   03   02   01   00   99   98   97   96

Typeset by Aarontype Limited, Bristol
Printed in Hong Kong

Contents

Part 2 PROPERTIES OF WOOD – INFLUENCE OF STRUCTURE 67

7 Appearance of Wood 69

8 Density of Wood 77

9 Moisture in Wood 81

10 Other Physical Properties of Wood 96

11 Strength, Elasticity and Toughness of Wood 102

Preface to the Seventh Edition

Dr H.E. Desch died in 1978 after a lifetime of service devoted to the acquisition and critical appraisal of knowledge on all aspects of timber technology, and to its subsequent dissemination by publication of this text together with its practical application through his consultancy. The first edition of this book was published in 1938 and the first revision in 1947. Since then it has been revised at regular intervals, continuing to fulfil the needs of students in timber technology for an authoritative and comprehensive text on the subject.

Dr Desch was about to embark on the sixth edition when he died, and I was subsequently approached to undertake that revision: this I was pleased to do and the sixth edition was published in 1981. Since then the knowledge of wood science has increased, new European Standards have replaced many British Standards, new grading techniques have become established, and new board materials have appeared on the market, to illustrate but a few of the many changes in wood technology. I was invited in 1993 to again revise the text: in agreeing to accept the challenge, I have again found this to be no light task. Indeed, the amount of new information that has had to be incorporated has resulted in an almost total rewrite of the text. The opportunity has been taken not only to rearrange certain topics to bring these into line with current thinking, but also to widen the scope of the book to cover areas of conversion, machining, and the application of paints and finishes, thereby making the text even more comprehensive; this in line with the many helpful comments received.

The text is divided into 23 chapters which are ascribed to five parts covering the Structure of Wood; Properties of Wood; Processing of Timber; Utilisation of Timber; and Timber in Service.

In Part 1, the chapters on structure have been updated and revamped. Separate chapters now deal with structure at the gross, cellular and molecular levels. Variability in structure is comprehensively covered in a further chapter, enlarged not only to deal with aspects previously covered in other chapters, but also to embrace new areas of variability in structure. Identification in structure has been enlarged to include section preparation. Regrettably, there have had to be losses as well as gains: feedback has indicated a much reduced need for timber descriptions and, since this subject is covered in many other handbooks, these have now been deleted.

In Part 2 of the text, the properties of wood are dealt with comprehensively in a set of five chapters, the last of which, on mechanical performance, has been enlarged to cover elastic behaviour, toughness and the use of structural-sized timber for strength tests.

Part 3, dealing with timber processing, covers many new areas not discussed in previous editions. Thus, the conversion of the tree to timber is described in terms of those properties and variables of wood that influence its conversion to usable timber. A second chapter covering the seasoning of timber describes not only traditional air- and kiln-drying, but also deals with solar and high temperature drying. A third chapter contains new information on the machining of timber with emphasis on health and safety. The grading of timber has undergone major changes since the previous revision and these are described: parallel changes have taken place in the derivation of grade stresses (and characteristic values) during this period and this subject is treated in much detail.

Part 4 is concerned with the utilisation of timber and board materials. New boards and structural composites have emerged since the last revision and the chapter on board materials has been extensively updated and enlarged. The British Standards on wood adhesives have been replaced with the new European Standards.

Part 5 is devoted to all aspects of timber in service, commencing with a new and generalised chapter on degradation, followed by a series of specialised chapters on fungal attack, insect attack and their eradication. This part concludes with preventative measures covering preservation and the application of finishes.

In view of this more comprehensive treatment of timber, it is no longer possible for any one individual to write authoritatively on every aspect of timber technology, embracing as it does structure, properties, conversion, utilisation and behaviour in service. Therefore, I am deeply indebted to a number of my colleagues in coming to my assistance and giving so willingly of their own free time to assist in a wide variety of ways, all of which are deeply appreciated.

I am particularly indebted to Mr Roger Berry in charge of the Biodeterioration Section, Building Research Establishment for completely revising Chapters 20 (insects) and 21 (eradication); to Dr Janice Carey of the BRE Biodeterioration Section for the revision of Chapter 19 (fungi); and to Dr Reginald Orsler in charge of the Preservation Section, BRE for the revision of Chapter 22 on the preservation of timber.

Several of my colleagues or former colleagues at BRE have spent many hours of their free time reading my first draft and making the most valuable of comments and suggestions: in particular I would like to thank Keith Maun, Head of the BRE Utilisation Section (Chapters 12 – log conversion, 13 – seasoning and 14 – machining); Clive Benham (Chapter 15 – grading); Gerald Moore (Chapter 6 – identification); Peter Bonfield (Chapters 9 – moisture, 11 – strength, 16 – utilisation and 18 – degradation); John Boxall (Chapter 23 – finishes); and John Brazier, former Head of Wood Properties section (Chapters 1 – introduction and 3 – cellular structure). I would also like to thank Mr E. van der Strataen for his guidance and valuable comments on the adhesives section of Chapter 16 (utilisation).

My very sincere gratitude is extended to Mrs Sue Phillips, Mrs Margaret Penney, Mrs Hilda Ridgewell and Mrs Jo Mundy for the many hours of their own time spent in word processing my scrawl, and in the subsequent updating of the text following receipt of comments. I am greatly indebted to both Mrs Jo Mundy and Dr Peter Bonfield who, between them, have read the entire draft text in their own time, thus reducing very considerably the number of errors in the published book.

I would also like to express my appreciation to the Chief Executive of BRE, Mr Roger Courtney, for permission to use a considerable number of illustrations from the former Princes Risborough Laboratory, now within the guardianship of the Building Research Establishment.

Lastly, I would like to extend my gratitude and appreciation to my wife, who has been so helpful, patient and understanding over the last nine months and who has spent many hours reading the proofs with me.

Princes Risborough, 1996 J.M.D.

1

Introduction

1.1 Consumption of timber

Timber has been used by man since the early days of recorded history, first as a means of constructing a shelter for himself, later as a hunting tool in the form of spear or bow, and later still as a widely used artefact of an industrialised society with uses ranging from truck sides and bottoms, electrical sockets and plug tops, tool-handles, sports equipment, boats, railway sleepers and musical instruments, to name but a few.

Why is timber such a versatile material? Its success lies partly in its attractive strength-to-weight ratio and its workability, and partly in its availability at a competitive price. Although many of its traditional uses have been replaced by man-made plastics and fibre-composites, timber still remains a competitive material in many applications.

Vast quantities are still used worldwide, as Figure 1.1 illustrates. Thus, timber production worldwide in 1988, the last year in which comprehensive statistical data are available, was almost $3.5 \times 10^9 \, \text{m}^3$, a colossal volume, and far greater than that of many other materials. Sadly, slightly over half of this volume was burned as fuel-wood in third-world countries. When consumption is examined by region, 46 per cent of this was produced in the tropical countries of the world ($1.57 \times 10^9 \, \text{m}^3$) and only 17 per cent of this amount was used by industry. Of the total volume of tropical wood produced, 93 per cent was hardwood; in terms of total world production, 56 per cent was hardwood. Worldwide, the trade in timber and wood products is about $£60 \times 10^9$ (£60 billion).

Timber is an important commodity not only worldwide but also in the UK. Thus, the UK import bill in 1989 for timber, panel products and pulp ($£6 \times 10^9$), is equivalent to 10 per cent of the worldwide trade value in these commodities. Pulp constitutes two-thirds of the value of UK imports and consequently the import value of timber and panel products is $£2 \times 10^9$, a far from inconsiderable import bill. This can be broken down into $£1.1 \times 10^9$ on softwood, $£0.3 \times 10^9$ on hardwoods and $£0.6 \times 10^9$ on panel products.

UK production is no longer insignificant with an annual value of sawn softwood of $£0.15 \times 10^9$ (about 14 per cent of the softwood imported value), and of sawn hardwood of $£0.06 \times 10^9$ (20 per cent of the hardwood imported value). The value for home-produced panel products was about $£0.245 \times 10^9$.

Taking UK imports and production together, the consumption of softwood in the UK is over $8 \times 10^6 \, \text{m}^3$, that of hardwoods over $1 \times 10^6 \, \text{m}^3$ and that of panel products $4.6 \times 10^6 \, \text{m}^3$.

An estimate of the relative proportions of softwood and hardwood used in each of the major areas of application is given in Chapter 16. Suffice it to say here that by far the greatest use is in construction; some 70 per cent of all sawn softwood used in the UK and 40 per cent of all sawn hardwood is used in this area.

As an introduction to the contents of this book, which spans the structure, properties and utilisation of wood, the remainder of this chapter is devoted to the tree and the classification of trees as a means of examining the nomenclature of timbers.

1.2 The tree

To a botanist, trees represent the climax of development in the plant kingdom. In nature there is continuous struggle between individuals for survival, and in the long run it is only the more

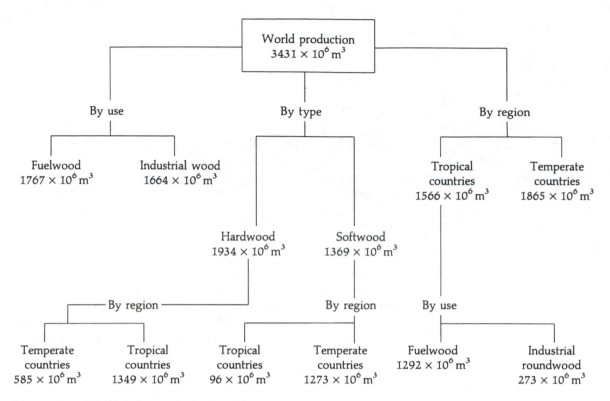

Figure 1.1   *World timber production in 1988*

efficient that prevail. In terms of its size, the tree is able to compete favourably in its demands for light, food and water; in order to develop to such large proportions it has increased its efficiency by developing groups of cells to perform specific functions, similar to what mankind has done, but to a lesser extent. This theme is developed in Chapter 3.

The forester makes use of the natural tendency of plants to compete against their neighbours by growing his trees just close enough to obtain the maximum volume of good-quality timber. The importance of the competition between individuals in producing clean, straight trees may readily be appreciated if the shape of a tree grown in park-land conditions be compared with one of the same species from a plantation: the former makes little height growth, with large branches near the ground, whereas the latter is tall and straight, and the bole is clear of branches to a

considerable height. From the economic standpoint, the plantation-grown tree produces a greater volume of better-quality timber than the park-land tree.

The tree habit, then, is a mode of growth assumed by certain plants to enable them to compete successfully with other plants in the struggle for air and light; this is essential to the development of the individual tree, and its subsequent duty of reproducing its kind.

1.3   Classification of trees

Commercial timbers fall into two main groups, the **softwoods** and **hardwoods**, and the trees that produce these two different classes of timber are themselves quite distinct. The former are **gymnosperms**, commonly referred to as conifers or cone-bearing plants, characteristically with needle or awl-shaped leaves and naked seeds; the latter represent one group of the **angiosperms** known as the **dicotyledons** which characteristically have broad leaves and seed enclosed in a seed case. The angiosperms have evolved to a higher state than the gymnosperms with a larger number of cell types each having a specific function. Although the division into softwoods and hardwoods is a convenient one for differentiating two broad groups of timber, there are a few timbers, for example pitch pine, among the softwoods that are actually harder than other timbers classed as hardwoods, for example balsa, lime and willow. Further, the divisions are not always applied correctly, particularly in the tropics. For example, native softwoods in such regions are usually soft hardwoods, that is they are broad leaved species with soft wood, although they are frequently referred to as softwoods.

Botanists found the need early for an orderly system of naming plants. They recognised that, although no two plants might be identical, minor variations between similar individuals did not alter the fact that several such individuals had many features in common, not shared by any other groups of plants, and these 'features in common' were reproduced in successive generations of such plants. This gave rise to the botanical concept that all plants could be separated into different **species**.

It was also observed that several species shared certain other 'features in common' – that is, they were more like one another than they were like other species. This gave rise to the second concept of a **genus**. Recognising the validity of these two concepts led botanists to adopt the binomial method for plant nomenclature, the first part of the **botanical name** indicating the genus to which the plant belongs, and the second part the species. These names are, by general consent, in Latin. Botanists have subsequently attempted to adopt a natural system of classification, based on evolutionary lines, arranging groups of similar genera in **families**, and bringing related families together into **orders**. The difficulty of reconciling all the complex factors that have to be considered has resulted in the

systems of classification being arbitrary, rather than natural.

Unfortunately, the position is not quite so simple as the foregoing paragraph suggests. A tendency of 19th-century botanists to name plants from inadequate material, the occurrence of actual errors of observation, and differences in interpretation of specific or generic characters, have all resulted in botanists not always being in agreement as to the correct botanical name of a plant: in consequence, some plants have been given more than one botanical name, and two or more different plants have been given the same name. Errors of observation arise through a failure to recognise the significance of some types of variation in morphological characters: 'immature' leaves, such as the leaves of seedlings or even saplings, are often much larger and very different in shape, from the leaves of a mature tree of the same species.

Internationally accepted rules have been drawn up for dealing with the types of mistakes that do arise in the naming of plants: the essence of these rules is that when errors are detected, the earliest name recorded in the literature must be revived and later names discarded but, if the earlier name should refer the plant to the wrong genus, then only the earlier specific name is retained. Transference of a plant to another genus may necessitate the selection of a new specific name if the original specific name has already been used for another species in that genus. There are also rules regarding the use of capitals in specific names: place names should not be capitalised, only vernacular names and the names of people.

1.4 Nomenclature of timbers

Timber names do not present as complex a problem as do botanical names, but there are nevertheless very real practical difficulties to be solved in selecting entirely satisfactory timber names. The precision essential in botanical work is seldom necessary in timber names, nor would it usually be practicable, because several botanically distinct species often provide a single commercial timber. Botanical names have the added disadvantages of being in a foreign language, often difficult to pronounce, and undesirably long. Thus, the timber familar to the trade as agba has the botanical name *Gossweilerodendron balsamiferum*.

In primitive communities, the problem is readily solved by the local inhabitants choosing words from their own tongue or dialects: these are **vernacular names**. However, since the local inhabitant is not as critical an observer as the scientific botanist, vernacular names are rarely as precise as botanical ones; they often refer to more than one species, and they may on occasions be applied to quite different species because of some superficial similarity in form between distinct plants. These objections are not of serious practical importance so long as there is little movement of timber from district to district, but the position is very different in a market drawing its supplies from many different localities or even countries. Even in a small country, dialects change from district to district, so that the vernacular name in one locality may be very different from that in another. This may lead to confusion in a distant, importing market. Nomenclature difficulties are further complicated by the deliberate hiding of the true identity of a timber under a misleading **trade name**, often that of some well-established timber, with the addition of a geographical or other qualifying adjective.

Some examples may assist the reader in clarifying the problem in his own mind. For example, the pines, firs, spruces and larches belong to one family the *Pinaceae*; the pines constitute one genus, *Pinus*, the true firs a second, *Abies*, spruces a third, *Picea*, and the larches a fourth, *Larix*. The different kinds of pine, such as Scots pine, Corsican pine, radiata pine or long-leaf pine, are separate species of the genus *Pinus*. This is a simple case in which trade practice follows botanical classification closely, although different countries have different trade names for *Pinus* timber; in Britain it is pine, whereas in France it is pin and in Germany Kiefer.

Where vernacular names are of the popular type, absurdities may occur. Thus, timber from *Pinus strobus* is sold locally in the USA as white pine, whereas when imported into the UK it is traded under the name yellow pine. Again, the standard trade name adopted for *Eucalyptus regnans*, F.v.M., is mountain ash in Australia and Tasmanian oak in the UK, but this timber has also been called swamp gum in Tasmania and Australian oak in Victoria; it is neither a true ash nor a true oak. Such anomalies are the outcome of European emigration: settlers name plants sometimes because of similarities of tree form or habit, sometimes because of colour similarities in the timbers, and sometimes because of other associated ideas. When the same species occurs over a wide area, several names based on these different concepts may come into being.

Similar timbers have different trade names when exported from different countries of origin. Thus, one group of species of *Shorea* is known as red meranti when exported from Peninsular Malaysia, Sarawak and Indonesia, as red seraya when exported from Sabah and as white lauan when exported from the Philippines. Yet another group of species of *Shorea* is known as dark-red meranti when exported from Peninsular Malaysia, Sarawak and Indonesia, as dark-red seraya when exported from Sabah, and as red lauan when exported from the Philippines.

The practice of borrowing names of familar timbers, and applying them to other and quite distinct woods, is at the root of much confusion in timber nomenclature. It frequently misleads the layman to the extent of causing him to use timbers for purposes for which they are unsuited. Alternatively, he may be induced to buy timbers that, were he more enlightened, he would not purchase, or, if he did consider them, he would not be prepared to pay the prices asked. The name 'mahogany' has been applied at some time or another to the timbers of more than two hundred distinct groups of timber. The original 'mahogany' of commerce was the so-called Spanish mahogany obtained from San Domingo and other West Indian islands then owned by Spain; Central American mahogany which is currently available is a very close relative, but is the timber of a different species of the same genus, *Swietenia*, but it is a very much milder timber to work than the original Spanish mahogany. African mahogany, on the other hand, is produced by several species of the genus *Khaya*; mahogany has also been applied to species of the genus *Entandrophragma* but this practice is to be discouraged. Both genera belong, however, to the same family as the American species of *Swietenia*. As recorded above, red lauan is applied to timber from several species of *Shorea* when exported from the Philippines. Regrettably, Philippine lauan is often offered for sale as Philippine mahogany, when the layman is not only confused but deceived. In consequence, timbers bearing the name 'mahogany' vary appreciably in their appearance and properties, and in the opinion of the author some have no real claim to be considered in the same class as true mahogany. Similar confusion has arisen through the widespread use and misuse of such names as walnut, ash, oak and teak, leading in some instances to lengthy and costly litigation.

Attempts have been made to standardise trade names. The British Standards Institution, for example, has published a list of **standard names** for timbers known to the UK trade (BS 7359). Publication of a list does not, of course, solve the problem of timber nomenclature, but it points the way. It has not yet been possible to compile a list based entirely on internationally acceptable rules; compromise has been necessary. The ideal solution would be the adoption of standard names on an international basis, whereby each trade timber would have a single unambiguous name. Persistence in the present confusion may well result in several potentially useful timbers being discredited, besides perpetuating deception of the laymen. It could also give rise to much expensive and acrimonious litigation. In the absence of general agreement regarding timber names, the existence of a difficult problem must be recognised: 'standard names' are used throughout in the text, but to assist the reader the botanical equivalents are given in Appendix I.

Reference

BS 7359: 1991 *British Standard Nomenclature of Commercial timbers including sources of supply*. BSI, London.

Part 1

STRUCTURE OF WOOD

2

Gross Structure

2.1   The tree

Almost all woody plants with which we are familiar have three principal parts: **roots**, **stems** and **leaves**. The characteristic that separates trees from other woody plants is their single main stem, commonly referred to as the **trunk** or **bole** (Figure 2.1).

Each of the tree parts is specially adapted to one or more particular functions (Figure 2.1). Thus the roots not only anchor the tree in the ground, but also absorb water and mineral salts in dilute solution. The stem or trunk of the tree not only supports the crown comprising branches and leaves, but also conducts these mineral solutions to it through the part that we call **wood**. The leaves comprising the crown absorb gases from the atmosphere which, together with the mineral solutions, are manufactured into glucose and other organic substances using energy obtained from sunlight. The substances that will sustain life in those living cells in the tree, as well as result in the production of new living cells, are passed down the inner layers of the outer part of the trunk in that region popularly known as the **bark**. At all levels in the trunk, these substances are passed into the woody part of the trunk where they are stored until required.

Consequently the woody part of the trunk has three principal functions to perform:

- support of the crown, which contains the manufacturing and reproductive elements;
- conduction upwards of dilute mineral solutions;
- storage of manufactured organic substances.

It will be demonstrated in the next chapter how these three functions are satisfied by different types of cells.

Being primarily interested in the woody part of the trunk, interest in the bark is somewhat superficial. In passing, however, it should be noted that this layer protects the wood from extremes of temperature and mechanical injury. The bark, being a conductor downwards of manufactured food materials, is often rich in chemical substances, such as tannin and dyes derived from the metabolic processes in the crown.

2.2   Growth of the tree

Growth of the tree occurs both outwards to produce with time a larger-diameter trunk, and upwards to produce a taller tree. Each form of growth is the responsibility of a different set of living cells.

Between the bark and the wood is a thin, delicate, living tissue known as the **cambium**, which forms a complete sheath in both the trunk and the branches (Figures 2.1 and 2.2). In the winter months, the cells comprising the cambium are dormant and consist of only a single layer. As growth commences in the spring, the cells of this single circumferential layer subdivide radially with the formation of new vertical primary walls to form a cambial zone some eight to ten cells in width. During the growing season, the cells of this cambial zone undergo further radial subdivision to produce what are known as daughter cells: some of these will remain as cambial cells, while others, to the outside of the zone, will develop into bark, or if on the inside, will form into wood.

In order to accommodate the increasing diameter of the trunk as the radial cambial division described above takes place, the cambium has to increase circumferentrially. This is achieved by the occasional development in a cell of a sloping

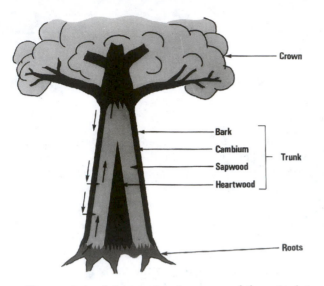

Figure 2.1 *Main parts of a tree and how food is manufactured and distributed (Building Research Establish-ment, © Crown Copyright)*

tangential wall and the subsequent elongation of the two new cells to form an overlapping pair of cells. Since this usually occurs in a preferred orientation it gives rise to the development of spiral grain, a feature that will be discussed under the general heading of variability (Chapter 5).

Increase in height is due to the subdivision of the cells that form the **apical meristem**. Cells are produced downwards, with the result that the apical meristem continues to rise. The new cells formed are of two types: those that form the soft tissue comprising the pith, and those that form vascular bundles: the latter contain cambial layers which later coalesce to form the circumferential cambium described in the previous paragraph.

2.3   Growth rings

The wood of trees grown under seasonal conditions consists of a series of concentric layers of tissue, called **growth rings** (Figures 2.2 and 2.3). Each growth layer comprises the wood produced by the cambium in a single growing season. The rings are actually layers of wood, extending the

full height of the tree, a new layer being added each growing season, like a glove, over the whole tree. Thus the wood nearest the outside of the bole is the youngest. In temperate regions, and certain tropical countries, the alternation each year of a growing season followed by a resting period, results in the growth rings being **annual rings**, thus providing a fairly accurate means of computing the age of a tree after it is felled. Double (or multiple) rings, consisting of two or more false rings, caused by serious interruptions to growth during the growing period, sometimes causes errors in such calculations. Where growing seasons are not well defined, as in many tropical regions, growth rings may be indistinct and they may not be annual.

Growth rings are apparent because the wood produced at the beginning of the growing season is different in character from that formed later in the season, and zones of **early wood** and **late wood** (the early wood was originally called **springwood**, and the late wood, **summer** – or **autumn** – **wood**) may be distinguished: where this is the case the early wood is softer and more porous than the late wood. However, in those rings near the centre of the tree there is a marked reduction in the contrast between the early and late wood of any one ring.

2.4   Sapwood and heartwood

The cross-section of the trunk can be divided into two distinct zones defined in terms of their physiological activity, that is how much or how little they contribute to the day-to-day growth of the tree. The **sapwood** comprising the outer part of the trunk is said to be physiologically active in that it is responsible for both the conduction upwards of mineral solutions and the storage of manufactured products. Conversely, the **heartwood** is physiologically inactive, having neither a conducting nor a storage role.

All cells in the heartwood are dead; in the sapwood, although the majority of cells are dead, the parenchyma cells of the ray (see Chapter 3) remain alive for some time in order to perform the task of storage.

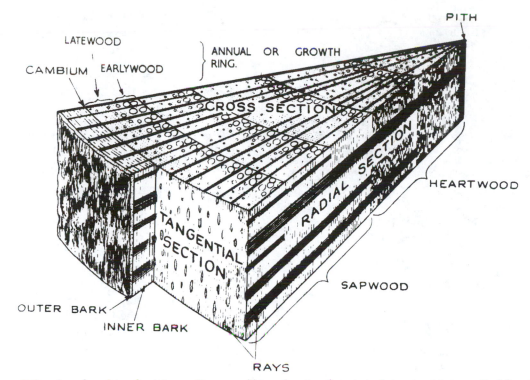

Figure 2.2   *A wedge of wood cut from a five-year-old tree showing the principal structural features (Building Research Establishment,* © *Crown Copyright)*

The sapwood, the outer of the two layers (see Figures 2.1, 2.2 and 2.3), forms a distinctive zone, typically 12.5 to 50 mm wide, but in a few species, particularly tropical species, the sapwood may be up to 200 mm or more wide, depending on the species and age of the tree, and the mode of growth of individual trees. Generally, the sapwood decreases in width as the tree gets older, though it is possible that the area of sapwood remains fairly constant with age. For a log 600 mm in diameter, a 50 mm wide sapwood would mean that the sapwood comprised 21 per cent of the log volume. Trees of the same age and species have a wider zone of sapwood when grown in the open than when grown under forest conditions in close competition with other trees.

The sapwood is usually lighter in colour than the heartwood and less durable and, when green, contains much more moisture. The line of demarcation between the two zones may be sharply defined or indefinite. In some species, such as Sitka spruce and common ash, there is no colour differentiation between the two; nevertheless a physiologically inactive heartwood exists.

As new growth layers are formed to the outside of the sapwood zone, so the heartwood advances to include former sapwood cells, bringing about a number of changes which are of importance in the utilisation of wood. Thus moisture content decreases significantly with the onset of heartwood formation while the degree of acidity increases generally slightly, but exceptionally considerably (see Chapter 4, section 4.6). However, there is no difference in either wood density (except where there is a high extractive content − see Chapter 4) or strength between sapwood and heartwood *when compared at the same moisture content.*

Colouration of the heartwood frequently occurs and is due to the deposition of small quantities of complex organic compounds

Figure 2.3 *Cross-section through the trunk of a Douglas fir tree showing the concentric annual growth rings comprising the darker-coloured late wood and lighter-coloured early wood. The darker-coloured region in the centre of the tree is known as the heartwood and the outer lighter region as the sapwood (Building Research Establishment, © Crown Copyright)*

Table 2.1   *Durability grades for heartwood*

Grade of durability	Approximate life in contact with the ground (years)	Examples
Very durable	More than 25	Teak, Iroko, Ekki
Durable	15–25	E. Oak, Utile, Yew
Moderately durable	10–15	Af. Walnut, Douglas Fir
Non-durable	5–10	Afara, Obeche, Redwood
Perishable	Less than 5	Ash, Beech, Birch

collectively known as **extractives**, but varying in composition among different species. Extractives are responsible for imparting the natural durability to the heartwood, and thus the heartwood durability will vary considerably among different species depending on the amount and composition of extractives present (see Chapter 4, section 4.5).

The natural durability of timber is assessed in the UK in terms of the life of 50 × 50 mm stakes planted in the ground, together with practical experience. A five-grade arbitrary classification has been derived (see Table 2.1) and the durability ratings for the heartwood of a wide range of species are listed in the BRE Digest 296 (1985). Sapwood on the other hand is devoid of extractives and is rich in plant food material that is attractive to certain wood-rotting fungi and insects, and consequently is always rated as 'perishable'.

The conducting tissue of wood usually undergoes modifications at the time of heartwood formation so that the free movement of liquids is interrupted. Further, various substances are deposited on the walls of most cells during transition to heartwood, which render them more or less impermeable to moisture movements. In consequence, heartwood is not so easily impregnated with preservatives or dyes as is sapwood. This may be of less practical importance than is apparent on the surface: heartwood possesses more natural resistance to fungal and insect attack than sapwood, and the reduced absorption of wood preservatives may still be sufficient to ensure that the more lightly treated heartwood will outlast the mechanical life of the treated timber. However, where service conditions impose no limits on the growth of fungi, it is not unusual for the heartwood to decay, while the outer, heavily impregnated sapwood remains quite sound.

In certain softwoods and hardwoods, **wetwood** may be formed. Two major types have been identified according to their origin of formation. The first is associated with a transitional stage in the formation of heartwood from previous sapwood. The width of this **transition zone** is restricted at the most to a few rings and its width in any one stem can vary not only between different seasons but also within

Figure 2.4   *Splay knot produced by cutting radially through the branch which has died when the cambial layer of the trunk was halfway along the current length of the knot. When the branch has broken off the growing trunk has enveloped the end of the branch (Building Research Establishment, © Crown Copyright)*

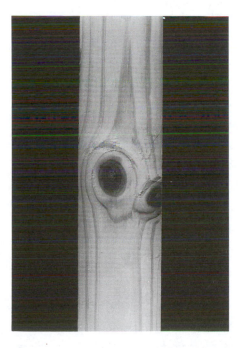

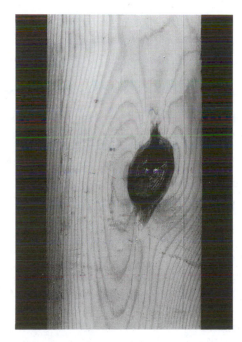

Figure 2.5   *'Live' or 'green' knot: note continuity in structure between the knot and the wood of the trunk (Building Research Establishment, © Crown Copyright)*

Figure 2.6   *'Dead' or 'black' knot: note the lack of continuity in structure between the knot and the wood of the trunk due to the presence of the bark (Building Research Establishment, © Crown Copyright)*

a growing season. As its name implies, its properties lie midway between those of sapwood and heartwood: its moisture content is always lower than sapwood and usually higher than that of heartwood, hence the name wetwood. However in a few timbers, this so-called wetwood has been found to have lower moisture content than the heartwood.

Unlike the transition zone which is of only academic interest, the second type of wetwood which is derived from previously formed normal heartwood has much more technical importance. These zones in the heartwood not only look wetter, but do have an appreciably higher moisture content: they are also usually of a darker colour than the remainder of the heartwood. There is some debate as to the cause of its formation, though there is some evidence to support the hypothesis that micro-organisms are responsible for its formation: these tend to produce a range of aliphatic acids or various gases (Ward and Zeikus, 1980; Hillis, 1987). Whatever the cause, the technical importance and commercial value of the heartwood are lowered.

2.5   Knots

It was stated earlier that the leaves in the crown are carried on branches. As a tree increases in diameter it gradually envelops the bases of branches; the portions of the branches enclosed within the wood of the trunk are called knots. A plank of wood cut radially along a knot is illustrated in Figure 2.4: such a section of a knot is called a **splayed** or **spike** knot. If the branches are alive at the time of their inclusion, their tissues are continuous with those of the main stem and the knots so formed are said to be **green**, **live** or **tight** knots (Figure 2.5).

When a branch dies, a stump remains, and this is gradually surrounded by the tissues of the expanding trunk. However, being dead, its tissues are not connected with the enveloping tissues of the main stem, and a **black**, **dead** or **loose** knot results (Figure 2.6); such knots fall out, either when timber is converted, or after it is seasoned and when it is being machined. The broken stubs of dead knots provide ready access for decay and, consequently, dead knots are frequently unsound. Returning to Figure 2.4, a plank cut at right angles to this section on the right-hand side of the figure would contain a green knot, while a plank cut at right angles from the left-hand side would contain a dead knot.

Knots vary in size from little more than a pinhead to many millimetres in diameter. They also vary in shape, according to the angle at which they are cut through in conversion, and five different projections of knots are recognised: **splay** or **spike**, **arris**, **edge**, **margin** and **face knots**. A splay knot, as described above and illustrated in Figure 2.4, is one cut more or less parallel to its long axis, appearing cone-shaped on one face and semi-circular on the adjacent face or edge. An arris knot, as the name suggests, is one that appears on two adjacent edges, the piece of included branch being inside the piece of wood. An edge knot is one appearing, either round or elliptical, on the edge of a board or plank, whereas a face knot is a similar knot, but on the face as opposed to the edge of a board or plank. A margin knot is one that appears on a face outside the middle half of the depth of the face near to, or breaking through an edge. Knots have an important bearing on the utilisation of timber; in many species they are the primary cause of lowering board quality; knots may also detract from the appearance of boards. The effect of knots on strength properties is discussed in Chapter 11; they may also adversely affect seasoning, machining and painting.

References

BRE (1985) Timbers: their natural durability and resistance to preservative treatment. *Digest 296*. Building Research Establishment (obtainable from BRE Bookshop).

Hillis W.E. (1987) *Heartwood and Tree Exudates*. Springer-Verlag, Berlin.

Ward J.C. and Zeikus J.G. (1980) Bacteriological, chemical and physical properties of wetwood in lining trees. In: Banch J. (Ed.), Natural variations of wood properties. *Mitt Bundesforschungsanst Forst-Holzwirtsch* **131**: 133–166.

3
Cellular Structure

3.1 Differentiation and types of cells

In Chapter 2, attention was drawn to the presence between the bark and the wood of living cells capable of subdividing to form new wood or bark cells. Attention was also drawn to the three functions that the trunk of a tree has to perform. Following their formation, the new cells on the woody side of the cambium undergo, over a period of up to three weeks, a series of changes which is known as **cell differentiation**. The cells tend to change shape and a secondary wall is formed, the structure of which is presented in Chapter 4. The cell dies and the degenerated cell contents are frequently found lining the cell wall; the cell is now ready to assume one or more of the three basic functions of conduction, support and storage.

The conifers, or softwoods, are characterised by having two types of cells to perform these three functions – while the broadleaves or hardwoods possess four types of cells to carry out the same three functions (see Table 3.1).

Those cells responsible for support (fibres and tracheids) and conduction (tracheids and vessels) are aligned vertically in the trunk and comprise from about 80 per cent of the wood volume in hardwoods to 95 per cent that in the softwoods.

Those cells responsible for storage (parenchyma) may be present both horizontally and vertically. They are certainly always present horizontally where they assume narrow bands of cells radiating outwards from the pith and known as **rays**. These are continuous outwards, because the cambial cells, from which they arise, produce only ray cells at each division. As a tree increases in girth, additional groups of specialised cambial cells are formed that produce only ray cells. In this way, the number of plates of ray tissue per unit length of circumference of a stem remains approximately the same, irrespective of the age of the tree. The number of rays per unit of circumference, however, varies appreciably in different species: from less than one to more than ten per millimetre of circumference. The use of the term 'medullary ray' should be avoided: **medulla** is an alternative term for pith. Hence, only those rays that originate in the first year's growth, or pith, are strictly medullary rays; away from the pith, such rays cannot be distinguished from those that originate later in the life of a tree and, therefore, are not true 'medullary rays'.

The rays are usually just visible to the naked eye on radial surfaces, where they appear as narrow, horizontal ribbons 0.05 to 0.5 mm wide (Figure 3.1b). They may appear discontinuous because the cut surface is rarely truly radial and the rays may run out of the section.

On transverse and tangential surfaces, the rays can usually be seen with the aid of a low-power lens. On transverse surfaces they appear as narrow lines radiating outwards, crossing the growth rings at right angles (Figure 3.1a), and on tangential surfaces (Figure 3.1c), where the rays themselves are seen in section, they appear as short, vertical, boat-shaped lines.

Parenchymatous cells can also be present running vertically in certain softwoods and many hardwoods. When these wood parenchyma cells are present in the softwoods they are generally in single isolated vertical series of cells known as a strand, but in the hardwoods, although they can again be present as single vertical strands, they are more frequently present in groups or bundles.

There is present then in each of the three principal planes (cross-section, longitudinal radial, longitudinal tangential) a different distribution

Table 3.1   *Functions and wall thicknesses of the various types of cells found in softwoods and hardwoods*

Cells	Soft-wood	Hard-wood	Function	Wall thickness
PARENCHYMA	✓	✓	Storage	
TRACHEIDS	✓	✓	Support conduction	
FIBRES		✓	Support	
VESSELS (pores)		✓	Conduction	

of cells: this is one of the main reasons for the difference in behaviour of wood among the three planes — a feature known as **anisotropy**: anisotropic behaviour in terms of strength, stiffness, moisture relationships and fracture methodology will be discussed in later chapters.

Individual cells can be studied by separating them from one another by the process known as maceration: this chemical treatment dissolves away the middle lamella, a non-structural compound which holds together the individual cells. The interrelationships between cells of any one type and between different cell types can be studied microscopically using thin sections (10 –15 μm) of wood cut using a base-sledge microtome as discussed in Chapter 6.

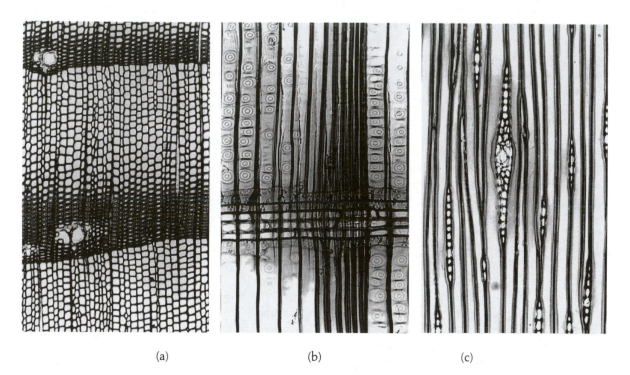

(a)   (b)   (c)

Figure 3.1   *(a) Transverse section of Scots pine (×75) showing one complete growth ring. Note the thin walls of the early-wood tracheids, the thick walls of the late-wood and the abrupt change from early- to late-wood. (b) Radial–longitudinal section of Scots pine (×150). Note the numerous and large pits in the early wood, and the few small pits in the late wood. (c) Tangential–longitudinal section of Scots pine (×150). Note the resin canal in one ray (fusiform ray) (Building Research Establishment, © Crown Copyright)*

3.2 Softwoods

3.2.1 *Conducting and supporting cells*

As can be seen in Table 3.1 the conducting and mechanical functions in softwoods are performed by a single type of cell known as a **tracheid**. These cells are hollow, needle-shaped and generally some 2.5–5.0 mm in length, though tracheids up to 10 mm in length are possible. The aspect ratio of the tracheid (length/breadth) is about 100/1, as illustrated in Figure 3.2.

The tracheids are packed closely together so that a cross-section through them resembles a honeycomb (Figure 3.1a). Those formed from the cambium early in the growing season have a large radial diameter and small wall thickness (about 2 μm), and consequently their function is primarily that of conduction: those tracheids comprising the late wood have a much smaller radial diameter (the tangential diameter remains approximately constant since all the cells in a radial file were produced from the same cambial initial) while the wall thickness is markedly increased (10 μm); they are also about 10 per cent longer than the early-wood tracheid. The function of the late-wood tracheid is therefore primarily support, though they may also play some part in conduction (Figure 3.1a).

This contrast in cell diameter and wall thickness between the two zones renders growth rings in most softwoods conspicuous to the

Figure 3.2 *Maceration of a piece of softwood: magnification ×20 (Building Research Establishment, © Crown Copyright)*

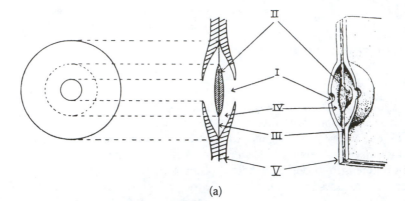

(a)

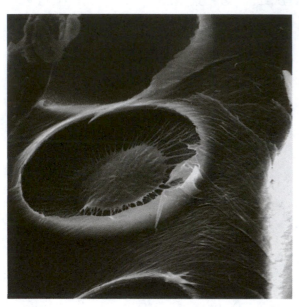

(b)

Figure 3.3 *(a) [Left and centre] – surface view and section through bordered pits in conducting cells; [right] – solid view of two pits cut in half (after a woodcut by Dr L. Chalk): I, pit opening; II, torus; III, margo strands formed from the primary wall; IV, pit cavity; V, secondary wall. (b) Bordered pits on the radial wall of a softwood. The arched dome on one side has been torn away during sample preparation to reveal the torus and supporting margo strands. Magnification ×3000 (scanning electron micrograph by Building Research Establishment, © Crown Copyright)*

naked eye; the early wood, containing a smaller proportion of wall substance, appears lighter in colour than the denser late wood. In some timbers, for example larch, Douglas fir and European redwood, the transition from thin- to thick-walled tracheids is abrupt, but in others, such as spruce and fir, it is gradual. In softwoods originating in tropical or sub-tropical climates, such as parana pine, rimu and podocarp, the growth ring cannot be subdivided into two zones and it is often difficult to determine the growth ring boundaries.

The quality of a softwood depends largely on the proportions of thin- to thick-walled tracheids, and on the contrast between the wood of these two zones. The higher the percentage of late wood, the stronger is the timber; moreover, marked differences in thickness of the walls of the early- and late-wood cells may cause the two zones to behave differently under tools and in service, and may give rise to painting problems, as in Douglas fir.

In popular language, tracheids are often called 'fibres', particularly in connection with wood-pulp and the paper industry, but this is incorrect, as true fibres occur only in hardwoods.

Since tracheids are needle-shaped, complete cells yet responsible for conduction, the transfer of moisture up the tree can be effected only by openings in the cell wall between overlapping tracheids. The inter-tracheid pit in a softwood is known as a **bordered-pit** and these are found predominantly on the radial walls of the early-wood tracheids since it is these cells, as noted above, that are primarily responsible for conduction. Within the early-wood tracheid, the bordered pits, although present along its entire length, tend to be concentrated towards the ends of the cells. Bordered pits are also present on the radial walls of late-wood tracheids but are smaller in size and much fewer in number (Figure 3.1b). In a number of softwoods, but certainly not all, bordered pits are also found on the tangential wall interconnecting the last tangential row of late-wood cells of any one year and the first row of early-wood cells of the following year (Figure 3.1c).

The bordered pit is about 15–20 μm in diameter and in cross-section is analogous to two saucers facing one another with the indented area in which the cup sits removed; between the 'saucers' is suspended a diaphragm. The detailed structure is illustrated in Figure 3.3; the secondary wall to the tracheid formed during differentiation is arched in adjacent tracheids to form the 'saucers'. Between these lies the thickened **torus** suspended from the edge of the pit cavity by strands known as **margo strands**. The torus can therefore respond to differences in liquid pressure in adjacent cells by moving to the tracheid having the lower pressure. In the process of wood drying, the retreating water meniscus generally pulls the torus to one side where it becomes firmly adpressed (Figure 3.4). In this state it is referred to as **aspirated** and markedly reduces the permeability of dried softwoods to treatment by preservative and fire retardants.

3.2.2 Storage cells

The storage tissue is known collectively as parenchyma; it consists of two kinds of cells that are essentially similar in details of structure, but that differ in their manner of distribution in the wood. These cells are brick-shaped, about 200 × 25 μm in size with the longer axis horizontal in the **ray-parenchyma** cells, and vertical in the **wood-parenchyma** cells. The cells have relatively thin walls with numerous pits (see below). They differ from tracheids in remaining alive for some years after their development is completed. This is because plant food is usually stored in some form other than that required by the growing cambium, and its conversion to a suitable state can only occur in a living cell. When no longer required for storage, the parenchyma cells die like any other cells of the secondary xylem.

The ray tissue occurs in narrow, horizontal bands or plates called **rays**, which radiate outwards from the centre of the tree to the bark, although on a small area from near the outside of a large tree they may appear as a series of parallel layers between several rows of tracheids, as in Figure 3.1a. The height of the ray varies considerably between adjacent rays. It also varies widely between different species

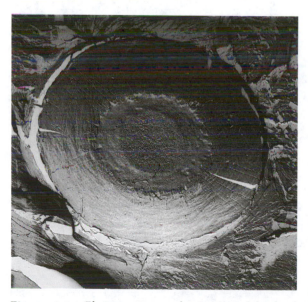

Figure 3.4　*Electron micrograph of softwood bordered pit with the torus in an aspirated state (×3000) (Building Research Establishment, © Crown Copyright)*

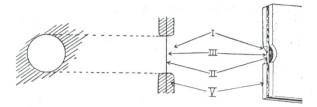

Figure 3.5   *Simple pit: (left) surface view and centre, a section through a pit in storage cells; (right) solid view of a simple pit cut in half – I, pit opening; II, primary wall; III, pit cavity; IV, secondary wall*

and it is possible to recognise timbers with very low rays (3–4 on average), such as juniper and cryptomeria, and those with very high rays (about 25 cells), such as fir. In most conifers the ray comprises a single row of cells, but in sequoia the ray is often biseriate.

In softwoods the wood-parenchyma tissue when present is sparse in amount, and usually visible only under the microscope. The strands are either scattered through the growth ring as in fir, western red cedar and podocarp, or in a layer at the end of the growth ring, as in juniper. In the last-mentioned case, the layer is usually visible to the naked eye on the end surface as a narrow line, lighter in colour than the surrounding tissue.

Passage of food material between parenchymatous cells, both axial, and within the ray is again by way of pits. However, the pits here are of a simple structure compared with the bordered pit between adjacent tracheids. This **simple** pit comprises a cylindrical opening through the two secondary walls, while the primary wall remains as a semi-permeable membrane (Figure 3.5).

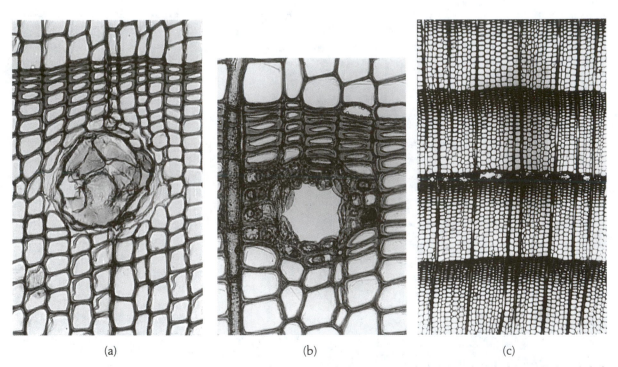

(a)   (b)   (c)

Figure 3.6   *(a) Transverse section of Scots pine (about ×200) showing a vertical resin canal with thin-walled epithelial cells. (b) Transverse section of larch (about ×200) showing a vertical resin canal with thick-walled epithelial cells. (c) Transverse section of cedar (×10) showing a tangential series of traumatic resin canals (Building Research Establishment, © Crown Copyright)*

3.2.3 *Resin-producing cells*

A characteristic feature of many, though certainly not all, softwood timbers is their resinous nature, which is often sufficient to give them a pronounced odour, and may cause freshly sawn timber to be 'tacky'. The resin in formed in parenchyma cells, and in some species these occur in special channels called **resin canals**, or **resin ducts**. These canals are not cells, but cavities in the wood, lined with an '**epithelium**' of parenchyma cells; these epithelial cells secrete resin into the canals.

Resin canals are present both vertically and horizontally; in the latter orientation they are present in some of the rays, when the ray is termed a **fusiform ray** (Figure 3.1c). Resin canals are a useful feature for distinguishing some timbers, since they are always present in some genera, such as larch, Douglas fir, pine (Figure 3.6a) and spruce, but they are normally absent in others, such as fir, sequoia and yew. Additionally, it is possible to separate the pines from spruce, larch and Douglas fir in that pine has thin-walled epithelial cells (Figure 3.6a) while the

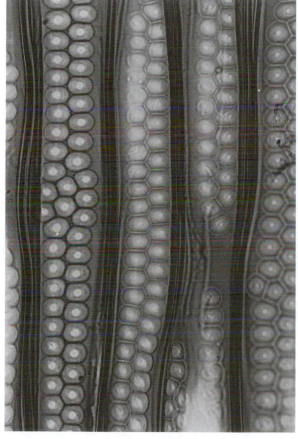

Figure 3.7 *Alternate arrangement of bordered pits characteristic of Parana pine. Usually the pit chambers assume a hexagonal arrangement owing to high packing density. Magnification ×350 (Building Research Establishment, © Crown Copyright)*

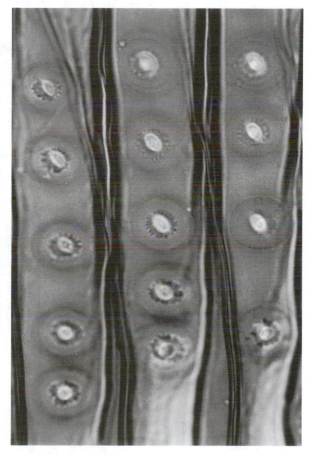

Figure 3.8 *High magnification view of the bordered pit in cedar showing the characteristic scalloped border of the torus. Magnification ×800 (Building Research Establishment, © Crown Copyright)*

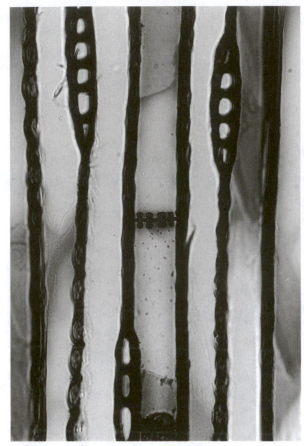

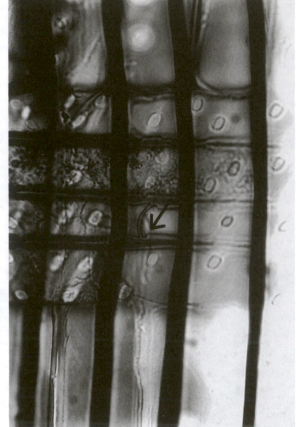

Figure 3.9 *Nodular thickening of the end walls of vertical parenchyma cells as seen on the tangential–longitudinal surface of taxodium. Magnification ×350 (Building Research Establishment, © Crown Copyright)*

Figure 3.10 *Indentures, or slight depressions of the horizontal wall at its junction with the vertical wall of ray parenchyma cells in western red cedar. Magnification ×750 (Building Research Establishment, © Crown Copyright)*

other three timbers have thick-walled epithelial cells (Figure 3.6b).

Resin canals are usually scattered across the growth ring though in a few timbers these can be restricted to the late wood. However, it is possible to find resin canals in tangential rows not only in those timbers that normally have resin canals, but also in timbers devoid of them.

These tangentially arranged resin canals are the result of wounding of the tree. The tree reacts by producing from the cambium **traumatic** resin canals, usually of irregular size and shape,

frequently touching one another and always in a tangential row (Figure 3.6c).

3.2.4   Variability in cellular features between species

Although there are basically only two types of cells in softwoods, the microscopic structure of these can vary systematically among different genera thereby forming the basis for identification at a microscopic level. The variability that occurs in the distribution of the bordered pits, the type of

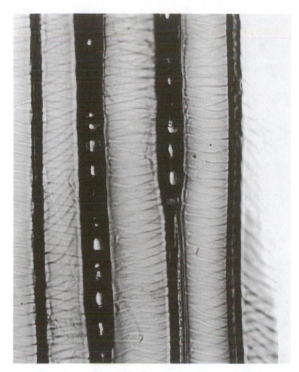

Figure 3.11 *Spiral thickening in the vertical tracheids of Douglas fir. Magnification ×250 (Building Research Establishment, © Crown Copyright)*

additional thickening to the cell wall and the type of pitting between tracheids and parenchyma are set out below.

Pitting on the vertical tracheid

The bordered pits of the tracheids typically occur in one or two rows, as seen on radial sections (Figure 3.1b), but, in the *Araucariaceae* particularly, the pits are distinctly angular in outline, and the pits in one row alternate with those in the rows above and below

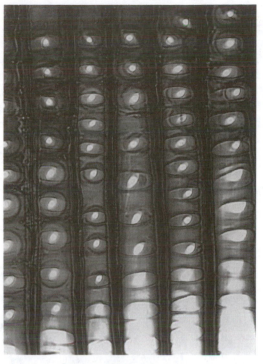

(a)

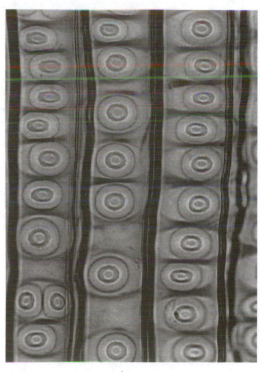

(b)

Figure 3.12 *(a) Horizontal thickening bars across the pit border, known as **callitroid thickening** and characteristic of* Callitris spp. *Magnification ×400. (b) Horizontal thickening bars above and below the bordered pits known as **Crassulae** or **Bars of Sanio**: common in most softwoods except the* Araucariaceae. *Magnification ×400 (Building Research Establishment, © Crown Copyright)*

instead of being arranged in parallel lines, one above the other (that is, *alternate pitting*) (Figure 3.7). Yet another variation in the distribution of pits in tracheids is their occurrence, mainly in rows of three, the pits in one row being immediately above and below those in the adjacent rows, that is, **multiseriate** and **opposite pitting** which is characteristic of taxodium (swamp cypress).

In cedar the margin of the torus, as seen in radial section, is regularly scalloped (**scalloped tori**), providing a reliable diagnostic feature for distinguishing this genus (Figure 3.8).

Pitting of the parenchyma cells

(a)   *Nodular walls*: the transverse or end walls of the wood parenchyma cells of some softwoods, such as taxodium (swamp cypress), are nodular or bead-like in appearance owing to the formation of many simple pits (Figure 3.9); in a few species the end walls of the ray cells may be similarly thickened.

(b)   *Indentures*: depressions in the horizontal walls of rays, in which the ends of the vertical walls stand, have been described as **indentures** (Figure 3.10). This feature has been observed in all families of softwoods,

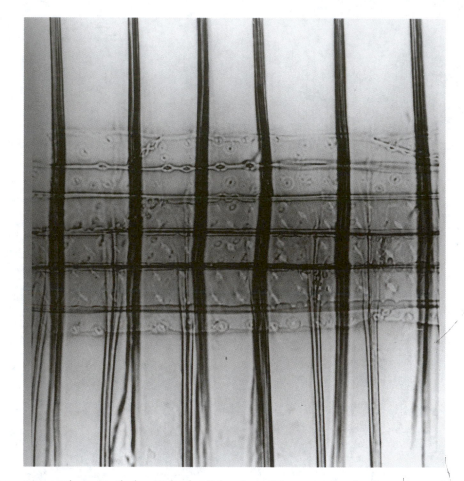

Figure 3.13   *View of the ray on the longitudinal radial surface of Norway spruce showing presence of ray tracheids with smooth margins and small bordered pits, and rows of ray parenchyma cells with piceoid pits in the cross-field. Semi-bordered pits are present linking the ray tracheids and ray parenchyma cells: small bordered pits link adjacent ray tracheids. Magnification ×700 (Building Research Establishment, © Crown Copyright)*

except the *Araucariaceae*, but is strongly developed only in some genera, such as western red cedar.

Wall thickening of the vertical tracheid

(a) *Spiral thickening* occurs as a characteristic feature in Douglas fir and yew, and is only rarely present in some other species. The spirals are inclined in one direction but, because of the depth of focus of the microscope, the spirals in the wall of the cell below may also be seen, producing a reticulate pattern on the wall of the tracheid above (Figure 3.11). The spirals are actual bands of thickening in the secondary wall. Care must be exercised to distinguish spiral thickening from checking of the cell wall as occurs in the development of compression wood (Chapter 5, section 5.4.1).

(b) *Callitroid thickenings*: pairs of thickening bars across the pit border occur in a few species, particularly in the genus *Callitris* (Figure 3.12a).

(c) *Crassulae* (formerly known as the 'Bars of Sanio'): these constitute concentrations of intercellular substance and appear as horizontal bars. They occur in the radial walls of all tracheids (except in the *Araucariaceae*) above and below the rows of bordered pits (Figure 3.12b).

Ray tracheid

The ray cells of some softwoods, spruce, larch, Douglas fir, pine, hemlock and usually cedar, are of two kinds – ray parenchyma and **ray tracheids**. The latter are not parenchyma cells, but mechanical tissue and physiologically inactive; they are equipped with bordered pits, which can usually be seen in section on the radial face; pitting between ray tracheid and ray-parenchyma cells is half-bordered (Figure 3.13). Ray tracheids are normally confined to the margins of rays when the outer wall is sinuous in shape, but in some species of *Pinus*, ray tracheids may also occur in the middle portions of a ray, and the low rays of the hard pines may consist wholly of ray tracheids.

The horizontal walls of the ray tracheids may be smooth walled as in the five-needled 'soft' pines, larch, cedar, Douglas fir and hemlock. The spruces are generally regarded as having smooth walls to the ray tracheids (Figure 3.13), even though, in some species extremely small teeth are scattered along the wall. In the remainder of the pines, the walls are thickened which can be of **dentate** or **reticulate** form, the latter characterised by overlapping teeth and occasional confluence of the teeth to form bridges (Figures 3.14 and 3.15). Dentate thickening is characteristic of most two-needled pines, while reticulate thickening is formed in the hard pines (three-needled plus contorta pine which has two needles).

Cross-field pitting

The area of wall contact between a ray cell and a vertical tracheid is referred to as a **cross-field**; the pitting occurring in a cross-field is **semibordered** and takes one or other of five more or less distinct forms. In the pines, the cross-field is occupied by one or two very large pits (**fenestrate type** – Figures 3.15 and 3.16a), or one to

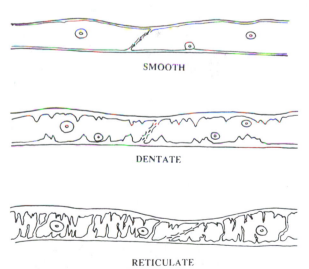

SMOOTH

DENTATE

RETICULATE

Figure 3.14 *Diagrammatic representation of the different types of thickening to be found in the horizontal walls of the ray tracheids (J.M. Dinwoodie)*

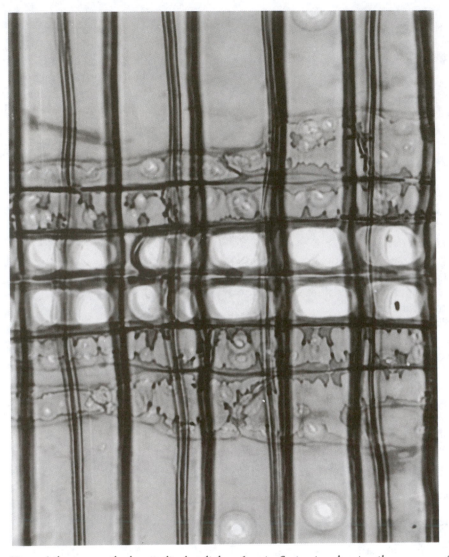

Figure 3.15   *View of the ray on the longitudinal radial surface in Scots pine showing the presence of ray tracheids with dentate margins and small bordered pits, and ray parenchyma cells with fenestrate pitting in the cross-field. Magnification ×700 (Building Research Establishment, © Crown Copyright)*

six similar but smaller pits (**pinoid** type — Figure 3.16b). The **piceoid** (sometimes called picoid) type refers to pits with very narrow apertures, sometimes extending beyond the margins of the pits (Figures 3.13 and 3.16c). This type of pit is to be found in many timbers including spruce, larch, Douglas fir and hemlock. In the **cupressoid** type, characteristic of juniper and cypress (*Cupressus* and *Chamaecyparis*), the apertures are included and rather narrower than the border (Figure 3.16d), while in the **taxodoid** type (as found in taxodium and western red cedar) the apertures, which are included, are ovoid to circular, and wider than the border (Figure 3.16e). The distinction between the last two types calls for careful observation of sections in proper focus.

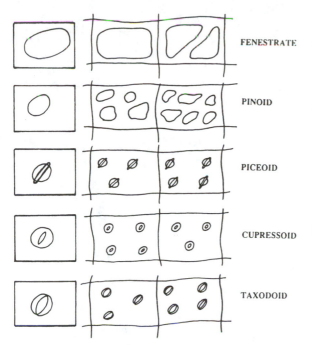

FENESTRATE

PINOID

PICEOID

CUPRESSOID

TAXODOID

Figure 3.16 *Types of cross-field pitting in plan view. The pit apertures in both the fenestrate and pinoid pits are straight-sided; in the other three types of pits, the walls are tapered and the openings in the top and bottom surfaces of the walls are superimposed*

3.3 Hardwoods

3.3.1 *Types of cells*

Whereas in softwoods both conducting and strengthening functions are undertaken by a single type of cell (the tracheid), in hardwoods there is a more distinct division of labour, and the conducting cells, called **vessels**, are quite different from the **fibres** that provide mechanical support. The presence of specialised conducting tissue provides a simple means of differentiating hardwoods from softwoods (Table 3.1 and Figure 3.17).

The cambial cells of hardwoods are shorter than those of softwoods, and so are the mature cells that arise from the division of these cambial cells. The maximum length rarely exceeds 2 mm, as compared with 10 mm attained by some softwood tracheids. In hardwoods, as in softwoods, the same cambial cell may give rise successively to wood cells, or bark tissue; the former will differentiate to form conducting, mechanical and storage tissue. As in softwoods, a special type of cambial cell gives rise to the ray cells.

3.3.2 *Conducting cells*

The counterpart in hardwoods of the thin-walled, conducting tracheids of softwoods are the **vessel members**, each about 0.2–0.5 mm in length and varying in diameter from 20 to 400 µm; these vessel members are illustrated in Figures 3.17 and 3.18. Figure 3.18a shows a vertical series of three fully-developed conducting cells, each of which is known as a vessel member; these series may extend for a considerable distance up the tree. When first formed, vessel members have end walls like other cells, but early in their development the cells swell and the end walls split and are dissolved away by enzymes, so that the vessel members form a continuous tube, like a drain-pipe, in the tree, known as a **vessel**. These are frequently visible on longitudinal surfaces of wood as fine-to-coarse scratches, often referred to as **vessel lines**. When examining a cross-section of a hardwood, the term **pore** is used to describe the cross-section of the vessel.

In Figure 3.18 it will be seen that the vessel members have no 'end' or transverse walls, but are open top and bottom. Where the open ends of the vessels comprise a single hole, the feature is known as a **simple perforation plate**. In some woods, however, only parts of the end wall may be dissolved away to produce a **multiple perforation plate**. The most common of these is the **scalariform perforation plate** in which horizontal bars remain in the form of a ladder (Figure 3.19). Such plates are always inclined at an oblique angle and always in the radial plane; they are characteristic of certain hardwoods, such as birch and alder, and can be seen with a hand lens on split radial surfaces if the vessels are not too small in cross-section.

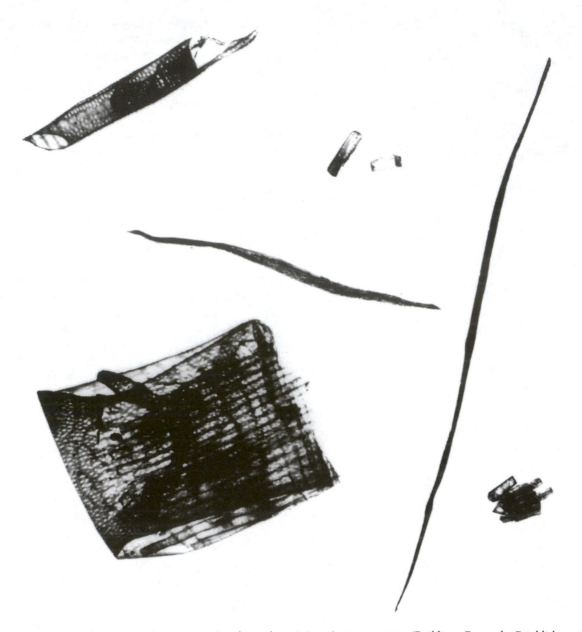

Figure 3.17 *Maceration of a piece of oak timber. Magnification ×120 (Building Research Establishment, © Crown Copyright)*

It will be clear that the conducting tissue of hardwoods is more effective in providing the water requirements of leaves than are tracheids in softwoods and this is necessitated by reason of the larger leaf area of broad-leaved species compared with that of conifer needles. Moreover, in addition to the open ends of the vessel members, pits occur in the longitudinal walls in great profusion.

The pits between the adjacent vessels are known as intervascular pits and are therefore fully bordered, similar in structure to the bordered

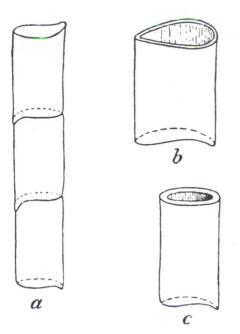

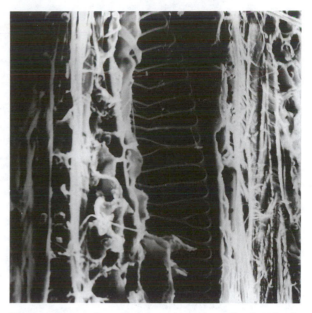

Figure 3.18   *Vessel members: a, vertical series of three vessel members; b, thin-walled early-wood vessel member; c, thick-walled late-wood vessel member (highly magnified)*

Figure 3.19   *Scanning electron micrograph of multiple perforation plate (scalariform perforation plate) at the end of a vessel member in Dillenia reticulata. Magnification ×300 (Building Research Establishment, © Crown Copyright)*

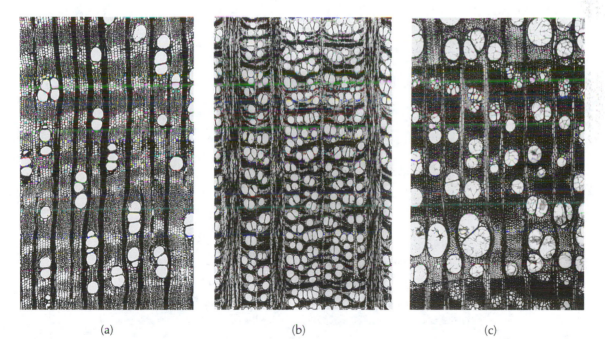

(a)   (b)   (c)

Figure 3.20   *(a) Transverse surface of sapele (×8) showing vessels in radial groups. (b) Transverse surface of Australian silky oak (×8) showing tangential groups of vessels. (c) Transverse surface of robinia (×8) showing clustered arrangement of vessels in late wood (Building Research Establishment © Crown Copyright)*

pit joining together two vertical softwood tracheids, but with elongated apertures and lacking a torus; they are considerably smaller in diameter (generally 5–12 µm) and, exceptionally, as in birch and box, are as small as 3 µm in diameter. Differences in distribution and shape of these pits occur: the pits may alternate as in poplar, be opposite one another as in tulip tree, or elongate considerably horizontally, when the pitting is called scalariform as occurs in magnolia. Small outgrowths from the secondary wall may develop into the pit cavity in certain timbers: these pits are known as **vestured pits** and can be observed in laburnum.

Pit pairs between vessels and tracheids are again fully bordered, but are usually different in appearance or size from those interconnecting two vessels side by side. Pit pairs between vessels and parenchyma are semi-bordered.

The walls of the vessels of certain hardwoods, such as sycamore, elm, lime and horse chestnut, have spiral thickening similar to that described earlier for yew and Douglas fir.

Some hardwoods, of which oak and sweet chestnut are examples, have tracheids in addition to vessels. These tracheids which are usually in association with the vessels, are similar in appearance to softwood tracheids, but are shorter (<1 mm).

Vessels can be distributed singly across the cross-section (Tasmanian oak), in radial groups (sapele), in tangential groups (Australian silky oak), or in clusters (robinia) (Figure 3.20).

The majority of hardwoods, whether of temperate or tropical origin, show little change in the size and distribution of vessels across the growth ring except for very late in the growth ring when a reduction in size occurs. Such woods are said to be **diffuse-porous**: examples are beech, lime mahogany and American whitewood (Figure 3.21a).

A few timbers, however, display the existence of two very different sizes of vessels across the growth ring. Thus, the large diameter vessels of the early wood contrast sharply with the much smaller-diameter vessels of the late wood. Woods with this distinct two-phase system are known as

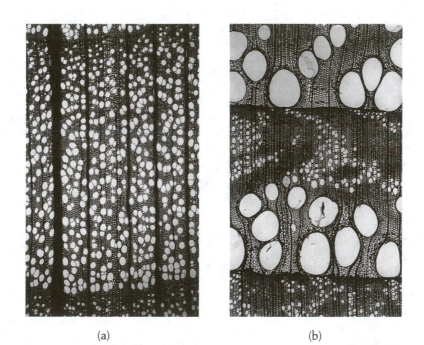

(a)   (b)

Figure 3.21   *(a) Transverse section of a diffuse-porous wood – beech (×8). (b) Transverse section of a ring-porous wood – sweet chestnut (×8) (Building Research Establishment, © Crown Copyright)*

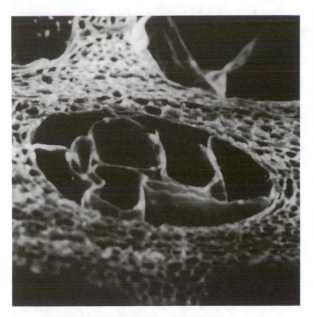

Figure 3.22 *Scanning electron micrograph of tyloses in a heartwood vessel of European oak. Magnification ×650 (Building Research Establishment, © Crown Copyright)*

ring-porous, and in this group oak, ash, elm, sweet chestnut, hickory and teak are the best known (Figure 3.21b).

The vessels of the heartwood do not conduct, being frequently blocked by solid deposits of a gummy type, or in some species by the development within the vessels of a structure known as **tyloses** (Figure 3.22). Even within a genus, some species will develop tyloses and others not.

In those timbers that have tyloses in the heartwood, a thin extra layer to the cell wall is formed in both the wood and ray parenchyma lying adjacent to the vessels; this layer has been called the **protection layer** (Schmid, 1965) and overlies the pit membranes. In the **transitional zone** between sapwood and heartwood (see Chapter 2), this protection layer is sucked through the pit into the adjacent vessel cavity under the negative pressure present in it, where it forms a balloon type structure (Wilson and White, 1986). Where these are common, the vessel is completely blocked to the transport of mineral solutions. Tyloses are present in the

heartwood of many, but certainly not all timbers; they are to be found, for example, in robinia and meranti. Closely related timbers may differ in the presence or absence of tyloses. Thus the white oaks develop tyloses – and the heartwood can therefore be used for the manufacture of sherry and wine barrels – while the red oaks do not form tyloses and are therefore unsuitable for barrel making.

3.3.3 Supporting cells

The mechanical tissue of hardwoods consists of wood **fibre**. These are narrow, spindle-shaped cells, not unlike the late-wood tracheid of softwoods, but they usually have more pointed ends and are shorter, being usually about 1 mm in length: as such, however, they can be from two to five times the length of the cambial initial from which they were originally formed, and quite unlike a vessel member which remains the same length as its original cambial initial. The increase in length of the fibre is due to its very marked elongation at the tips. The shape of the fibre is, therefore, needle-like with very sharp tapering ends: the ratio of length-to-width at mid-length is about 100 to 1 (Figure 3.17).

Fibre wall thickness varies considerably among the different species, being very thin in balsa and so thick in lignum vitae as to occlude the cell cavity. Wall thickness directly affects the density of wood, and through density its mechanical performance.

Pitting in the walls of the fibres is much less frequent than with vessels. Bordered pits are absent and the pits, though often slit-like in appearance, are basically simple pits (Figure 3.5); they are not confined to a particular wall, though they are more frequent on the radial wall.

Pits with very small borders are present, however, in fibres of some woods and these fibres are called **fibre-tracheids**.

In some timbers, such as teak and gaboon, the cavities of the fibres are divided into small compartments by thin horizontal partitions; such fibres are called **septate fibres**. The reason for the partitioning is not known, but such fibres are more common in species with little parenchyma.

Fibres are sometimes arranged in very regular radial rows (as seen in cross-section); such an arrangement is a characteristic and constant feature of several timbers, such as Central American cedar (*Cedrela* spp).

3.3.4 *Storage cells*

In hardwoods, the storage tissue is essentially similar to that of softwoods, but it is frequently more abundantly developed, and it displays greater variety in distribution and arrangement. In consequence, both the axial wood parenchyma and the horizontal rays are among the most useful features for distinguishing between different hardwoods.

Wood parenchyma

Three distinct types of distribution may be differentiated: **apotracheal parenchyma**, which is parenchyma lying independent of the vessels; **paratracheal parenchyma**, which is parenchyma associated with the vessels; and **banded parenchyma**, which is parenchyma in tangentially or obliquely oriented bands; these may be mainly independent of the vessels, associated with the vessels, or both (Figure 3.23 and Wheeler *et al.*, 1989).

Apotracheal axial parenchyma (axial parenchyma not associated with the vessels). This can be subdivided according to its distribution (Figure 3.24).

(a) *Axial parenchyma diffuse*: Consists of single strands, or pairs of strands, distributed irregularly among the fibres, as in pear and box; this distribution of parenchyma is, as a rule, distinct only under the microscope. However, when the individual strands are of sufficiently large cross-section, they are discernible with a hand lens as indistinct, light-coloured dots, as in opepe. In woods with numerous parenchyma strands of rather large cross-section, the end surface may appear characteristically speckled, as in punah, or opepe.

(b) *Axial parenchyma diffuse-in-aggregates*: Consists of strands grouped into short discontinuous tangential or oblique lines, as in lime and obeche.

Paratracheal axial parenchyma (axial parenchyma associated with the vessels or vascular tracheid). The distribution of parenchyma relative to the vessels is a complete spectrum, but for convenience it can be subdivided into six types

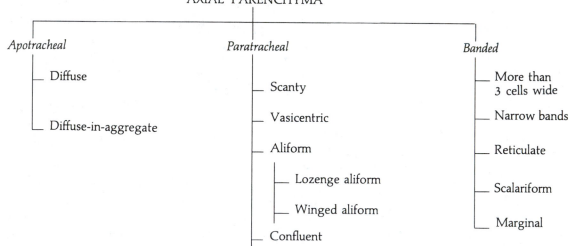

Figure 3.23 *The different forms of distribution of axial parenchyma in hardwoods (graphical presentation of information in Wheeler et al., 1989)*

WOOD PARENCHYMA

APOTRACHEAL

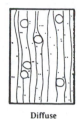

Diffuse

Diffuse-in-aggregate

PARATRACHEAL

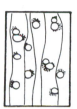

Scanty

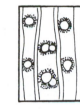

Vasicentric

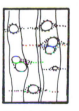

Winged aliform

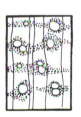

Confluent

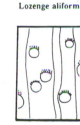

Unilateral

BANDED

Wide bands

Narrow bands

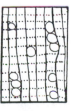

Reticulate

Scalariform

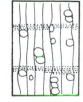

Marginal

(Figure 3.24). Quite frequently, any one timber will have more than one type present, though usually only one type predominates.

(c) *Axia parenchyma scanty paratracheal*: Consists of occasional parenchyma cells associated with the vessels or an incomplete sheath around the vessels, as in some of the African mahoganies.

(d) *Axial parenchyma vasicentric*: Consists of a complete sheath of parenchyma cells around a solitary vessel or a vessel multiple; as found in ash and idigbo.

(e) *Axial parenchyma aliform*: Consists of parenchyma cells either surrounding or to one side of the vessel with lateral extension. Two subdivisions occur — where the overall shape of the parenchyma tissue is like a diamond, the parenchyma is known as *lozenge-aliform*, as in afzelia and albizia; where the lateral extensions are more elongated and narrow, the parenchyma is known as *winged-aliform*; as found in ramin.

(f) *Axial parenchyma confluent*: Consists of coalescing vasicentric or aliform parenchyma surrounding or to one side of two or more vessels; as found in iroko.

(g) *Axial parenchyma unilateral paratracheal*: Consists of paratracheal parenchyma forming semi-circular hoods or caps only on one side of the vessels and which can extend tangentially or obliquely in an aliform or confluent or banded pattern; as found in araracanga.

Banded parenchyma (axial parenchyma in bands which may be independent of the vessels, definitely associated with the vessels, or both; these bands may be wavy, diagonal, straight, continuous or discontinuous). Five types here have been described (Figure 3.24).

(h) *Axial parenchyma in wide bands of more than three cells*: as found in omu and ekki.

Figure 3.24 *Diagrammatic illustration of the different types of axial parenchyma. This is portrayed as black dots for purposes of reproduction, but in a wood section or wood surface the parenchyma would appear as a lightly coloured tissue (see Figures 3.20 and 3.21) (J.M. Dinwoodie)*

(i) *Axial parenchyma in narrow bands or lines up to three cells wide*: as found in jelutong.

(j) *Axial parenchyma reticulate*: Consists of parenchyma in continuous tangential lines of approximately the same width as the rays, *regularly* spaced and forming a network with them; the distance between the rays is approximately equal to the distance between the parenchyma bands; as found in makore and nyatoh.

(k) *Axial parenchyma scalariform*: Consists of parenchyma in fairly regularly spaced fine lines or bands arranged horizontally (tangentially) or in arcs, appreciably narrower than the rays and with them producing a ladder-like appearance in cross-section – the distance between the rays is greater than the distance between the parenchyma bands; as found in lancewood.

(l) *Axial parenchyma in marginal or in seemingly marginal bands*: Consists of parenchyma bands which form a more or less continuous layer of variable width at the margins of a growth ring, or are irregularly zonate; as found in American mahogany.

Ray parenchyma

In softwoods, ray tissue is sparsely developed and typically only one cell wide in the tangential direction, that is the rays are **uniseriate**, but in hardwoods there is a considerable variation in both size and number of the rays. Some hardwoods, such as willow and poplar, have only uniseriate rays, but in the majority the rays are **multiseriate**, that is more than one cell wide. In some timbers, the rays are comparatively uniform in size; they may be relatively small and not easily visible to the naked eye, as in birch, or they may be broad and high, and conspicuous to the naked eye, as in plane. In other woods, rays of two distinct sizes occur: very large rays in association with uniseriate ones, as in oak. In a few species, groups of small rays occur in aggregations that appear to the unaided eye, or at low magnifications, as single large rays; these are known as **aggregate** rays. The apparently broad rays of hornbeam, hazel and alder are of this type.

Very broad rays give rise to the handsome 'silver figure' of quarter-sawn timber of true oaks and Australian silky oak. The presence of broad rays is also an indication that the timbers will split readily in a radial direction, an important property for certain specialised purposes, for example the best-quality 'tight' barrel staves.

Rays are sometimes arranged in regular storeys or tiers that appear on tangential surfaces as wavy, parallel, horizontal lines, known as **ripple marks**; these are characteristic of some woods, such as mansonia, and are usually present in others, such as American mahogany.

In most timbers with storeyed rays, the wood parenchyma tissue, and sometimes the fibres, are also storeyed. In a few, however, the wood parenchyma or fibres, or both, are storeyed but not the rays.

In some timbers all the rays cells may be procumbent (longest dimension radial as seen in a radial section – as in maples and albizia), or all the ray cells may be square or vertically elongated, or a mixture of both. The individual ray cells may be either more or less similar in size and shape, in which case the rays are said to be **homocellular**, or distinctly variable, in which case the rays are **heterocellular**. Another variant of a ray is one that contains **sheath** cells; these upright cells form a more-or-less complete outer layer to the ray, one cell in thickness; the sheath cells are much larger than the other ray cells. One timber that contains sheath cells is yellow sterculia.

3.3.5 *Deposits*

The storage tissue of many timbers contains crystals, usually of calcium oxalate (Figure 3.25). These may be confined, in different species, to the wood parenchyma or rays, or they may occur in both tissues. More rarely, these cells contain deposits of silica, for example the ray cells of white meranti, Queensland walnut, keruing (apitong, gurjun, and yang) and parinari (Figure 3.26). Deposits of silica have an important bearing on the working qualities of timbers; an appreciable amount of silica in a wood renders

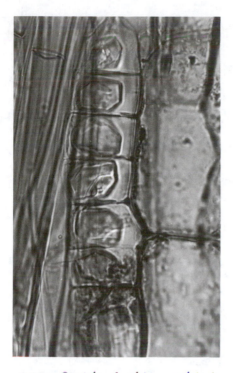

Figure 3.25 *Crystals of calcium oxalate in scented guarea; magnification ×400 (Building Research Establishment, © Crown Copyright)*

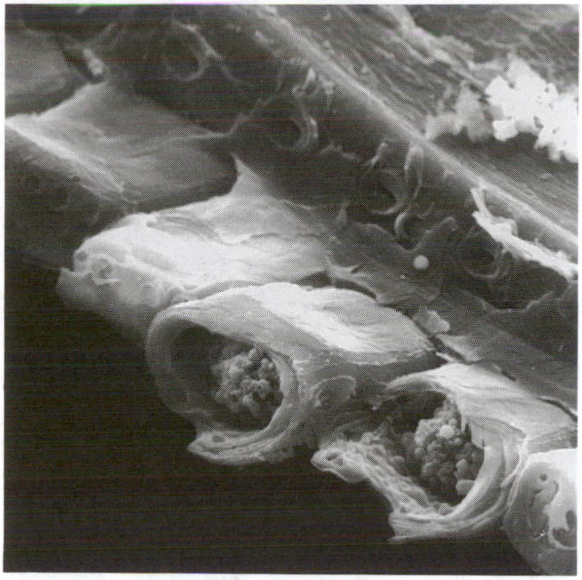

Figure 3.26 *Inclusions in cells. Silica grains in the ray parenchyma cells of* Parinari *species; magnification ×750 (scanning electron micrograph by the Building Research Establishment, © Crown Copyright)*

ordinary machine tools and feed speeds uneconomic in the conversion of logs to sawn timber. The particles of silica have an abrasive effect on the saw-teeth, producing rapid blunting of cutting edges and heating of the saw. Experiments indicate that saws of thicker gauge and wider gullet than normal, and teeth tipped with tungsten carbide, are suitable for the conversion of timbers with high silica content.

Occasionally large lumps of mineral are found in internal cracks in wood. One timber that appears to be prone to such development is iroko; the large lumps of dense calcium carbonate are referred to as 'stone'.

3.3.6 Resin canals or gum ducts

Normal resin canals or **gum ducts** are comparatively infrequent in hardwoods, although they are a constant feature of certain families, for example meranti (lauan), the keruing group and mersawa, of which the most important commercially is the *Dipterocarpaceae*. They may occur either as vertical canals in the wood, or horizontally in the rays, or, more rarely, both in the same species. The vertical canals may occur in tangential series, producing the appearance of growth-ring boundaries or they may be distributed in short tangential series throughout the wood, or scattered singly through the wood as in mersawa, resak and agba. The contents of the canals of the *Dipterocarpaceae* usually consist of white or yellow, solid, dammar deposits, but in *Dipterocarpus* spp the deposits are viscous oleo-resins, which tend to ooze over sawn surfaces even after the wood is thoroughly seasoned, causing difficulties in painting and other finishing processes. The white deposits, particularly when the canals are in more-or-less continuous tangential series, are often conspicuous to the naked eye on all surfaces, appearing as prominent white lines, erroneously called 'mineral streaks' in the trade.

Intercellular canals or gum ducts are produced as a result of wounding in many hardwoods and such canals are said to be **traumatic**: they may be distinguished from the normal type because they are invariably in tangential series, are irregular in outline, contain no epithelial cells and usually have dark-coloured, more-or-less viscous, gum-like deposits. Further, traumatic canals are usually larger than the vessels, and typically wider tangentially. In some species, such as African walnut, traumatic canals are sufficiently frequent in occurrence to be regarded almost as a characteristic feature of the timber.

3.3.7   *Latex tubes*

Special cells, containing latex and called **latex tubes**, occur in the ray tissue of certain timbers. They are usually invisible to the naked eye, but where they can be detected they are a helpful feature in identification. In a few timbers, such as jelutong and mujwa, specialised parenchymatous tissue, containing numerous latex tubes, develops from leaf-traces and continues outwards during the subsequent growth of the bole; such **latex traces**, as seen on tangential surfaces, are up to 20 mm high and lens-shaped in section. As the leaf-traces occur in whorls, the latex traces are found in tangential series up the tree at intervals of 0.6–0.8 metres, disfiguring long lengths of timber, and rendering it unsuitable for many purposes. Long splits can develop from the latex traces during seasoning.

References

Schmid R. (1965) The fine structure of pits in hardwoods. In Coté W.A. Jnr (Ed.), *Cellular Ultrastructure of Woody Plants*. Syracuse University Press, Syracuse, New York, pp 291–304.

Wheeler E.A., Baas P. and Gasson P.E. (1989) IAWA list of microscopic features for hardwood identification. *Bulletin (n.s.) of the International Association of Wood Anatomists* **10(3)**, 219–332.

Wilson K. and White D.J.B. (1986) *The Anatomy of Wood: Its Diversity and Variability*. Stobart & Son Ltd, London.

4

Molecular Structure

Previous chapters have dealt with the gross structure of wood as seen by the naked eye, and the cellular structure as can be observed with the aid of a microscope or hand lens. This chapter looks at the structure of wood at an even finer level by enlisting techniques such as the electron microscope, chemical analysis and X-ray diffraction analysis. These have identified the basic building blocks in wood and have revealed how these fit together to produce the high level of physical and mechanical performance that wood possesses.

The principal components of the structure of wood are **cellulose**, **hemicellulose** and **lignin**. Additional to these are very small quantities of minerals which are required for the basic metabolism of the tree that is carried out in all living cells.

In many timbers highly complex organic compounds are formed in the heartwood and, because they can be easily removed from the wood without altering its structure, these compounds have been given the collective name of **extractives**.

4.1   Principal chemical constituents

4.1.1   Cellulose

Between 40 and 50 per cent of the dry mass of wood is in the form of cellulose. In Chapter 2, it was mentioned that, within the crown of the tree, glucose is produced by the action of photosynthesis. The glucose units ($C_6H_{12}O_6$) in the cambial zone bond themselves together into long chains to form a molecule of cellulose. Water is removed from the glucose in this process and so each individual unit is really an anhydroglucose unit ($C_6H_{10}O_5$). Every alternate unit is rotated through $180°$ and generally about 8000 of

these link up longitudinally to form the cellulose molecule which is in the form of a long, straight and very thin chain or filament with the structure $(C_6H_{10}O_5)_n$ (Figure 4.1).

It should be noted from this figure that the anhydrogluclose unit, which is not flat, is in the form of a six-sided ring consisting of five carbon atoms and one oxygen atom, the former carrying side groups which, as will be discussed later, play an important role in intra- and inter-molecular bonding. These anhydroglucose units are linked together in the 1,4 positions, that is, it is the first and fourth carbon atoms after the oxygen atom moving clockwise around the ring (in its unrotated form) that combine with adjacent anhydroglucose units to form the long-chain molecule of cellulose.

There are actually two chemical forms of glucose, named α and β depending on the position of the $-OH$ group on carbon 1 relative to the ring. When this group lies below the ring on the opposite side to the group on carbon 4, the unit is known as β-glucose; it is the β-glucose units that combine through a condensation reaction to form the cellulose molecule.

Over the greater part of their length, the cellulose molecules lie parallel to one another in a particular pattern to form a crystal, a repeating unit of which is known as the **unit cell** (within the box in Figure 4.2). In natural, as distinct from man-made cellulose, the form of crystalline cellulose is known as cellulose I. Over the years, there have been many attempts to model cellulose I and in the literature a whole series of contrasting models will be found.

One of the more recent, and one which has gained worldwide acceptance, is that proposed by Gardner and Blackwell (1974). Using X-ray diffraction methods, these authors proposed an eight-chain unit cell with all the chains running

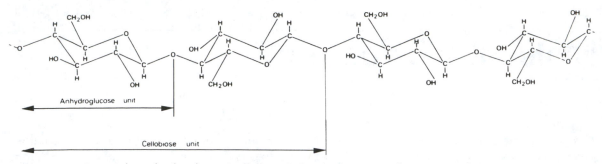

Figure 4.1   *Part of the molecular chain of cellulose which is built up from glucose units, alternate units being rotated through 180° (Building Research Establishment, © Crown Copyright)*

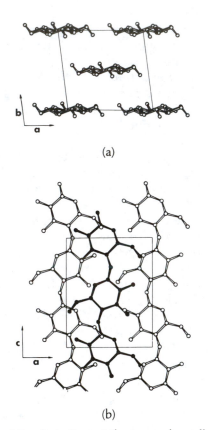

(a)

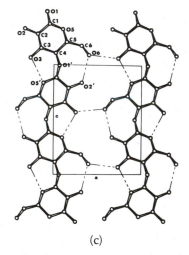

(b)   (c)

Figure 4.2   *Projections of the proposed parallel two-chain model for cellulose. (a) Unit cell viewed perpendicular to the ab plane (along the fibre axis). (b) Perpendicular to the ac plane. Note that the central chain has the same orientation as the four other chains, and is staggered longitudinally with respect to them by an amount equal to b/4. (c) Projection of the (020) plane only, showing the hydrogen bonding network. Each glucose residue forms two intramolecular bonds (03–H...05, and 02–OH...06) and one intermolecular bond (06–H...03) (From K.H. Gardner and J. Blackwell (1974) Biopolymers 13, 1975–2001, by permission of John Wiley and Sons, Inc)*

in the same direction (contrary to certain earlier models). Forty-one reflections were observed in their X-ray diffractions and these were indexed using a monoclinic unit cell having dimension $a = 1.634$ nm, $b = 1.572$ nm, c (cell axis) = 1.038 nm, with $\beta = 97°$; this unit cell therefore comprises a number of whole chains or parts of chains totalling eight in number.

All but three of the forty-one reflections found could be indexed using a simpler two-chain unit cell almost identical to an earlier model (Meyer and Misch, 1937) except that this had adjacent chains running in opposite directions. Because these three reflections were so weak, Gardner and Blackwell take a two-chain unit cell ($a = 0.817$ nm, $b = 0.786$ nm, $c = 1.038$ nm) as an adequate approximation to the real structure. This two-chain model comprising one whole chain and four quarter-chains is illustrated in Figure 4.2, and it should be noted that the centre chain is staggered relative to its neighbours by a distance of $0.266c$ ($= 0.276$ nm) (Figure 4.2b).

Although Gardner and Blackwell's work was carried out on the cellulose of an alga, it is difficult to see why the structure of cellulose in wood should be different: indeed the Gardner and Blackwell model is now widely accepted for cellulose I in timber.

Both primary and secondary bonding is present within this structure of cellulose I. Primary (covalent) bonding is limited in its presence to within the glucose rings and in joining these together to give a long and very strong molecular chain along its length (Figure 4.1). Secondary bonding comprising both fairly strong hydrogen bonds and much weaker van der Waals' forces is present in specific areas: thus hydrogen bonds occur both within the cellulose molecule and between different molecules linking these together in sheets within a single plane (ac or 020 plane, see Figure 4.2c). Van der Waals forces link together these different sheets within the cb plane.

It should be noted that the −OH groups of the cellulose molecules which give rise to hydrogen

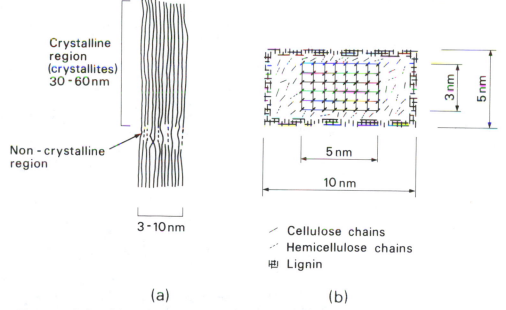

Crystalline region (crystallites) 30 - 60 nm

Non - crystalline region

3 - 10 nm

5 nm

10 nm

3 nm

5 nm

Cellulose chains
Hemicellulose chains
Lignin

(a)

(b)

Figure 4.3 *The microfibril in longitudinal view (a) and in cross-section (b), showing the disposition of the various chemical constituents and the presence of crystalline and non-crystalline regions (illustration (a) by Building Research Establishment, © Crown Copyright, illustration (b) adapted from Preston, 1974)*

bonding are also highly attractive to moisture, an area of considerable technical importance which will be discussed further in later chapters.

Since the length of the cellulose molecule (about 5000 nm) is considerably greater than the length of crystalline regions (average 60 nm), it follows that any one molecule will pass through regions of high crystallinity known as **crystallites** as well as regions of low or no crystalline structure (Figure 4.3a) in which the molecules are only in loose association with each other. However, the majority of molecules emerging from one crystalline will pass into the next, generating a high degree of longitudinal association to form a unit of indefinite length known as the **microfibril** (see section 4.2 and Figure 4.3); the degree of crystallinity along its length will vary therefore, but on average, about 70 per cent of the cellulose present in wood is crystalline.

4.1.2   Hemicelluloses

Between about 25 and 40 per cent of the dry mass of wood comprises a series of hemicelluloses. These, like cellulose, are carbohydrates built up of sugar units, but unlike cellulose in the type of units forming them. Hemicelluloses, therefore, comprise mixtures of polysaccharides manufactured in wood from basic sugars such as mannose, galactose and xylose. The most important result of this composition is that they are generally low in degree of crystallinity and comprise relatively few (150–200) sugar units in each molecule.

The hemicelluloses present in softwoods are markedly different in composition from those in hardwoods and generally the proportion of hemicellulose present is greater in hardwoods compared with softwoods.

4.1.3   Lignin

Lignin is a highly complex non-crystalline molecule comprising a large number of phenyl-propane units. The molecular weight has been determined on extracted material to be as high as 11 000, and it has also been shown to be of differing composition in the hardwoods and softwoods. Unfortunately, because of their size and complexity, the detailed composition of the lignins have not yet been fully resolved.

4.2   The microfibril

The structure of the microfibril in the wood cell wall can be conveniently regarded as a fibre composite, one of nature's fibre composites, and analogous to the man-made fibre composites such as glass reinforced plastic or carbon-fibre reinforced plastic.

As in the case of the man-made fibre composites where we can recognise long, thin fibres embedded usually in a particular pattern in a synthetic resin, so in wood we can identify 'fibre' and 'matrix' constituents. In very broad terms in both the natural and synthetic fibre composites, the 'fibre' constituent is usually regarded as conferring strength, particularly tensile strength, to the composite, while the primary role of the matrix is to confer not only lateral stiffness to the composite, but also to transfer stress from fibre to fibre and to inhibit crack growth, thereby improving toughness.

The 'fibre' constituent in our fibre composite model of the microfibrillar structure is cellulose present in a crystalline state, while the 'matrix' comprises the hemicelluloses, lignin and cellulose present in a non-crystalline state. Once again, many models of the structure of the microfibril have been proposed over the years, but that which has received general acceptance was that proposed by Preston (1974) and this is illustrated in Figure 4.3.

The microfibril which is of indefinite length is believed to be some 10×5 nm in cross-section. Within it there is a crystalline core some 5×3 nm in cross-section comprising 48 molecular chains of cellulose: this crystalline core comprises the 'fibre' in our analogy with man-made fibre composites with its inherent high tensile strength imparted by its covalently bonded cellulose chains.

Surrounding this core is first an area of low crystallinity comprising hemicelluloses and non-crystalline cellulose, to the outside of which is an amorphous layer (non-crystalline) comprising the lignin. The combined layers of hemicellulose and lignin are therefore analogous to the 'matrix' in man-made fibre composites.

Whereas X-ray diffraction studies have indicated that the crystalline core is non-reactive to the presence of moisture, the matrix is certainly reactive to its presence.

4.3 The cell wall

The cell wall is made up of millions of microfibrils and can be conveniently subdivided into a number of different layers dependent upon the arrangement of these microfibrils.

The original wall laid down during cambial division, and referred to in an earlier section as the primary wall, is characterised by being very thin with a random arrangement of microfibrils (see Figure 4.4). The secondary wall which is laid down after cell division can be subdivided into three layers. The outermost, known as the S_1, is thin, comprises less than 10 per cent wall thickness, and has microfibrils lying parallel to one another in two distinct spirals, one right-handed the other left-handed, but both with a pitch of from 50° to 70° to the vertical axis (Figures 4.4 and 4.5).

The middle layer, known as the S_2 layer, comprises some 85 per cent of the wall thickness and has microfibrils lying parallel to each other in a spiral with a pitch to the vertical axis of from 10° to 30° (Figures 4.4 and 4.5). The innermost layer, the S_3, comprises only 1 per cent of the wall

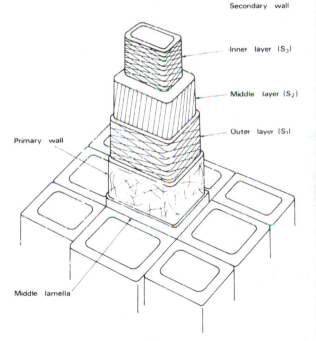

Figure 4.4 *Simplified structure of the cell wall showing the angle of orientation of the microfibrils in each of the major wall layers (Building Research Establishment, © Crown Copyright)*

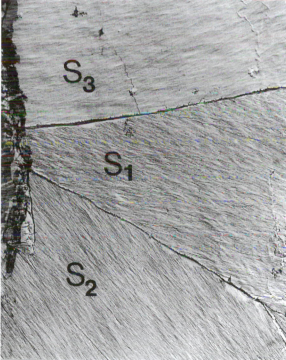

Figure 4.5 *Transmission electron micrograph of the different layers of the cell wall which can be exposed during sample preparation; magnification ×10 000 (Building Research Establishment, © Crown Copyright)*

thickness and has a similar arrangement of microfibrils to the S_1 layer (Figures 4.4 and 4.5). In some species the S_3 layer is overlaid with a thin warty layer (Figure 4.6).

The performance of wood is closely associated with the microfibrillar angle of the S_2 layer, and it is possible to relate much of the variation in strength, stiffness, dimensional instability in the presence of moisture, and fracture morphology to variations in this angle.

Measurement of microfibrillar angle is usually fairly laborious, irrespective as to whether iodide staining, polarisation microscopy, or X-ray diffraction techniques are used; different techniques tend to yield slightly different values (Saiki *et al.*, 1989). Considerable debate occurs on the method of analysing data from X-ray diffractions (see, for example Paakkari and Serimaa, 1984). An approximate value of micro-fibrillar angle in the S_2 layer may be obtained in certain conifers by measuring the angle of the slit in the piceoid semi-bordered pits relative to the principal axis of the tracheid.

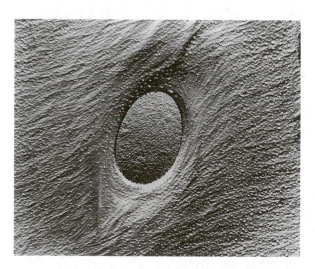

Figure 4.6 *Transmission electron micrograph of the warty layer in Western hemlock. Note how the underlying microfibrils sweep around the aperture to the bordered pit thereby reinforcing the structure; magnification ×2700 (Building Research Establishment, © Crown Copyright)*

4.4   Mineral content

Minerals are present in wood either as an integral part of substances manufactured by the living cells, or as extraneous substances brought up in suspension and deposited in the wood. Examples of the latter, such as silica aggregates in the ray cells of certain tropical species, and calcium carbonate ('stone') deposits within the cavities of cells have already been described in the previous chapter.

Elements such as calcium, sodium, potassium, phosphorus and magnesium are all components of new tissue laid down in the tree: magnesium is required for the manufacture of chlorophyll in the leaves while the other three elements are required in the formation of the nucleus of new cells. The actual mass of these inorganic components is small, ranging on the basis of percentage mass per dry weight of wood from 0.1–1.0 per cent in temperate woods to a maximum of 5 per cent in tropical woods; heartwood will always have lower values than sapwood owing to the absence of living cells in the former (Hillis, 1987).

4.5   Extractives

This term embraces a large number of compounds often of complex chemical structure, which can be extracted from the wood (or bark) with both polar and non-polar solvents. Frequently, the type of extractive is characteristic to a particular family or genus, such as the tropolones in the Cuppressaceae, and pinosylvins in Pinaceae. However, it must be appreciated that different extractives may be present in different cell types in the heartwood of one tree, or that different extractives may be present in the same cell type in different parts of the same tree. Among the more common types are the terpenes and numerous phenolic compounds: a comprehensive list and their location is given by Hillis (1987).

Many of these extractives impart coloration as well as conferring natural durability. As a very general rule, it can be stated that the darker the coloration of the heartwood, the higher will be its natural durability. Natural durability will be further discussed in Chapters 18 and 19.

The heartwood of many timbers has a characteristic odour, which is apparent when these woods are worked in a fairly fresh condition, but which usually disappears as the wood dries out. Perhaps the most outstanding examples are the characteristic resinous odour of the pines, the spicy aroma of sandalwood and Central American cedar (*Guarea cedrata*) and the camphor-like odour of Formosan camphorwood. Certain Australian timbers of the *Acacia* group possess an odour not unlike violets, coachwood is reminiscent of new-mown hay, West Indian satinwood of coconut oil, and Queensland walnut has an objectionable foetid odour that disappears as the wood dries out. The junipers, cypresses and western red cedar have distinctive variants of a spicy nature, though much less strong than that of sandalwood or Central American cedar.

4.6 Acidity

With very rare exceptions, wood is acidic, the level of acidity being considerably higher in the heartwood relative to the sapwood. Like many other properties of wood, the level of acidity varies considerably among the different woods, but values of pH as low as 3.0 have been recorded in the heartwood of oak, sweet chestnut, eucalypt and western red cedar. However, the majority of woods appear to have a heartwood pH of from 4.5 to 5.5. In the sapwood, the pH is generally at least 1.0 pH higher than the corresponding heartwod with pH values usually of the order 5.5 to 6.5, but occasionally as low as 4.0.

Because of its acidity, wood can cause corrosion of metals either by direct contact or, in confined spaces, by the emission of corrosive vapour; the propensity to corrode is greatly increased at high relative humidities. The principal corroding agent is volatile acetic acid which forms part of the structure of wood constituting 2–5 per cent by mass of the dry wood. Hydrolysis to free acetic acid occurs in the presence of moisture at a rate varying from one species of wood to another.

Small quantities of formic and other types of acids are also formed, but their effects can be neglected in comparison with those of acetic acid.

References

Gardner K.H. and Blackwell J. (1974) The structure of native cellulose. *Biopolymers* **13**: 1975–2001.

Hillis W.E. (1987) *Heartwood and Tree Exudates*. Springer-Verlag, Berlin.

Meyer K.H. and Misch L. (1937) Position des atomes dans le nouveau modèle spatial de la cellulose. *Helvetica Chimica Acta* **20**: 232–244.

Paakkari T. and Serimaa R. (1984) A study of the structure of wood cells by x-ray diffraction. *Wood Sci Technol* **18**: 79–85.

Preston R.D. (1974) *The Physical Biology of Plant Cell Walls*. Chapman & Hall, London.

Saiki H, Xu Y. and Fujita M. (1989) The fibrillar orientation and microscopic measurement of the fibril angles in young tracheid walls of sugi (*Cryptomeria japonica*). *Mokuzai Gakkaishi* **35(9)**: 786–792.

5

Variability in Structure

In the utilisation of timber, perhaps the single most important factor detracting from its outstanding performance as a material is its variability. In all applications of timber, whether it is in furniture manufacture or housing construction, large quantities are rejected on the grounds that they are different in appearance or behave differently in machining. It is most unlikely that two pieces of timber are identical in both appearance and performance: within a batch of timber pieces, there will be a wide spectrum in what we could loosely call quality. At some arbitrary point, a line is drawn above which the pieces though variable are acceptable, and below which the pieces are unacceptable for a particular use.

There are very many factors that induce this large degree of variability in timber, but it is convenient to group them into four broad groups under the headings:

- genetical causes
- systematic variability
- environmental reasons
- presence of defects.

5.1 Genetical causes of variability

Although it is widely accepted that variation in colour and strength occurs in timbers from different families, it is often not realised how extensive this variability can be. At the extreme end of the range, a sample of lignum vitae could be ten times as strong as a sample of balsa from the same age of tree. While this is admittedly the extreme that could occur, it is not uncommon among a group of timbers used for a particular purpose to find a range of strengths of two to three times.

Another manifestation of genetics in determining variability is in the comparison on site of

adjacent trees of the same species and same age. Samples taken from the same relative position in each tree could have differences in strength values of between 10 and 50 per cent.

5.2 Systematic sources of variability

Within a tree there are systematic trends in cell length, density, microfibrillar angle and grain angle. Such patterns are present not only in a horizontal, but also a vertical plane.

Looking first at these horizontal patterns in cell behaviour, the length and diameter of cells, and the density of the wood will increase progressively outwards from the pith until they appear to reach a 'maximum' some 12 to 15 rings from the pith, after which further increase will occur but at a very slow rate. Such systematic patterns occur irrespective of rate of growth, though marked changes in growth rate may raise or lower the general slope of the line. The increase in tracheid length with increasing ring number from the pith is illustrated in Figure 5.1.

The pattern of variation for both microfibrillar angle and spiral grain will show the reverse from that for tracheid length, starting high and rapidly decreasing before apparently levelling out; for spiral grain this pattern is discussed in more detail in Chapter 7, section 7.2.1.3 and illustrated in Figure 7.2.

It will be appreciated that near the core of a tree, the cells are short and of low density, while the slope of grain is very high. All these factors contribute to the formation of timber that is both weak and will show a propensity to warp when dried. It will be discussed in a later chapter how this core or **juvenile** wood should be rejected during conversion: unfortunately the laws of economics prevail

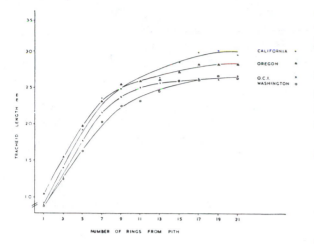

Figure 5.1   *The relationship between the mean length of the springwood tracheids in six trees and the number of rings from the pith in each of the four provenances of Sitka spruce from Radnor Forest (Building Research Establishment, © Crown Copyright)*

over common sense and quantities of juvenile wood appear on site or in factories causing problems if used, and high waste if rejected.

Systematic variation also occurs upward in the tree when rings of the same age after formation are compared at different levels. Thus tracheid length, tracheid diameter and wood density all decrease upward in the tree.

Variability due to systematic sources is additional to that due to genetical reasons.

5.3   Environmental influences

Any environmental factor, whether climatic or silvicultural, as occurs when the forester carries out a thinning, will affect the rate of growth and, through that, the density and length of cells produced: this in turn will influence the strength of the timber.

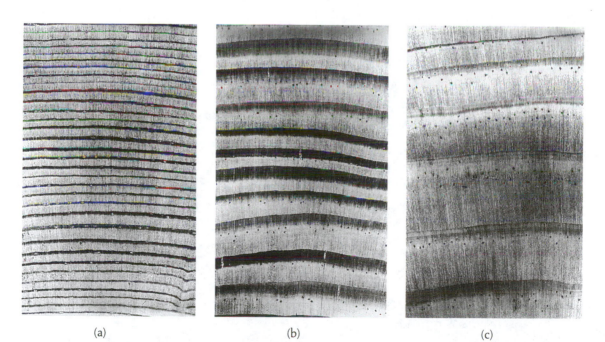

(a)   (b)   (c)

Figure 5.2   *Transverse sections of Scots pine (×3): (a) Very slow grown (44 rings per 25 mm); (b) medium grown (10 rings per 25 mm); (c) fast grown (5 rings per 25 mm) (Building Research Establishment, © Crown Copyright)*

In softwoods, and the ring-porous hardwoods, variations in ring width are associated with variations in the proportion of late wood to early wood (Figures 5.2 and 5.3). In the diffuse-porous woods, however, in which the wood produced in a single growing season is not differentiated into early and late wood, variations in ring width have no effect on cell type distribution. In all three types of wood, extremely narrow (Figures 5.2a and 5.3a) and extremely broad rings are an indication of exceptionally weak timber; probably in all species there is an optimum rate of growth for the production of the strongest timber, but the rate differs with the species.

In softwoods with a ring width between the two extremes, it is found that the amount of late wood remains fairly constant and hence the wider the ring width, the wider is the layer of early wood (Figures 5.2b and 5.2c). Thus the percentage of late wood decreases with increasing ring width and consequently both density and strength will decrease with increasing ring width. Where strength or toughness is involved, softwoods with between 7 and 20 rings per 25 mm should be used.

In ring-porous hardwoods the opposite relationship occurs. Here the width of the early wood remains fairly constant and increasing ring width results in increasing width and hence percentage of late wood (Figures 5.3a and 5.3b). Density, strength and toughness will therefore increase with increasing ring width up to about 4 rings per 25 mm. For most practical purposes, ring porous timbers having from 4 to 14 rings should be used.

The weakness of the very slowly grown softwoods and ring-porous hardwoods (Figures 5.2a and 5.3a) is explained by the very narrow layers of late wood that their rings contain. This is frequently due to the growth of the tree at high latitudes or altitudes where a short growing season restricts the amount of late wood formed. The weakness of rapidly grown softwoods, on the other hand, is explained by the very wide layers of early wood. In hardwoods grown faster than the optimum limits, however,

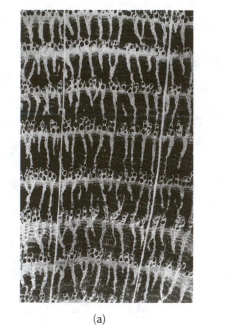

(a)

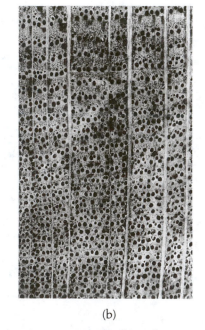

(b)

Figure 5.3   *Transverse section of oak (×4): (a) slow grown (40 rings per 25 mm); (b) medium grown (9 rings per 25 mm) (Building Research Establishment, © Crown Copyright)*

although there is an increase in the proportion of late wood, the individual fibres are abnormally thin-walled, and the timber is, in consequence, weaker than less rapidly grown material in which the late-wood fibres have thicker walls.

Environmental factors influencing variability are additional to variability due to genetical and systematic effects. Thus, as the width of ring tends to decrease towards the bark, the effect of ring width on density tends to be additional to the systematic trend as far as softwoods are concerned, and density increases towards the bark, whereas in the case of ring-porous hardwoods, the effect of decreasing ring width is to reduce density and hence detract from the systematic patterns such that density may decrease slightly in the outer rings. In diffuse-porous hardwoods, ring width has little effect on density, subject to the reservations set out earlier, and the general systematic pattern prevails.

5.4 Variability due to presence of natural defects

Because wood comes from the living tree, it is particularly prone to the development of defects as the tree responds to a wide range of external agencies. While it is the aim in commercial forestry to eliminate this source of variability in wood, in practice this is impossible to achieve owing in part to the vagaries of nature.

Perhaps the natural defect with the greatest impact of the use of wood is the development in the tree of abnormal tissue called **reaction wood**, a special type of tissue formed in leaning trees and in heavy branches. When formed in softwoods it is called **compression wood**, and in hardwoods it is known as **tension wood**.

5.4.1 Compression wood

Compression wood is induced to form in softwoods where there is an uneven distribution in the growth-controlling substance auxin around its circumference. Therefore, wherever a softwood tree is growing out of the vertical,

compression wood will develop on the underside, that is the side of the tree that is actively in compression and where there is an increase in the amount of auxin. Compression wood development therefore occurs in softwood trees that have been partially blown over, in trees on the windward side of exposed plantations where trees assume an inclination in growth due to the force of the prevailing wind, in the lower part of trees planted on a slope where initial growth tends to be perpendicular to the slope before correcting itself to a truly vertical plane, and on the underside of heavy branches where these are allowed to develop as in parkland trees. The extent of development of compression wood is dependent on the degree of disorientation of the dividing cambial cells: consequently, where this occurs in softwoods, the compression wood may be of limited or extensive development. Much softwood timber is found to contain a mild form of compression wood.

Figure 5.4 *Cross-section of softwood log showing marked development of compression wood (the dark, wide-ringed portion shown in the upper part of the section) (Building Research Establishment, © Crown Copyright)*

Compression wood is therefore associated with elliptical cross-sections of softwood trees and in most species it may be recognised by its relatively dark red-brown colour, reflecting the increased proportion of lignin relative to cellulose. When developed only moderately it is usually restricted to only the late wood of each annual ring, thereby giving rise to an increase in colour contrast between the early and late wood. When fully developed, the early wood is also affected and darkens in colour appreciably; it is also no longer possible to distinguish between the early- and late-wood zones (Figure 5.4).

In well-developed compression wood, the cells become circular in cross-section, giving rise to small intercellular spaces (Figure 5.5a). Longitudinal splits or checks arise in the secondary wall running at an angle indicative of the microfibrillar angle of the S_2 layer (Figure 5.5b); this angle is characteristically higher (up to $45°$) in compression wood compared with normal wood.

The principal technical feature of compression wood is its abnormally high longitudinal shrinkage due to the higher microfibrillar angle. Whereas normal wood shrinks 0.1–0.2 per cent longitudinally in drying from the green to the oven-dry condition, compression wood may shrink as much as 5 per cent and commonly 0.3–1 per cent (Figure 5.6). Consequently, boards and planks containing some compression wood are liable to bow in seasoning. The abnormal wood is also exceptionally dense owing to the higher proportion of lignin, but the extra weight is not accompanied by a proportional increase in strength; in particular, compression wood has relatively low bending strength and lacks toughness.

5.4.2   Tension wood

Tension wood develops in hardwoods again from an uneven circumferential distribution of

(a)

(b)

Figure 5.5   *(a) Transverse section of compression wood showing circular cross-section of cells and development of intercellular spaces. Magnification ×250. (b) Longitudinal section of compression wood showing development of checks in the cell wall. Magnification ×250 (Building Research Establishment, © Crown Copyright)*

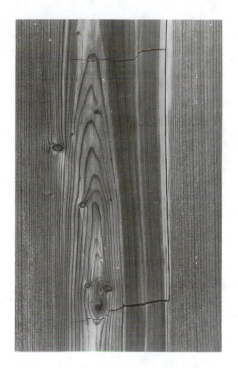

Figure 5.6 *Transverse fractures in Norway spruce due to the abnormal longitudinal shrinkage of compression wood. The photograph shows the darker colour and denser character of compression wood (Building Research Establishment, © Crown Copyright)*

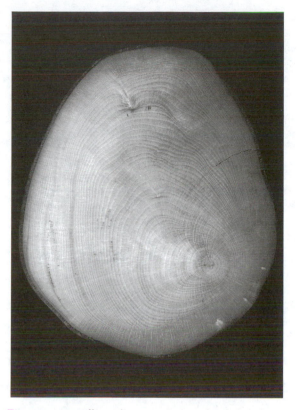

Figure 5.7 *Elliptical cross-section of beech log showing tension wood in upper half*

the growth controlling substance auxin, but unlike compression wood which develops because of an unusual increase in the amount of auxin, tension wood develops on the upper side of leaning trees and heavy branches where the amount of auxin is unusually low. As with softwoods, leaning hardwoods are usually the result of exposure to wind, or the effect of another tree leaning against them. Once again it is possible to have all shades of tension wood from mild to severe development.

The cross-section of the hardwood tree will be elliptical with the longest radius on the upper side of the leaning trunk or heavy branch (Figure 5.7).

Tension wood is paler than normal wood, a reflection of the decreased proportion of lignin to cellulose present; it is more lustrous when viewed by obliquely reflected light.

The increased proportion of cellulose present in tension wood is a manifestation not of increased amounts throughout the wall layers, but of the presence of an additional layer to an otherwise normal cell wall: this extra layer comprises only cellulose and is usually referred to as the gelatinous or **G layer** (Figure 5.8). Tension wood differs technically from normal wood of equal density in being exceptionally weak in compression parallel to the grain. It is, however, slightly stronger in tension and toughness than normal wood of the same density. As with compression wood, tension wood has abnormally high longitudinal shrinkage, probably a reflection of the presence of the G layer; radial shrinkage is normal and tangential shrinkage rather greater than normal. In sawing, planing and turning, the surface of tension wood tends to be woolly since the fibres tend to be pulled out rather than cut cleanly, a manifestation of its higher-than-normal cellulose content (Figure 5.9).

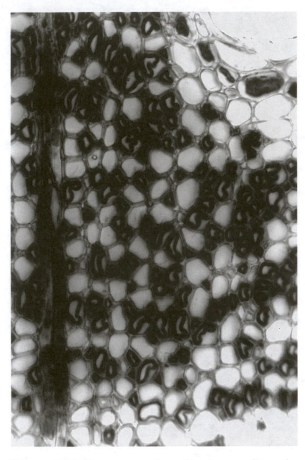

Figure 5.8 *Transverse section of tension wood in agba (×300) showing formation of gelatinous layer (G-layer) in the fibres. In some cells this layer has become detached from the secondary cell wall during section cutting (Building Research Establishment, © Crown Copyright)*

5.4.3   Pith flecks

Patches of abnormal parenchymatous tissue, called **pith flecks**, occur in some timbers, as a result of the tunnelling of the cambium by the larvae of certain insects. Pith flecks are usually wider tangentially than radially, and extend considerable distances vertically; their inner faces follow the outline of the cambial sheath and their outer faces are irregular in outline. They are a common feature of some timbers, such as alder, birch, maple and sycamore.

5.4.4   Bark pockets

Pockets of bark are sometimes enclosed in the wood of the main stem. They result from injury to the cambium. Growth ceases locally until the adjacent cambium has completed the occlusion of the damaged area, resulting in portions of bark becoming embedded in the wood. Such pockets obviously constitute a defect, the seriousness of which depends on the size of the pocket and the extent to which decay may have developed in the vicinity (Figure 5.10).

5.4.5   Resin streaks (or pitch streaks)

Resin streaks or pitch streaks are narrow brown streaks extending along the grain, and fading out gradually, that occur in spruce, Douglas fir and

Figure 5.9   *Board of African mahogany showing rough surface and concentric zones of well-developed tension wood on the end grain (Building Research Establishment, © Crown Copyright)*

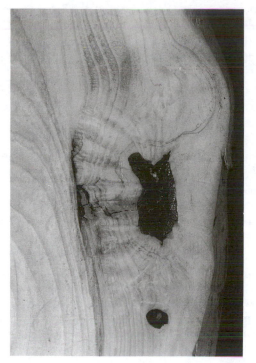

Figure 5.10 *A bark pocket and distorted grain in Norway spruce following wounding (Building Research Establishment, © Crown Copyright)*

Figure 5.11 *Resin streaks in maritime pine (Building Research Establishment, © Crown Copyright)*

other softwoods (Figure 5.11). They are caused by local accumulations of resin in the tracheids. Resin streaks may sometimes be confused with discoloration caused by incipient decay, but they can be distinguished because the darkened wood is not soft or otherwise affected; the strength properties are not influenced in any way.

5.4.6 Pitch pockets, seams or shakes

Serious injury to the cambial cells may result in the formation of **pitch pockets**. These vary in size from about 3 mm to several millimetres wide tangentially, and up to 300 mm or more longitudinally. They are saucer-shaped, with the concave face towards the pith, and up to 2.5 cm or more at their greatest depth radially. They contain liquid resin which flows out readily when the pockets are sawn through. Openings of the wood along the growth

rings may also become more-or-less filled with pitch in a liquid or granulated state: these are known as **pitch seams** or **pitch shakes**.

Pitch pockets, blisters, seams, or shakes are for the most part defects of softwood species, but similar defects also occur in one hardwood family, the *Dipterocarpaceae* (for example lauan, meranti, seraya, keruing and gurjun). In this family, however, defects of the types referred to are much smaller than typical, corresponding defects in softwoods, and their contents are usually solidified dammars or oleo-resins.

A resin pocket in plank of Sitka spruce is illustrated in Figure 5.12.

5.4.7 Gum veins

Gum veins are traumatic canals that occur in some wood; they are usually filled with dark-coloured

Figure 5.12   *A resin pocket in a plank of Sitka spruce: the contained resin has spilled out over the sawn surface (Building Research Establishment, © Crown Copyright)*

deposits. In some timbers, such as jarrah, they are usually infrequent in occurrence, but timber from fire-swept forest may contain gum veins in such considerable numbers that they constitute definite defects; in other timbers, such as African walnut, they are so frequently present as to constitute a characteristic feature of the wood, which may enhance its appearance.

5.4.8   Strawberry mark

Strawberry mark is a minor defect sometimes encountered in Sitka spruce. The discoloration is caused by accumulation of resin and consists of a red-brown zone up to 3 cm wide and 5 cm high, running radially through the wood. In truly radial faces, the defect appears as a bar of darker-coloured wood running across the piece but, if the cut surface is oblique, the discoloration

usually appears more as a blotch. Unless the accumulation of resin is accompanied by enlargement of the rays to many times their normal size, the strength properties of the wood are unaffected and the defect is no more than a minor blemish.

5.4.9   Mineral streaks

Mineral streaks are a type of localised discoloration of timber, in the form of streaks or patches usually darker than the natural colour; these do not impair the strength of the piece. Mineral streaks have been found in rock maple, sycamore and wych-elm. The term has also been applied somewhat incorrectly to the light-coloured streaks occurring in timbers of the family *Dipterocarpaceae*. These streaks are really the resin canals in longitudinal section, which, because of their white or yellow contents, show up against the red or brown background of the wood.

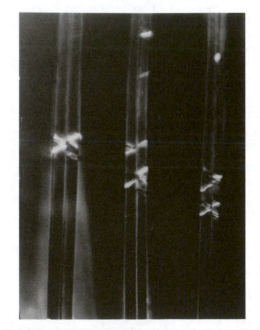

Figure 5.13   *Kinks (slip-planes) formed in the cell wall during longitudinal compressive stressing: observed under polarised light (magnification ×800) (Building Research Establishment, © Crown Copyright)*

5.4.10 Compression creases, kinks and natural compression failures

Compression creases are horizontal series of **kinks** in the cell wall induced by localised high levels of compressive stress (Dinwoodie, 1968). These kinks in the cell wall are frequently referred to as slip-planes though they are not comparable in structure to the crystallographic slip-plane. In timber, the microfibrils are deformed in the form of a 'Z' and as a result of the change in crystal orientation in the centre of the Z it is possible to see these kinks quite clearly under polarised light (Figure 5.13). A group of creases lying parallel to one another and in close proximity is usually referred to as a **compression failure**.

High localised stress can be induced in a tree as a result of severe buffeting in gale-force winds, or when a tree is felled across some obstacle on the ground. In the former case, subsequent growth layers are formed over the crease and the only outward appearance is a slight swelling of the stem due to the development of some traumatic

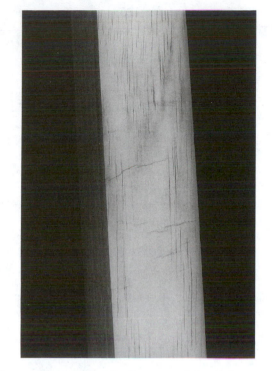

Figure 5.14   *Natural compression failures showing up on a planed surface illuminated with light at a low angle (Building Research Establishment, © Crown Copyright)*

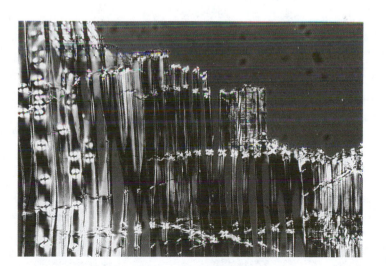

Figure 5.15 *Failure of timber containing compression creases: note how the fracture has run transversely along the creases: polarised light (magnification ×100) (Building Research Establishment, © Crown Copyright)*

tissue. It should be noted that both these sources of natural compression failures can occur in any species and are to be found towards the outer part of the stem, unlike the smaller creases associated with brittleheart (described below) which occur in the centre of trees of only certain species. Natural compression failures were formerly called 'thunder-shakes'.

The natural compression failure can usually be seen on planed surfaces, especially when illuminated by light at a low angle to the board surface (Figure 5.14); however, on sawn surfaces they are almost impossible to see and frequently timber containing creases is used in service, often leading to fatal accidents, such as in scaffold boards. Although the tensile and bending strengths are reduced only slightly, the impact resistance (toughness) is reduced appreciably, frequently by as much as a half (Dinwoodie, 1978). Failure will occur along the line of the natural compression failure, producing a very 'short' or brittle-like fracture surface (Figure 5.15). (See also Chapter 18, section 18.7.)

Figure 5.16 *Creases in a piece of brittleheart: note fairly regular distribution over an extensive area (Building Research Establishment, © Crown Copyright)*

5.4.11   Brittleheart

When a tree is growing, the new cells formed from the cambium (see Chapter 2) are laid down under tension and this induces a compressive stress on the wood in the centre of the tree. This effect is cumulative, and in very large trees high compressive stresses occur; in many species, the natural compression strength of the wood is high enough to withstand such stresses, but in certain low-density hardwoods, such as meranti, African walnut and agba, this is not the case, and compression failure occurs in the centre of the tree, extending upwards and outwards for about one-third of the way. This zone is known as **brittleheart**, though terms such as 'spongy' or 'punky-heart' have also been used; further information on the development of brittleheart is to be found in Chapter 12, section 12.1.2.

The kinks formed in the cell wall are identical to those described for natural compression failure, but the creases tend to be finer in width and distributed more regularly in the timber throughout the central region (Figure 5.16); thus on the basis of size, distribution and location, it is possible to say whether the compression kinks are part of a natural compression failure or brittleheart.

All the strength properties of brittleheart are lower than those of normal wood; this is more a reflection of the very low density of core wood than the development of creases as such, with the possible exception of impact resistance. The very marked reduction in this property can be related primarily to the development of creases, though lower density is also a contributory factor.

Failure in brittleheart occurs preferentially along the line of the creases, giving rise to a very short and brittle-like fracture (Figure 5.16).

References

Dinwoodie J.M. (1968) Failure in timber. Part 1. Microscopic changes in cell-wall structure associated with compression failure. *J. Inst Wood Sci* **21**: 37–53.

Dinwodie J.M. (1978) Failure in timber. Part 3. The effect of longitudinal compression on some mechanical properties. *Wood Sci & Technol* **12**: 271–285.

6

Identification of Timbers

The identification of timbers may, at first sight, appear to be a comparatively simple matter; however, when it is realised that there are tens of thousands of woody species in the world, it will be appreciated that in some cases correct identification may be exceedingly difficult. Actually it is not always possible to arrive at the correct specific name from the examination of a single sample of wood, although it is usually possible to narrow down the identification to a group of related species, and this may be sufficient for most practical purposes. Moreover, although there are so many species that produce woody stems, only a small proportion grow to timber size. Even so, the number of species producing commercial timber runs into several hundreds. The characters available for distinguishing woods are not numerous, and identifications should be based on an examination of features that are known to be reliable, rather than on the more obvious characters, such as colour and weight, that tend to be far from consistent.

6.1   Use of gross characteristics

The average timber-user handles relatively few timbers and can usually recognise those with which he is familiar by a fairly cursory glance combined with an appreciation of the mass of the block in his hand; he is not, however, in a position to name timbers with which he is not familiar, not only because he is not aware of their gross characteristics, but also because the larger the range of timbers, the greater the probability that the same combination of gross features can apply to more than one timber.

Thus there is a naive belief that identification of timber is relatively simple and can be achieved by assessing:

- the colour of the sample
- the mass, and hence for its size an estimate of density
- whether or not the wood has a distinctive smell
- whether it is a softwood or hardwood by the absence or presence of vessels
- the texture of the wood as revealed by the size of the various cells on the planed surface (see Chapter 7)
- the presence of very conspicuous rays (see Chapter 7).

Such an approach has its place where interest is confined to a small number of timbers, or where for example the question is raised as to whether or not a sample of timber is oak or not. However, in view of the magnitude of variability that can occur in timber, as was described in considerable detail in Chapter 5, it will be appreciated that the very arbitrary mode of timber identification described above will lead to error where large numbers of timbers are involved, and, more importantly, will not stand up to the rigours of legal cross-examination.

6.2   Identification of hardwoods using a hand lens

A much higher degree of assurance can be given to identification of hardwoods made on the basis of cellular structure assessed with the use of a ×10 hand lens. Such a method however, is not appropriate to softwoods where identification can be achieved only with the use of a microscope.

In passing, it should be appreciated that whereas it is usual for the magnifications of microscopes and textbook illustrations to be expressed in terms of *linear* magnifications, pocket lenses are usually inscribed in terms of area magnification. Thus a ×10 hand lens has a linear magnification of √10, or just over 3. For examination of vessel size and distribution, therefore, a ×10 lens is the minimum size that should be used.

Before using the hand lens, the surface of the block to be identified must be cut cleanly with a new razor blade to produce a smooth, flat and clean surface devoid of cell damage in each of the three principal planes (cross-section, longitudinal-radial and longitudinal-tangential – see Chapter 2). In preparing the cross-section, the cut should be made radially in a direction from bark towards the pith. In each of the three surfaces, the razor cut must be made in a single stroke and consequently the prepared surfaces should rarely exceed 5 mm × 5 mm.

In every case, the first point to decide is whether or not vessels are present: this may not be as easy as it appears, since the resin canals in a softwood can be easily mistaken for solitary vessels in a hardwood at this level of magnification. Next to decide is the type of growth rings (or their absence), the size and arrangement of the vessels, and then the distribution of the wood parenchyma relative to the vessels (see Chapter 3). Ray type and size are also important in identification. In addition to these cellular features, recourse should be made to such physical features as colour, mass and smell, but in view of what has been said previously on variability, such physical features should merely be used to confirm an identification already made on the cellular characteristics.

6.3 Identification using a microscope

6.3.1 *Slide preparation*

The method described below is taken from part of the Building Research Advisory Service Timber Information Leaflet No. 52 (1974) which has been withdrawn, as reference is made in it to certain chemicals now deemed to be hazardous.

Selection of material

The material for microscopic examination should be chosen as representative of the timber; this may best be done after inspection of a cleanly cut transverse surface of the timber. For sectioning, a block of convenient size (13 mm cube) should then be cut out with sides truly transverse, tangential and radial. It should be noted that wood within twenty rings or so of the pith shows variation in structure corresponding with a period of 'adolescence' in the growth of the tree, and accordingly, if possible, such wood should be avoided if it is desired to examine only the typical adult structure. The inclusion of both sapwood and heartwood in the same block is to be avoided since it may prove impracticable to achieve uniform softening.

When several blocks are being prepared together, each should be distinctively marked, for example, by removing one or more corners, or by notching; a record should be kept of these markings.

Softening and other treatments preparatory to sectioning

Freshly cut green wood, especially sapwood, can sometimes be sectioned without any special treatment; the surfaces of the block should be kept wet with water or alcohol. Seasoned timber generally requires some kind of softening treatment to bring it into a suitable condition for sectioning. The degree of softening required depends on the kind of timber and can best be decided by experience.

The simplest method, which is generally suitable for fairly soft timbers such as spruce and poplar, is to immerse the block in a vessel of gently boiling water until it becomes waterlogged and sinks. (The waterlogging may be hastened by repeatedly removing the block from the boiling water and dipping it in cold water.) Heat plays an important part in the softening process; the block should be taken from the water and sectioned while it is still hot. A longer period of treatment in boiling water will soften the wood still further.

An alternative method of waterlogging, which is to be preferred for fragile material, is to place the block in water in a vessel connected with a vacuum pump and exhaust the air. This is repeated several times, if necessary, until the block is thoroughly waterlogged. The wood can then be softened, if necessary, by one of the methods described below.

Some moderately hard woods can be softened satisfactorily by soaking in a mixture of equal parts of glycerine and alcohol (methylated spirit is suitable). After waterlogging, the blocks should be transferred direct to the mixture, and should remain immersed for about three days. This treatment is usually adequate even for fairly hard woods such as Canadian yellow birch or oak.

When blocks intended for sectioning have to be stored for any length of time, they may conveniently be given a slow softening treatment by immersion in a glycerine and alcohol mixture containing not more than 10 per cent of glycerine.

Section cutting

Apparatus. With practice, satisfactory sections can be cut freehand, but large sections should not be attempted by this method (Hoadley, 1990). The block may be held in the hand, or in a vice or clamp, thus leaving both hands free to manipulate the razor.

Several inexpensive forms of hand microtome are available; they consist essentially of a vice to hold the block, a screw adjustment to raise the block small distances, and a glass or metal surface to act as guide for the razor or microtome knife.

For the preparation of first-class sections of large size and uniform thickness, a microtome of the heavy 'sledge' type is necessary (see Figure 6.1), fitted with special wedge-shaped knives, 160 mm in length (Wilson and White, 1986).

The microtome knife and its sharpening. Good sections devoid of tearing to the cell walls are possible only with really sharp knives with the correct bevel edge. The most consistent quality of cutting edge is obtained by using an automated knife-sharpening machine which holds the knife against an oscillating and rotating plate to which is added a small quantity of diamond paste and lubricating fluid.

Where an automated knife-sharpening machine is not available, it is possible to sharpen the knives

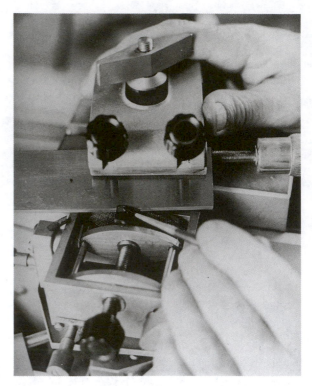

Figure 6.1   *Cutting the section of wood on a 'sledge'-type microtome; note the use of the brush to lift off the section from the knife; this may also help to prevent the section from curling (Building Research Establishment, © Crown Copyright)*

by hand, though it will take much practice before a high quality of edge is consistently produced. A piece of quarter-plate glass 300 × 160 × 6 mm is required, and this should sit in a metal tray. Each knife must have its own marked back bevel which is fitted to it prior to sharpening, together with a handle which should be thin in order to allow it to pass over the glass plate. The knife edge has to be well bedded with the glass plate before first-class results can be obtained, and this state should be achieved after the first few sharpenings.

A small quantity of grinding powder is shaken on to the glass plate and made into a slurry with 20 per cent alcohol. A metal tray to catch any overflow is advised. Aluminium oxide optical smoothing powder No. 95, supplied under the trade name 'Aloxite', has been found satisfactory for this purpose. The knife is pushed forward with even pressure at an angle of about 45° with the cutting edge leading, turned over when at the top of the plate, and brought back down the plate at 90° to the upward push. At the bottom, the knife is again turned over and the cycle started again. The glass plate should be kept wet throughout by adding more alcohol from time to time, but it should be unnecessary to add more grinding powder. About 5 minutes' grinding is normally required to restore the edge of a knife that has become dulled (but not nicked or otherwise seriously damaged). The same glass plate should be used for the final polishing, after washing it and also the tray, knife and bevel in running water, until all traces of the grinding powder have been removed. After drying, the glass plate should be placed in exactly the same position for polishing as for grinding, and a mark scribed in one corner on the underside of the plate helps to ensure this. For polishing, a small quantity of liquid silver polish is poured on the glass plate and the knife movements carried out as for grinding, the process being completed in less than 5 minutes. No stropping is required. The glass plate is rewetted as necessary (with polishing liquid, not alcohol).

General method of sectioning. The block is clamped in the microtome so that the softest tissues (either rays or sizeable bands of parenchyma) are parallel to the travel of the knife. For hardwoods, the knife should meet the late wood of each ring first, and for softwoods the early wood. Any preliminary trimming of the surface of the block should be done with the heel of the knife and the remainder of the edge reserved solely for section cutting. The knife is adjusted so that it makes an angle of 10° with the surface of the block in the vertical plane, while the blade makes an angle of 10° with the line of motion. Where it is essential to avoid mechanical damage to the fine structure of the cell walls, this angle should be reduced to 5° or less (Dinwoodie, 1966).

To commence sectioning, the knife, flooded with alcohol, is moved in its slide with a steady, firm movement of the right hand, while a brush held in the left hand is placed gently on the block to prevent the section curling (see Figure 6.1); no attempt should be made to push the section on to the knife, as this may cause tearing. By means of the brush, the section is then carefully transferred to a dish of alcohol. Each time the knife is moved forward over the block, a section is cut and, on the return stroke, the block is automatically raised by a predetermined amount (say 0.01 mm or 10 microns) for the next cut. The strong tendency to curl shown by sections of some woods can often be overcome by wetting the knife with a mixture of equal parts of glycerine and alcohol, and allowing each section to remain on the blade for a few seconds after cutting; the section is then transferred to a dish containing the same mixture. Another method which is often effective in flattening a tightly curled section is to draw it on to a section-lifter and press it with a needle previously heated in a flame.

Sectioning of plywoods and veneers. Small specimens such as portions of plywood and veneers may be sectioned, after softening in the usual way, by holding them between pieces of wood of similar hardness. Transverse and radial sections of rotary-cut veneers can usually be prepared in this way; to obtain tangential sections (parallel to the face of the veneer), a strip of the veneer is bent round a small block of wood so that the material may be securely clamped in the microtome with a tangential surface uppermost.

Staining

The general procedure is as follows – for the chemicals listed below, the manufacturers' safety instructions must be followed:

1. Cover the sections with 1 per cent aqueous Safranine for 5 minutes.
2. Drain off the Safranine and wash with at least three changes of distilled water.
3. Wash twice with 97 per cent alcohol.
4. Cover the sections with 1 per cent Fast Green in clove oil and alcohol (1:3) for 2 minutes.
5. Drain off the Fast Green and wash with at least three changes of 97 per cent alcohol. (If absolute alcohol is obtainable, one or two washes will suffice.)

6. Transfer the sections to clove oil for 5 minutes and then to cedarwood oil for 1 minute; mount in Canada balsam. (If absolute alcohol is used in stage 5, xylol may safely be used here in place of the cedarwood oil.)

This combination of stains differentiates between lignified (stained red) and unlignified (stained green) structures, and has superseded the older safranine and haematoxylin combination.

For the detection of fungal hyphae in sections of wood, the following procedure is recommended:

1. Place a longitudinal section on a slide and stain in 1 per cent aqueous Safranine for about one minute, or rather less.
2. Wash with distilled water, leaving the section slightly overstained.
3. Stain with Picro-aniline Blue (prepared by mixing saturated aqueous solutions of aniline blue (1 part) and picric acid (4 parts), and warm over a flame until just on the point of simmering). Picric acid is both toxic and explosive, and appropriate safety precautions must be taken.
4. Wash with distilled water.
5. Wash in two changes each of 70 per cent and 97 per cent alcohol.
6. Clear the section in clove oil, wash in cedarwood oil and mount in Canada balsam.

(The alternative procedure indicated under stage 6 of the preceding staining method may also be substituted here.)

In the resulting preparation, the lignified walls will be stained red and any fungus mycelium will be stained a clear blue.

Mounting

(a) *Canada balsam mounts.* These are permanent mounts and are prepared as follows. Transverse, tangential and radial sections are transferred from cedarwood oil or xylol and placed in this order on a clean 75 × 25 mm glass slide. Two or three drops of Canada balsam dissolved in xylene and of a treacle-like consistency are then placed on the sections, and a warm coverslip of appropriate size (such as 40 by

20 mm) is lowered on to them. Xylene is classed as 'harmful' and appropriate safety precautions must be taken. The slide is gently heated until bubbles appear, and pressure applied to squeeze out surplus mountant. Unless the slide is likely to be required for critical examination at high magnifications or for photomicrography, the preparation may be completed at this stage merely by cleaning away surplus balsam with a cloth dipped in xylene. Before being stored, the slides should be left lying flat for several days until the balsam becomes quite hard. For critical work, where it is essential that the sections should lie perfectly flat in a balsam layer of uniform thickness, greater care in drying and pressing is necessary. Thus, after the initial pressing, the preparation is left for about a fortnight to dry. Then a piece of blotting paper is placed over the coverslip and a thin piece of wood clamped on top by means of two spring clothes pegs or a 'bulldog' clip. The slides are kept in an oven or over a hot radiator at 60°C for 12 hours, and then allowed to cool. When hard, the surplus balsam is carefully scraped away and the edge of the coverslip brushed with xylene. The slide is finally polished with alcohol and labelled.

(b) *Glycerine jelly mounts.* Temporary mounts may be made in glycerine jelly. If the sections have not previously been stained, a little 1 per cent aqueous Safranine (or other aqueous liquid stain) may be added and thoroughly mixed with the warm molten jelly. Sections are placed in line on the slide, a few drops of molten jelly added, a warm coverslip placed gently over the sections, and the slide heated until bubbles appear; pressure is then applied until the jelly has cooled and set. Sections containing alcohol should be well washed in water before mounting. Such mounts can be made more or less permanent by sealing the cover with gold size of Brunswick black.

6.3.2   *Microscopic examination*

It is necessary for the microscope, whether monocular or binocular, to be equipped with × 10 eyepiece(s) and two objectives with magnification of about × 10 and × 40.

Initial study of the cross-section will reveal whether the timber is a hardwood or softwood.

If the former, much of the subsequent examination can probably be carried out using the objective lens of the lower magnification. The features to be examined are again those which were set out in the previous section (6.2).

However, when dealing with the identification of softwoods, recourse will have to be made to the objective lens of higher magnification; this is especially the case in deciding whether ray tracheids are present, and, if so, the type of wall thickening they have; the type of semi-bordered cross-field pitting present; and whether the torus has a scalloped border (see Chapter 3). The lower powered lens can be used to determine whether or not resin canals are present, and, if so, whether or not the epithelial cell walls are thickened; the arrangement of the bordered pits along the radial wall of the tracheid; and whether helical thickening of the tracheid walls is present (see Chapter 3).

6.4 Keys to assist identification

6.4.1 Dichotomous keys

The commonest form of key in general use in botanical and entomological work is the **dichotomous key**, whereby successive pairs of mutually exclusive conditions are so arranged that, by a process of elimination, one is led step-by-step to the identity of the specimen. Such keys are suitable for a restricted number of timbers, plants, or insects: they become unmanageable if their construction is attempted for too many different individuals, because it is frequently necessary to use features that are subject to considerable variation within a single species, and it becomes increasingly difficult to find pairs of characteristics that are mutually exclusive. In some circumstances, it is often necessary to include the same timber, plant, or insect in more than one section of a key to ensure covering varia-

tion within a species, or the lack of really satisfactory, mutually exclusive, characteristics.

Given the above limitations of a dichotomous key, it is still possible to prepare a useful and reliable key for a limited number of timbers, and an example of one is given below. This has been constructed using the information set out in Table 6.1, following examination of the cellular properties using a hand lens as described in section 6.2. In using the key, the operator works progressively from the top, selecting at each stage the appropriate description from the alternatives provided (Wilson and White, 1986).

1	Wood ring-porous	2
1	Wood not ring-porous	5
2	Pore clusters present (parenchyma absent or indistinct)	Elm
2	Pore clusters absent	3
3	Rays wider than vessels (parenchyma in fine lines)	Oak
3	Rays not wider than vessels	4
4	Wood white (no odour)	Ash
4	Wood brown (distinctive odour)	Teak
5	Vessels exclusively solitary	6
5	Vessels not exclusively solitary	7
6	Normal vertical canals present	Gurjun
6	Normal vertical canals absent	Opepe
7	Scalariform perforation plates distinct	American whitewood
7	Scalariform perforation plates indistinct or absent	8
8	Ripple marks distinct	9
8	Ripple marks indistinct or absent	10
9	Confluent parenchyma present	Sapele
9	Confluent parenchyma absent	Central American mahogany
10	Vessels in radial groups	11
10	Vessels not in radial groups	12
11	Wood red-brown	Makore
11	Wood yellow or black, or streaked brown-black	Ebony

12	Normal vertical canals present	13
12	Normal vertical canals absent	14
13	Canals in short tangential series	Gurjun
13	Canals not in short tangential series	Meranti
14	Tissue other than rays storeyed	Obeche
14	Tissue other than rays not storeyed	15
15	Parenchyma in broad conspicuous bands	Ekki
15	Parenchyma not in broad conspicuous bands	16
16	Parenchyma distinct to naked eye	17
16	Parenchyma not distinct to naked eye	20
17	Wood white or yellow, terminal parenchyma indistinct	Idigbo
17	Wood not white, terminal parenchyma prominent	18
18	Parenchyma apparently terminal only	Central American mahogany
18	Parenchyma not apparently terminal only	19
19	Confluent parenchyma present	Iroko
19	Confluent parenchyma absent	Dahoma
20	Wood walnut-brown	African walnut
20	Wood not walnut-brown	21
21	Wood pink-brown to red-brown, rays just visible to naked eye	African mahogany
21	Wood light pink, rays distinct only with lens	Gaboon

Timbers are so numerous, and the differences between many are so small, that it is impossible to construct a workable dichotomous key to embrace all the timbers in the world. Several good keys exist that are restricted to the timbers of particular countries or localities. All dichotomous keys have the disadvantage that additional timbers cannot be included in the key without reconstruction of large sections, if not the greater part, of the key.

6.4.2 *Multiple entry keys*

The Forest Products Research Laboratory, now the Building Research Establishment, adapted the **Paramount sorting** system to timber identification using the **multiple entry key**. Special cards, patented by Messrs Copeland–Chatterson Co, Ltd, containing punched holes along their four sides, are employed; each hole is used for one feature (Figure 6.2). Every timber to be included in the key requires a separate card and, if certain features are variable in different samples of the same timber, two or more cards may be necessary, exceptionally as many as eight cards. The card is completed for each timber by clipping away the edges of the punched holes to give V-shaped slots where the features are present. When all the cards are prepared, they are sorted so that all are arranged the same way round and the key is ready for use. To facilitate rapid sorting, it is convenient to cut one corner of each card on the splay, for example the top right-hand corner in the cards patented by Messrs Copeland–Chatterson.

The Forest Products Research Laboratory prepared the data for a set of cards for more than 400 commercial hardwoods, and this information has been published as FPR Bulletin No. 25 (Anon, 1960). Similar data for microscopic features of hardwoods have been published as FPR Bulletin No. 46 (Brazier and Franklin, 1961), while the data for the microscopic features of softwood have been published as FPR Bulletin No. 22 (Phillips, 1960). Blank cards for all these keys were available from HM Stationery Office. The definitions of the different anatomical features given in the Bulletins should be studied, as certain features are used in a slightly different sense from the definitions given in Chapter 3. A set of photomicrographs of the majority of the timbers figuring in these keys, originally published by HMSO (*Photomicrographs of World Woods* by Anne Miles), is now available from the BRE Bookshop.

To identify a timber, any feature present in the specimen is selected and a steel needle threaded through the punched hole for the selected feature. The cards for those timbers in which the

Table 6.1   *Tabulated list of hardwood features visible to the naked eye or with the aid of a hand lens (the number of features used may, of course, be considerably enlarged)*

Name of timber	Vessels							Parenchyma											Other features		Rays					Physical properties		
	Exclusively solitary	Radial multiples	Pore clusters	Scalariform perforations	Distinct to naked eye	Distinct only with lens	Barely visible with lens	Absent or indistinct	Distinct to naked eye	Terminal prominent	Apparently terminal only	Vasicentric	Aliform	Confluent	Broad bands	Broad conspicuous bands	Fine lines	Reticulate	Normal vertical canals	Ring porous	Storeyed (not rays)	> ½ width of vessels	Wider than vessels	Storeyed	Aggregate rays	White	Yellow or brown	Red or purple
Ash	−	−	−	−	+	−	−	−	−	+	−	+	±	±	−	−	−	−	−	+	−	−	−	−	−	+	−	−
Dahoma	−	−	−	−	+	−	−	−	+	+	−	+	+	+	−	−	−	−	−	−	−	−	−	−	−	−	+	−
Ebony	−	+	−	−	−	+	−	−	−	+	−	−	−	+	−	−	+	−	−	−	−	−	−	−	−	−	+	+
Ekki	−	−	−	−	+	−	−	−	+	+	−	−	−	+	−	−	+	+	−	−	−	−	−	−	−	−	+	+
Elm	−	−	+	−	+	−	−	+	−	+	−	−	−	−	−	−	−	−	−	+	−	−	−	−	−	−	+	−
Gaboon	−	−	−	+	+	−	−	±	−	−	−	+	−	−	−	−	−	−	−	−	−	−	−	−	−	+	−	−
Gurjun	+	−	−	−	+	−	−	−	−	−	−	×	−	−	−	−	−	−	+	−	−	−	−	−	−	+	−	−
Idigbo	−	−	−	−	+	−	−	−	+	×	−	+	+	+	−	−	−	−	−	−	−	−	−	−	−	+	+	−
Iroko	−	−	−	−	+	−	−	−	+	+	−	+	+	+	−	−	−	−	−	−	−	−	−	−	−	+	+	−
Mahogany—																												
Central American	−	−	−	−	+	−	−	−	+	+	−	×	−	−	−	−	−	−	−	−	−	−	±	−	−	−	−	+
African	−	−	−	−	+	−	−	+	−	+	−	×	−	−	−	−	−	−	−	−	−	−	−	−	−	−	−	+
Makoré	−	+	−	−	−	+	−	−	−	−	−	−	−	−	−	−	+	−	−	−	−	−	−	−	−	−	−	+
Meranti	−	−	−	−	+	−	−	−	−	+	−	+	±	±	−	−	−	−	−	+	−	−	−	−	−	−	+	−
Obeche	−	−	−	−	+	−	−	−	+	−	−	+	−	−	−	−	−	−	−	+	−	−	−	−	−	+	+	−
Opepe	+	−	−	−	+	−	−	+	−	−	−	+	−	−	−	−	−	−	−	+	−	+	+	−	−	−	+	−
Oak	−	−	−	+	−	−	−	−	−	−	−	−	−	−	−	−	+	±	−	+	−	+	+	−	−	±	±	
Sapele	−	−	−	−	+	−	−	+	+	−	−	+	+	+	±	−	−	−	−	−	−	+	−	−	−	−	+	−
Teak	−	−	−	−	+	−	−	−	±	−	×	−	+	−	−	−	−	−	+	−	−	−	−	−	−	−	+	−
Walnut—African	−	−	−	−	+	−	−	±	−	−	−	+	−	−	−	−	−	−	−	−	−	−	−	−	−	−	+	−
Whitewood—American	−	−	+	−	−	−	+	−	−	+	+	−	−	−	−	−	−	−	−	−	−	−	−	−	−	+	−	+

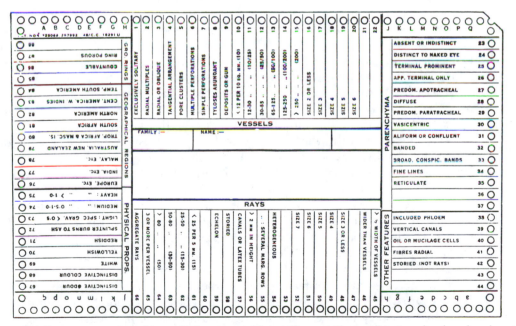

Figure 6.2   *A Paramount card used in multiple entry card keys. The card illustrated is for a key based on lens features (courtesy of Messrs Copeland–Chatterson Co. Ltd)*

selected feature occurs drop out as the needle is shaken. These cards are again sorted so that they are all the same way round, a second feature is selected, the needle threaded, and the shaking process repeated. The cards that drop out on the second occasion are again sorted and the process continued with different features in turn until only one or a few cards remain: *provided a card for the specimen to be identified has been prepared*, it will be among those finally eliminated. With some timbers, the card-key system results in the elimination of all cards but one, although more often a selection has to be made from two or three cards; correct identification can usually be arrived at by matching the unknown timber with authenticated specimens of the few alternative choices. Indeed, a final decision on a timber's identity, irrespective of whether one or more cards are left, should always be confirmed by matching of the unknown block with authenticated samples. In working with a card-key of this type, three precautions must be observed: (1) in shaking the pack, care must be taken to ensure that all the cards free to drop out do, in fact, drop out; (2) care must be taken that the correct pack is used after each sorting, that is if a feature is being used positively, the cards that drop out are used for the next operation, and vice versa; and (3) the cards must be kept in good condition, that is should any of the punched holes become torn, a new card must be made out.

The two special advantages of the card-key system are: (1) the simplicity with which new timbers can be added to the key (all that is required is an additional card); and (2) any sequence can be adopted that promises speedy identification.

It is usual to confine keys to features visible only with the aid of a microscope, or to features visible to the naked eye or with the aid of a low-power hand lens, but there is no reason why a key should not combine both classes of features, although such keys can, of course, only be used in the laboratory. In a lens key it is desirable to record features as they appear, whether or not the observations are in accord with the facts. For

example, the line of tissue bordering a growth ring in beech is not terminal parenchyma but it is convenient to record it as such because, with a lens, it is rather difficult to establish that it is not true terminal parenchyma. Actually, the line consists of a few rows of radially flattened fibres, wanting in diffuse parenchyma; this zone contrasts with the remainder of the background comprising fibre and diffuse parenchyma to give the effect of a line of distinctive tissue.

The successful use of keys necessitates some experience in the examination of small samples of wood, and this can only be obtained by practice. Readers who are anxious to be in a position to identify only the few common timbers in everyday use would be well advised to make their own keys from a study of a collection of authenticated samples of timbers. As a preliminary to a concerted attack on the problem of identification, there is no better method than the preparation of scale drawings of the end surfaces of different woods. The procedure is to prepare a clean-cut portion of the end surface, and to mark on this a square with 1 cm sides. Next, a sheet of paper with a square of 5 cm sides is required. The details visible in the marked square can then be transferred to the paper, more or less to correct scale. It is usually helpful to commence with the rays, and then to draw in the vessels, and subsequently the other details. The method is admittedly laborious, but the preparation of thirty or forty drawings fixes the distinctive patterns of the woods in the mind, and is of great help in mastering the technique of timber identification.

In attempting to identify timber, it is just not possible always to determine the specific identity of even a common commercial timber from extremely small fragments, such as sawdust. For example, for certain timbers it is essential to have transverse, radial and tangential sections from the same fragment to be in a position to establish the identity of the fragment, and no degree of expertise or of elaborate laboratory equipment will surmount the problem of too small sections of these different surfaces being available for purposes of identification.

6.4.3 *Computer-aided wood identification*

The use of multiple entry cards using a needle to make a physical selection of cards was described in the previous section. The data from these perforated cards can be easily computerised, and sorting by computer has many advantages over mechanical card sorting. Not only is the process much quicker by computer, but it also imparts a very important element of flexibility where a specified number of mismatches in feature descriptors between an unknown sample and the database can be permitted. This is important in dealing with such variable material as wood, but cannot be achieved when using mechanical sorting.

One of the few databases that is available with computerised sorting is that developed by the North Carolina State University (Wheeler *et al.*, 1986) using data from existing multiple card systems originated by Princes Risborough Laboratory (formerly the Forest Products Research Laboratory and now the Building Research Establishment), the Oxford Forestry Institute and the Centre Technique Forestier Tropical (CTFT) of France.

6.5 Cross-checking of identification

Having identified the sample of wood by hand lens or microscope using one of the keys, this identity must be checked against:

(1) a list of anatomical features recorded for that wood, or a published set of micrographs to see if these agree with the features found by examination;
(2) a description of the colour, texture and density of the wood, to see if these features agree with those of the sample.

Anatomical descriptions of woods are to be found in Hoadley (1990), Wilson and White (1986), and Bulletins 22, 25 and 46 of the former Princes Risborough or Forestry Products

Research Laboratory. Photomicrographs of the end-grain of over 450 woods were published originally by HMSO and are now available from the BRE Bookshop (Miles, 1978).

Detailed descriptions of the physical properties of various wood can be found in Hoadley (1990), the *Handbook of Softwoods* (Anon, 1977) and the *Handbook of Hardwoods* (Farmer, 1972), originally published by HMSO and now available only from the BRE Bookshop.

References

Anon (1960) Identification of hardwoods – a lens key. *Bulletin No. 25 of the Forest Products Research Laboratory*.

Anon (1977) *A Handbook of Softwoods*. HMSO: available only from the Building Research Establishment Bookshop.

Brazier J.D. and Franklin G.L. (1961) Identification of hardwoods – a microscope key. *Bulletin No. 46 of the Forest Products Research Laboratory*.

Dinwoodie J.M. (1966) Induction of cell wall dislocations (slip planes) during the preparation of microscope sections of wood. *Nature (Lond)* Oct. 29, pp 212, 525–527.

Farmer R.H. (1972) *A Handbook of Hardwoods*, 2nd edn. HMSO: available only from the Building Research Establishment Bookshop.

Hoadley R.B. (1990) *Identifying Wood: Accurate Results with Simple Tools*. The Taunton Press, Newtown, USA.

Miles A. (1978) *Photomicrographs of World Woods*. HMSO: available only from the Building Research Establishment Bookshop.

Phillips E.W.J. (1960) Identification of softwoods by their microscopic structure. *Bulletin No. 22 of the Forest Products Research Laboratory* (available only from the Building Research Establishment Bookshop).

Wheeler E.A., Pearson R.G., La Pasha C.A., Zack T. and Hatley W. (1986) Computer-aided wood identification. *Bull. 474, Dept of Wood and Paper Science, North Carolina State University, Raleigh, NC, USA*.

Wilson K. and White D.J.B. (1986) *The Anatomy of Wood: Its Diversity and Variability*. Stobart & Son Ltd, London.

Part 2

PROPERTIES OF WOOD
– INFLUENCE OF STRUCTURE

7

Appearance of Wood

Although vast quantities of timber are used in construction, and much of this is softwood, there are still large quantities of wood used primarily on the grounds of its aesthetic appeal, for example, for furniture, door skins, wall panelling and certain sports goods. The decorative appearance of many timbers is due to the **texture**, or **figure**, or **colour** of the wood, and often to a combination of two, if not all three of these characteristics.

7.1   Texture

The texture of a piece of wood relates to the size of the cells and their arrangement. Texture cannot be quantified and recourse has to be made to the use of such qualitative terms as coarse, fine, uneven and even. The differentiation between **coarse** and **fine** texture is made on the dimensions of the vessels, and the width and abundance of the rays. Timbers in which the vessels are large, such as oak, keruing, or the rays wide (oak again), are said to be coarse textured, but when the vessels are small, such as box, and the rays narrow, the wood is said to be fine textured. On this arbitrary subdivision, all softwoods will fall into the category of fine textured.

Where the arrangement of the types and sizes of cells remains constant throughout the growing season, as occurs in beech, box, birch and other diffuse-porous timbers, the wood is said to be **even textured**. The term can also be applied to those softwoods in which wall thickness does not change across the growth ring, such as Parana pine and white pine. Where there is contrast in cell size across the growth ring, as in teak, or contrast in wall thickness across the growth ring, as in larch or Douglas fir, the wood is said to be of **uneven texture**.

7.2   Figure

The term figure relates to the ornamental markings produced on the longitudinal surface of wood, as a result of either its inherent structure, or its induced structure following some external interference. The four principal structural features controlling figure are grain, growth rings, knots and rays.

7.2.1   Grain

The following types of grain can be distinguished, some of which give rise to important types of figure.

Straight grain
In straight-grained timber, the fibres or tracheids are more-or-less parallel to the vertical axis of the tree. In addition to being a contributory factor in strength, straight-grained timber makes for ease of machining and reduces waste. On the other hand, it does not give rise to ornamental figure.

Irregular grain
Timber in which the fibres are at varying, and irregular, inclinations to the vertical axis in the log, is said to have irregular grain. It is frequently restricted to limited areas in the region of knots or swollen butts. It is a very common defect and, when excessive, seriously reduces strength, besides accentuating difficulties in machining. Irregular grain, however, often gives rise to an attractive figure (Figure 7.1a, b, c). Pronounced irregularities in the direction of the fibres, resulting from knoll-like elevations in the annual rings, produce **blister figure**, the features of which can be either small (Figure 7.1a) or very large when the figure is referred to as quilted (Figure 7.1b).

The valuable and attractive **bird's eye figure** (resulting from conical depressions, as opposed to the elevations in blister figure) seen on the finished tangential surfaces of selected material of a few species, such as maple, is held to be the result of temporary injury to the cambium (Figure 7.1c).

Spiral grain

This is produced when the fibres follow a spiral course in the living tree. The twist may be left- or right-handed. The inclination of the fibres may vary at different heights in the trunk and, at any one height, the inclination may vary at different distances from the pith, it being highest in the centre where angles of 14° have been recorded, decreasing to zero in mature wood. The cause of spiral gain is not definitely known, but there is evidence that it is a hereditary characteristic of individual trees, and acting through the preferred orientation of the pseudo-transverse tangential division of the cambial initials (see Chapter 2). Although not always readily visible, spiral grain may often be detected from the direction of the surface seasoning checks (Figure 7.2), often visible, for example, on telegraph poles; it is quite common in softwoods, where it always takes the form of a left-hand spiral (Figure 7.3). Spiral grain reduces the strength of timber and is, therefore, a serious defect in timber for important structural work.

Interlocked grain

Interlocked grain, or **interlocked fibre** as it is often called, results from the fibres of successive growth layers being inclined in opposite directions, producing the familiar figure known as **ribbon** or **stripe figure** (Figures 7.4a, b) on radial cut surfaces. Interlocked grain is relatively uncommon in temperate woods, but it is a characteristic feature of a large proportion of tropical timbers. As far as is known, it does not appreciably affect the strength of timber, but it may cause serious twisting during seasoning and, if pronounced, makes the wood difficult to split radially (Figure 7.4b). There is also the added disadvantage that such timber 'picks up', particularly when being planed on the radial plane, leaving a very rough finish. In timbers with heavily interlocked grain, that is when the grain angle

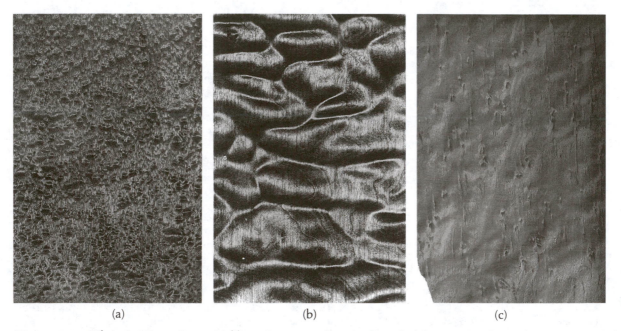

(a)   (b)   (c)

Figure 7.1   *Effect of grain on figure: (a) blister figure in mahogany; (b) quilted figure in Pacific maple — one form of blister figure; (c) bird's eye figure in maple (Building Research Establishment, © Crown Copyright)*

Figure 7.2 *Marked spiral grain in trunk of plum tree as indicated by splitting in drying (Building Research Establishment, © Crown Copyright)*

between the fibres and the vertical axis of the tree exceeds 20°, and the successive changes in inclination of the fibres occur at intervals of 6–12 mm radially, sawing difficulties may be very great; the fibres tend to pull out and wrap themselves round the saw-teeth until the saw becomes buried in the log. With timbers in this class, for example keledang, sepul and terentang, interlocked grain can cause as much trouble in conversion as does the high silica content of other timbers. A reasonably smooth surface can, however, be obtained in sawing and planing with modern machines, employing more set than is ordinarily required, a suitable cutting angle and a modified rate of feed (Chapter 12).

Wavy grain
When the direction of the fibres is constantly changing, so that a line drawn parallel to them appears as a wavy line on a longitudinal radial surface, the grain is said to be **wavy**. This type of grain gives rise to a series of diagonal, or more or

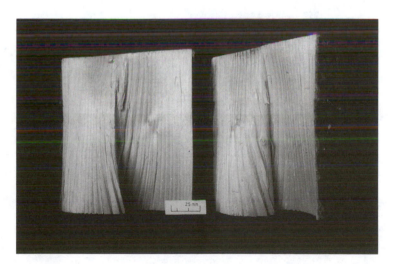

Figure 7.3 *Section of spruce trunk split to reveal systematic pattern in spiral grain: grain angle is at maximum at the pith and decreases progressively from the pith (Building Research Establishment, © Crown Copyright)*

(a)

(b)

Figure 7.4   *Relation of stripe or ribbon figure to interlocked grain: (a) stripe or ribbon figure on the radial surface in African mahogany; (b) block split radially to show interlocked grain (Building Research Establishment, © Crown Copyright)*

less horizontal, darker or lighter stripes on longitudinal surfaces, because of variations in the reflection of light from the surface of the fibres: this is called **fiddle-back figure** (Figure 7.5a). Wood with wavy grain presents a corrugated surface when split, as shown in Figure 7.5b. The importance of this type of grain lies in its decorative value, and any reductions in strength are of no consequence. Wavy grain may occur, together with interlocked grain, in one piece of timber, giving rise to a broken 'ripple' on quarter-sawn surfaces, called **roe** or **mottle figure** (Figure 7.6).

7.2.2 Growth rings

In the many hardwoods and softwoods where there is variability across the growth ring in either the distribution of cell types or in the thickness of the cell wall, distinct patterns of concentric arcs will appear on machined longitudinal–tangential surfaces as the rings are cut through (Figure 7.7a, b). The use of ash as a high-class panelling wood is due in part to its light coloration, and in part to the pleasing patterns of growth rings formed by the large-diameter early-wood vessels, as the wood is

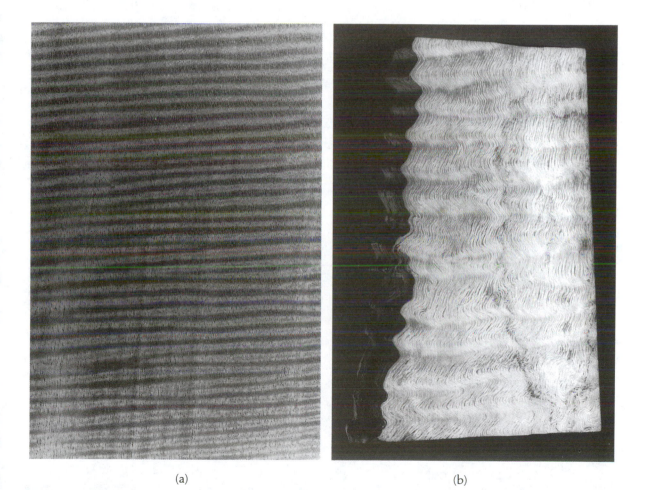

(a)   (b)

Figure 7.5 *Relation of fiddle-back figure to wavy grain: (a) fiddle-back figure in walnut; (b) split block of oak showing wavy grain (Building Research Establishment, © Crown Copyright)*

rotary peeled to produce high-quality veneer with a tangential surface.

7.2.3   Knots

A few trees respond to a marked increase in light levels by producing from the trunk a mass of twigs known as epicormic shoots. Left on the tree, these will eventually be engulfed by the enlarging trunk to produce a feature called a **burr**. When cut through, these burrs produce a beautiful figure comprising the distorted growth rings of the large number of small knots (Figure 7.8). Burr walnut has always been highly prized for top-class furniture manufacture.

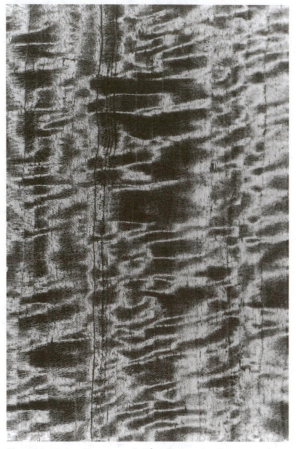

Figure 7.6 *Roe or mottle figure in Japanese horse chestnut (Building Research Establishment, © Crown Copyright)*

7.2.4   Rays

When woods with deep, wide rays are cut through on a radial plane (quarter-sawn), a beautiful figure can be obtained on the quarter-grain, as the ray is exposed against the vertical fibres. Two good examples are European oak (Figure 7.9) and 'silky' oak from Australia.

7.3   Colour

Distinctive colours in wood are caused largely by various extractives present in the cell walls of the heartwood (see Chapter 3). A whole range of colours can be found, from the pale yellow of boxwood, the orange of opepe or agba, the pale brown of oak, the greyish brown of European walnut, the dark brown of keruing, the reddish brown of true mahogany, the deep red of padauk, the purple of purpleheart, the green of greenheart and the black of ebony.

Colour may be used to assist in the identification of woods (Chapter 6), but great caution has to be exercised, not only because of the variation in colour that can be obtained among trees of the some species, but also, and possibly more important, because the colour of most woods changes on exposure to light. Under sunlight, most timbers will lighten appreciably, and there exists many examples of 'bleached' mahogany furniture that has stood near a window. A few timbers darken on exposure, the best known being Douglas fir, commonly sold under the trade name of Oregon pine.

It should be noted in passing that, in furniture manufacture, the pale sapwood of timbers having coloured heartwood is frequently artificially stained to match the colour of the heartwood, in order to increase the outturn from a log.

The moist heat employed during kiln seasoning darkens many woods, so much so that some are steamed purposely to alter the colour, for example the sapwood of beech and walnut. Colour changes are also effected by chemical means, for instance, liming lightens the colour and fuming (with ammonia gas) darkens it, removing the pink or red shades. Bleaching of wood with hydrogen peroxide is also feasible.

7.4 Lustre

Lustre depends on the ability of the cell walls to reflect light. Some timbers, such as East Indian satinwood, lauan and sapele, possess this property in a high degree, but others, such as hornbeam are comparatively dull. As a general rule, quarter-sawn surfaces are more lustrous than flat sawn, and if stripe, fiddle-back, or roe figure is present, the figure is considerably enhanced in timbers possessing a natural lustre; lustre is an important asset in a cabinet timber. From a practical viewpoint, the capacity for taking a good polish is quite as important, and the two do not necessarily go hand in hand.

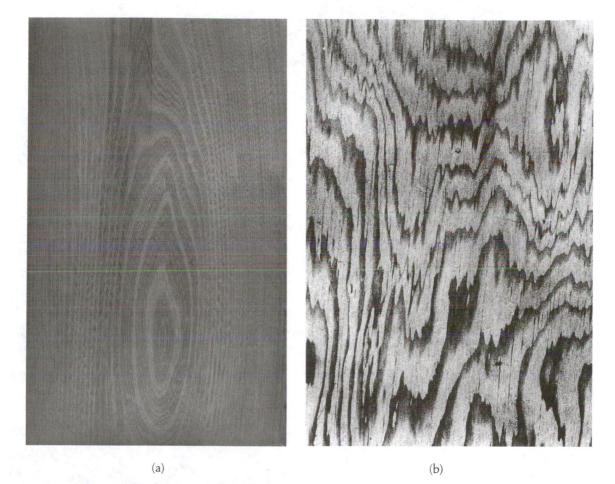

(a)   (b)

Figure 7.7 *(a) Figure due to cutting in a tangential plane the growth rings of a ring-porous timber. (b) Rotary peeled Douglas fir showing contrasting early (light) and late wood (dark) on the tangential surface (Building Research Establishment,* © *Crown Copyright)*

Figure 7.8   *Burr figure in alerce arising from cutting across a large number of small epicormic shoots (Building Research Establishment, © Crown Copyright)*

Figure 7.9   *The silver figure produced in cutting European oak in a truly radial plane (Building Research Establishment, © Crown Copyright)*

8

Density of Wood

8.1 Definition of density

Perhaps the single most important property controlling the mechanical performance of a piece of wood is its density. Density is the ratio of mass to volume

$$\rho = \frac{m}{v}$$

where ρ = density in kg/m^3
m = mass in kg
v = volume in m^3.

Care has to be exercised in the application of the equation and the interpretation of results since the density of a piece of wood is determined not only by the amount of wood substance present, but also by the presence of both extractives and moisture. Although in the majority of timbers extractives are usually absent or present in small amounts (<3 per cent dry mass of wood) such that they can be ignored in the determination of density, in a number of cases extractive contents of up to 10 per cent have been recorded and, in order to obtain an accurate measure of density, these must first be removed.

The presence of moisture in wood not only increases the mass of the timber, but also increases its volume. Consequently, in order to obtain an accurate measure of density, determination of mass and volume must be carried out at the same moisture content.

Generally, in a laboratory both mass and volume are determined at zero moisture content following drying in an oven at 105°C until constant mass is obtained. Frequently density is required at 12 per cent moisture content, the level at which most timbers are in equilibrium with a relative humidity in the atmosphere of 65 per cent. Either the determination of density

can be carried out at this moisture content, or the value at zero moisture content can be corrected for changes in mass and volume at the higher moisture content (Dinwoodie, 1993). As a very approximate rule of thumb, the density of wood increases by approximately 0.5 per cent for every 1.0 per cent increase in moisture content up to a moisture content of 30 per cent. Above this value, density will increase very rapidly at an accelerating rate with increasing moisture content, and it is not possible to provide a general rule of thumb over this range of moisture contents.

8.2 Determination of density

8.2.1 Using a dry measure of volume

To achieve a high degree of accuracy the sample of wood should be squared with a minimum dimension of 25 mm. Its volume is determined by calculation based on the direct measurement of length, width and thickness, and in order to maintain the high degree of accuracy these measurements should be taken using vernier or digital callipers.

Mass is assessed by use of a balance and should be carried out and recorded to an accuracy of at least 1 per cent.

The ratio of mass to the calculated volume will provide a measure of density at the moisture content of the sample when measured.

8.2.2 Using volume displacement

For smaller blocks of wood than that used above, or for blocks of irregular shape that for one reason or another cannot be cut to produce a squared section, it is necessary to determine volume by using a displacement technique.

The block of wood is first weighed, and as wood is a porous material, it is necessary to coat the block prior to immersion with an impervious material such as paraffin wax. This is done by dipping the block in a bath of melted wax and quickly removing it: when set, the excess wax is scraped off. A beaker of water is placed on a balance and, on old balances, counterbalanced by the addition of weights to the opposite pan, or on new balances, the original mass is tared to zero. The block of wood, suspended by a needle clamped in a stand, is then gently lowered into the beaker and completely immersed in the water without it either touching the sides of the beaker or any of the water running over the top of the beaker. Weights are then added to the opposite pan to restore equilibrium, or in new scales, the new mass is read off. Since the density of water is 1.000 g/cm^3 at 40°C, the mass in grams added is equal to the volume of the test block in cubic centimetres.

It must be appreciated that the ratio of

$$\frac{\text{mass of block (g)}}{\text{mass added to (restore) balance (g)}}$$

is NOT the density of the block, but rather its **relative density** or **specific gravity**.

The relative density (specific gravity) of a material is merely its density in comparison with a standard density, usually that of pure water since the mass of 1 cm^3 of water is 1 g at 4°C. Consequently, provided the oven dry mass and volume of the wood block are determined, density and relative density are numerically equal and it is therefore possible to express the relative value as a true density with units of kg/m^3. However, where the sample of wood is not at zero moisture content it is not possible to obtain a true density using immersion techniques, though it is possible mathematically to correct the relative density reading at some moisture content to the relative density value for zero per cent.

8.3   Variation in density of wood

A piece of perfectly dry wood is composed of both the solid material comprising the cell walls and the cell cavities which contain air and small quantities of gum and other substances. The relative density (specific gravity) of the solid material of the cell walls has been found to be similar in all timbers, and is in the neighbourhood of 1.5; that is, the cell walls are about one and a half times as heavy as water, and a cubic metre of solid wood, without cell cavities and intercellular spaces, would weigh roughly 1500 kilograms. Different timbers, however, vary in mass from about 160 to 1250 kilograms per cubic metre; this variation is caused by differences in the ratio of cell wall to cell cavity.

This ratio is controlled by both the relative proportions of the thinner-walled vessel and parenchyma cells and thicker-walled fibres (Chapter 3) and the extent of development of the secondary walls of the fibres; generally both factors operate to produce this very wide range in density among timbers. Thus the almost tenfold increase in density of lignum vitae compared with balsa is due in part to a higher proportion of vessels in the balsa, but mainly to the presence of only a thin wall in the fibres in balsa and an extremely thick wall in the fibres in lignum vitae.

Variation in density among timbers is clearly illustrated in Figure 8.1 from which it should be noted that the range in densities of the softwood timbers is much lower than that for the hardwood timbers and, equally as important, the density range of the softwoods is encompassed by that of the hardwoods; that is, some of the hardwoods are less dense that the softwoods.

In addition to the range in density that occurs among timbers of different species, there is considerable variation in density between different samples of the same species: this variation can be as high as fourfold and occurs between timber from different trees, and in timber from different parts of any one tree. As discussed in Chapter 5, variation in the former is influenced by such factors as rate of growth, site conditions and genetic composition. In the latter, systematic patterns of variation occur; as a general rule the heaviest wood is found at the base of the tree, and there is a gradual decrease in density in samples from successively higher levels in the trunk. At any given height in the trunk there is

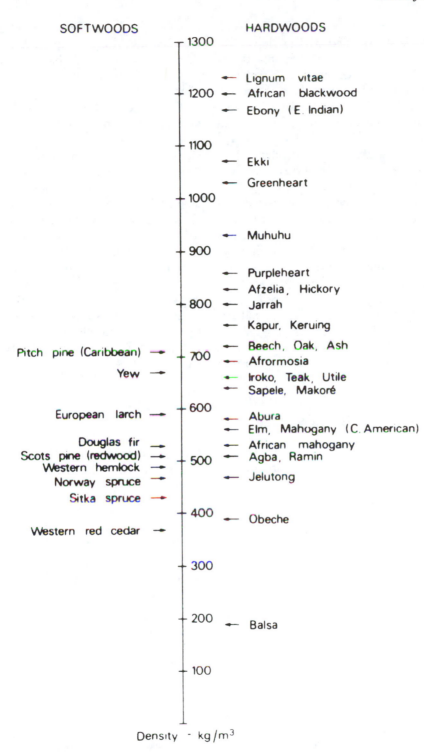

Figure 8.1 *Density values at 12 per cent moisture content for some common hardwoods and softwoods (Building Research Establishment, © Crown Copyright)*

usually a general increase in density outwards from the pith, fairly marked in the rings near the pith, but slowing down considerably thereafter. Superimposed on this systematic variation outwards from the pith is the effect of growth rate. It was shown in Chapter 5, section 5.3, how the composition and hence density of the growth ring was influenced by the rate of growth, and how this effect was different in conifers, ring-porous and diffuse-porous hardwoods.

As the width of ring tends to decrease towards the bark, the effect of ring width on density tends to be additional to the systematic trend as far as softwoods are concerned and density increases towards the bark, whereas in the case of ring-porous hardwoods, the effect of decreasing ring width is to reduce density and hence detract from the systematic pattern such that density may decrease slightly in the outer rings. In diffuse-porous hardwoods, ring width has little effect on density subject to the reservations set out in Chapter 5 and the general systematic pattern prevails.

8.4   Practical significance of density

The mean density of a timber is of practical interest because it is the best single criterion of strength. This generalisation, however, requires qualification. Mean density is of limited value in representing the strength properties of individual pieces of wood, because of the influence of other factors discussed in more detail in Chapter 11.

Two other factors modify the importance of density as a criterion of strength, namely the arrangement of the individual cells and the physico-chemical composition of the cell walls. If, for example, the parenchyma is distributed in broad layers, these may constitute planes of weakness along which the timber will shear, despite a relatively high density for the sample as a whole. It has now been established that the physico-chemical composition of the cell wall is the major influence in determining the strength properties of individual pieces of wood; in particular, the angle at which the microfibrils in the middle layer of the secondary wall (S_2) are lying, and the degree of lignification of the cell walls have a direct bearing on most strength properties and stiffness.

Reference

Dinwoodie J.M. (1993) In Illston J.M. (Ed.), *Construction Materials — their Nature and Behaviour*. Chapman & Hall, London.

9

Moisture in Wood

One of the most important variables influencing the performance of wood is its moisture content. The amount of water present not only influences its strength, stiffness and mode of failure, but it also affects its dimensions, its susceptibility to fungal attack, its workability as well as its ability to accept adhesives and finishes. For wood to perform well, its moisture content must be reduced to a level corresponding to at least 12 per cent of its oven-dry mass.

9.1   Amount and location of moisture

Some idea of the considerable amount of moisture in the timber of living trees and newly felled logs can be obtained from Figure 9.1: the actual amount however, will vary considerably among trees of different species. In some it is only about 40 per cent of the over-dry mass of the wood, while in others it may exceed 200 per cent of the mass, a value commonly found in freshly felled Sitka spruce. In most species there is usually a marked difference in the moisture content of sapwood and heartwood; this is particularly the case with softwoods (see Chapter 2). In long-leaf pitch pine, for example, the moisture content of the sapwood usually exceeds 100 per cent while the heartwood ranges only 30–40 per cent. Moisture content may also vary with height in the tree: thus butt logs of sequoia and western red cedar often sink in water, although the upper logs float. In species with a marked difference between the moisture content of sapwood and heartwood the position may, however, be reversed, because the upper logs contain a higher percentage of sapwood of high moisture content.

There is no evidence in support of the widely held opinion that timber felled in winter, when the sap is said to be 'down', is drier than that felled in summer, when the sap is said to be 'up'. In fact, such evidence as is available suggests that there is either no difference, or the moisture content is, if anything, somewhat higher in the winter months. In the light of these figures the prejudice against summer felling cannot be sustained on the grounds usually advanced. There is some justification on scientific grounds, however, for preferring winter felling. Chief of these is the reduced activity of insects and fungi during the cool months, and the lower degrade in seasoning resulting from a slowing down of drying from the ends of logs. Moreover, the risk of damage to the bole is probably greater when felling heavy-crowned trees in full leaf.

It was described in Chapter 3 how wood is made up of hollow cells, while in Chapter 4 the basic chemical constituents were described. When a tree is felled, water is present in two modes: first, it is present within the cell cavities when it is described as **free** water. The removal of this free water during drying has no effect whatsoever on either the mechanical performance of the wood or its dimensions.

The second mode in which water occurs in wood is when it is chemically bonded to constituents of the cell wall. In describing the cell-wall structure of woody tissue in Chapter 4 it was explained that the walls consist of several concentric layers, and that the layers are composed of microfibrils. The water in the cell walls is chemically bound (by hydrogen bonding) to the hydroxyl groups of the matrix constituents forming the sheath to the microfibril, and this type of water is therefore referred to as **bound** water. In most timbers the walls can hold about 25 to 30 per cent of their dry mass.

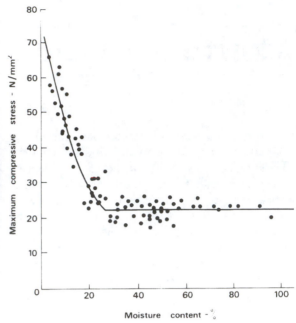

Figure **9.2** *Relationship between longitudinal compressive strength and moisture content; the inflexion in the graph corresponds to the theoretical state where there is no free water and maximum bound water present (the fibre saturation point) (Building Research Establishment, © Crown Copyright)*

Figure 9.1 *The amount of water in wood. Two pieces of oak (750 × 70 × 70 mm), the one on the left cut from a newly felled log and containing 2½ pints of water, the one on the right having been dried to a moisture content of 12 per cent still has one-third of a pint of water in it (Building Research Establishment, © Crown Copyright)*

When freshly felled or **green** wood is dried, it is the free water which is first removed because of the lower energy levels required. In theory it should be possible to remove all the free water before removing the bound water. In practice, however, it is impossible to remove all the water in the cell cavities without removing some from the cell walls, but as a convenient concept it is possible to imagine a theoretical situation where the cavities are empty and the cell walls are saturated; this idealised state is known as the **fibre saturation point**.

To appreciate how strength is influenced by the removal of the different forms of water, reference should be made to Figure 9.2. This clearly illustrates that as the moisture content of the wood is reduced artificially from 110 per cent to about 27 per cent, there is no change in the longitudinal compression strength of the wood because it is free water which is being removed at this stage. Below 27 per cent, bound water is removed from within the cell walls, resulting in the microfibrils coming into closer juxtaposition and increasing the strength of the cell wall (and hence of the timber). In this case, 27 per cent corresponds to the fibre saturation point; however, the value of the fibre saturation point will be influenced by density, the value decreasing with increasing density (see Skaar, 1988).

The fibre saturation point, therefore, corresponds to the moisture content of the timber when placed in a relative humidity of 100 per cent and it is usual for the moisture content of hardwoods under this condition to be about 1 to 2 per cent higher than for softwoods. The hygroscopic nature of wood will be discussed more fully in section 9.3.

9.2 Determination of moisture content

The moisture content (mc) of a piece of wood is defined as the mass of water in the piece expressed as a percentage of the oven-dry mass of the wood:

$$\text{mc (\%)} = \frac{\text{Mass of water in wood (g)}}{\text{Oven-dry mass of wood (g)}} \times 100$$

Consequently, it is possible to have wood with more than 100 per cent moisture content, as occurs in the sapwood of newly felled trees.

In the determination of moisture content, sample pieces of wood which cannot be assessed for moisture content immediately must be tightly and securely wrapped in a waterproof covering such as a polythene sheet or bag: this is particularly important where samples are being sent through the post to a testing laboratory.

There are several ways of determining the moisture content of wood, but by far the most satisfactory for most purposes is the oven-dry method.

9.2.1 Oven-dry method

The accuracy of the technique decreases with decreasing size of sample. Consequently, as large a sample as possible should be prepared commensurate with the amount available, the size of the oven, the capacity of the balance and the time available, since large pieces will take longer to dry than small pieces. Wherever

possible more than one sample should be used, and where the moisture content of a large stack of timber is being assessed between three and ten samples should be taken at random.

Once the test sample has been cut out, rapidity of weighing is essential to minimise the chance of the sample picking up or losing moisture, thereby introducing appreciable errors in the calculated moisture content. After the initial weighing, the samples should be transferred to the drying oven. This should be run at a temperature of 60°C for the first few hours to prevent the moisture in the centre of the samples from being sealed in as a result of case-hardening (see Chapter 13, section 13.6.2). The temperature can be raised to 105°C afterwards, and the samples left in the oven overnight. They should be re-weighed first thing on the following morning, and again two hours later. Rapidity of weighing is of particular importance when the samples are oven dry, as, in this state, they will absorb moisture in a very short space of time. If the difference between the last two weighings is less than 0.2 per cent, the lower may be taken as the oven-dry weight. If, however, the second weighing is greater than 0.2 per cent, drying must be continued for further periods until the difference is less than 0.2 per cent. Drying in an oven does not expel all the moisture, but the small discrepancy – the last one per cent or so – is not of practical importance.

The moisture content of the sample is therefore:

$$\text{mc (\%)} = \frac{m_{\text{init}} - m_{\text{od}}}{m_{\text{od}}} \times 100$$

where

m_{init} = initial mass of sample (g)

m_{od} = oven-dry mass of sample (g).

This method is the most precise way of determining the amount of moisture present; it is generally accepted for basic laboratory work and as a standard for calibrating other methods. It does, however, have a number of limitations:

(a) it is time-consuming,
(b) the sample is changed,

(c) it will be inaccurate where the sample contains either naturally occurring or artificially added volatile materials, such as oil-based preservatives. Thus resinous samples of pine, for example, will produce erroneous results when moisture content determinations are carried out by oven-drying. Much of the resin will be lost on drying and in the calculation will be treated as moisture, thereby giving moisture content values which are far too high. Moisture content determinations of wood containing volatile substances must be carried out using the distillation method described below.

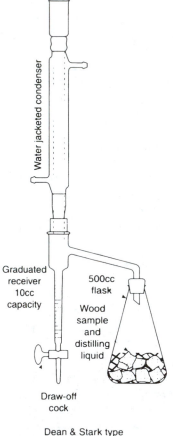

Water jacketed condenser

Graduated receiver 10cc capacity

500cc flask

Wood sample and distilling liquid

Draw-off cock

Dean & Stark type distillation apparatus

Figure 9.3 *Distillation method of moisture-content determination*

9.2.2 Distillation method

In this method, water is removed from the wood sample in a closed system in which the water is collected and measured separately, while the volatile materials, instead of being evaporated, are actually dissolved out of the wood sample by an organic solvent during the water extraction process.

A sample of about 50 grams of chips, borings, or sawdust is placed in a flask containing an organic solvent. The apparatus employed is illustrated in Figure 9.3. It consists of a flask, with suitable heating arrangements, and a reflux condenser discharging into a graduated trap which collects the condensed water from the wood and returns the solvent to the flask. Distillation is continued until no more water collects in the trap. The volume of water collected is read directly in cubic centimetres. As one cubic centimetre of water weighs one gram, the mass of water in the sample is obtained automatically. With samples containing only natural oils or resins, the percentage moisture content is arrived at simply as follows:

$$mc\ (\%) = \frac{m_{water}}{m_{init} - m_{water}} \times 100$$

where

$$m_{water} = \text{mass of water collected (g)}$$

$$m_{init} = \text{initial mass of wood sample.}$$

Samples of impregnated wood contain, in addition to wood substance and water, an unknown mass of preservative. This mass must next be determined; this is done by extraction of the preservative, with a suitable solvent, from the liquid remaining in the flask at the end of the initial distillation process. The percentage moisture content of the treated wood can then be calculated:

$$mc\ (\%) = \frac{m_{water}}{m_{init} - (m_p + m_{water})} \times 100$$

where

$$m_{\text{water}} = \text{mass of water collected (g)}$$

$$m_{\text{init}} = \text{initial mass of sample (g)}$$

$$m_{\text{p}} = \text{mass of preservative (g)}.$$

The distillation method probably gives slightly more accurate results than the oven-drying method for any wood, and has the additional advantage that the time required is only three to four hours. However, the delicate apparatus required, and the risk from fire that heating organic solvents entails, render it suitable for use only in properly equipped laboratories. As with oven-drying, the sample is changed with the distillation method.

9.2.3 Moisture meters

Reservations expressed above about the length of time and the changes to the wood sample in both the oven-dry and distillation methods do not apply when moisture meters are used to determine the moisture content of timber.

These instruments determine moisture content indirectly by measuring some electrical property that varies proportionately with changes in moisture content.

Three types of electric moisture meters are available: (1) the conductance (or resistance) type, which uses the relationship between moisture content and direct current conductance; (2) the capacitance type, which employs the relationship between moisture content and the dielectric constant of the wood; and (3) the power-loss type, which uses the relationship between moisture content and the dielectric loss factor of the wood (James, 1988).

Resistance meters

Although these meters are based on the principle of electrical conductance they are popularly known as resistance meters (Figure 9.4). Technically, this is quite acceptable since resistance = 1/conductance.

Electrical conductance is the property of a material that permits the flow of electrical current through it. In wood, the direct current conductance varies greatly with moisture content and below the fibre saturation point an approximately

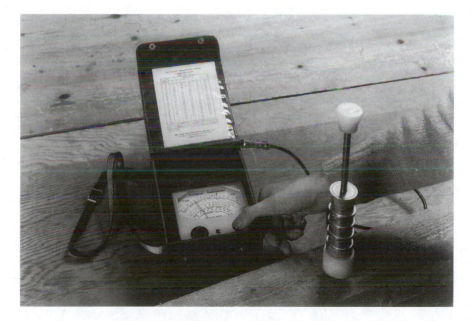

Figure 9.4 *Electrical moisture meter (resistance type) fitted with long probes which have been inserted into the wood block in the foreground (Building Research Establishment, © Crown Copyright)*

linear relationship exists between the logarithm of conductance and the logarithm of moisture content. This relationship forms the basis for this type of meter.

It should be noted that conductance varies with the temperature, increasing with a decrease in temperature, and decreasing with a rise in temperature. Moreover, the conductance for any given moisture content and temperature is not constant for different species. Hence, corrections have to be made for temperature and species. Fortunately, this is not difficult to do with sufficient accuracy for practical purposes; instruments are usually supplied by the manufacturers with the necessary correction data. Conductance (resistance) also varies with angle of the grain and for consistent results conductance should always be measured along the grain.

The conductance (resistance) type of instrument has electrodes in the form of spikes, which are driven into the wood and moisture content is read directly from one of several scales, each of which relates to different group of timbers. Generally, moisture content values for board products derived using this type of meter are not very reliable.

Besides the question of cost, and the fact that a different scale, or correction factor, has to be used for each species, resistance moisture meters have other limitations. In the first place, the figure read off the scale is the moisture content of the piece to the depth of penetration of the electrodes. In the early stages of drying, the surface moisture content of a board or plank is likely to be very different from that of interior, and with thick timbers this is likely to be so always, hence the meter reading gives too low a figure for the piece as a whole. On the other hand, a surface film of water left by a shower of rain will cause the reading to be too high. In the second place, the instruments are delicate, and require careful handling and considerable technical knowledge to maintain them in an efficient condition. In addition, the reading on the meter will be influenced by temperature: it will also be affected quite considerably by the presence of salts, as can occur with the use of certain water-borne preservatives, fire retardants or through exposure to sea water when timber is carried as deck cargo. However, the greatest drawback to the use of this type of meter is that their working range is limited to between about 7 and 27 per cent.

Moisture meters, however, have several distinct advantages: they give results almost instantaneously, and when appropriately used will give readings within one to two per cent of the true values. Moreover, they provide the only practicable means of determining the moisture content of finished woodwork *in situ* without damaging the wood. Greater accuracy can be obtained by taking several readings and averaging them or by using longer than normal electrodes, for example 12 to 25 mm long, but, so equipped, the relatively large diameter of the longer electrodes may be an objection — the diameter of a 25 mm long electrode leaves a far from inconspicuous hole. Moisture meters are particularly suitable where comparative rather than absolute figures are required. Since the reading of a conductance (resistance) type meter is determined by the wettest wood that contacts both electrodes, and since there is usually a moisture gradient in a piece of wood with moisture content increasing with depth, it is possible to determine such gradients by taking readings at increasing depths.

Capacitance meters
This type of meter uses the relationship between moisture content and the dielectric constant of the wood, which is a measure of how much electric potential energy is stored in the material when it is placed in a given electric field.

Capacitance meters have a pad which is pressed against the surface to be measured and hence no pin holes are left behind. However, they do require a flat surface and are sensitive to the applied pressure. The reading obtained applies only to the wood which is near the surface and they are calibrated on the basis of an average wood density, thereby giving erroneous results in timbers of very low or very high density. They are more expensive than resistance type meters, yet have the same limitations in range of moisture content, though the minimum value is slightly lower than with the resistance type meter. Very few capacitance meters are to be found in use.

Power-loss meters

These use the relationship between moisture content and the loss factor. The wood is penetrated by an electric field and the power absorbed (loss factor) by the wood can be measured and used as an indication of mixture content.

These meters are similar in performance to the capacitance-type meter and are not found in use in the UK though they are used in North America.

9.3   Equilibrium moisture content

Wood is **hygroscopic**, that is, the number of water molecules (bound water) at the sorption sites in the matrix constituents of the cell wall increase or decrease so that the vapour pressure of the wood is in equilibrium with that of the air surrounding it. The particular moisture content of the wood that is in equilibrium with a given vapour pressure (or relative humidity) is known as the **equilibrium moisture content**, frequently expressed as **emc**.

The value of the emc varies slightly with species and considerably with temperature. Inter-specific differences are probably related to a density effect and a number of studies have shown a negative relationship between a specific emc (such as the fibre saturation point) and specific gravity. However this has been contested by Chafe (1991) who recorded a positive relationship between these variables on samples of *Eucalyptus regnans* at three different emc values.

The factor with the greatest influence on the emc, however, is the past history of moisture levels in the wood. The emc at constant temperature will increase as the relative humidity of the surrounding air increases, and decreases with decreasing relative humidity. However, the interesting point is that the emc at a given relative

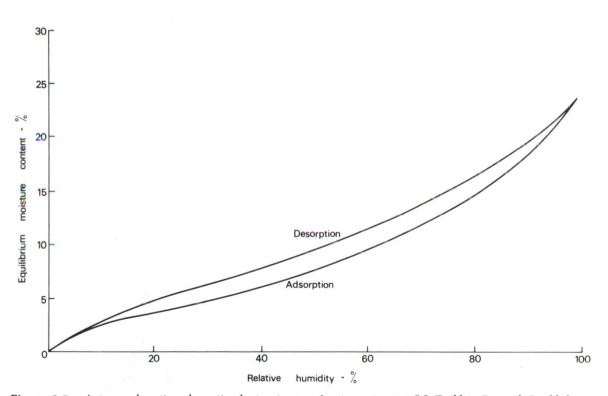

Figure 9.5   *Average adsorption–desorption hysteresis curve for six species at 40°C (Building Research Establishment,* © *Crown Copyright)*

humidity is not constant, but depends on whether the level of moisture in the timber has increased or decreased to reach equilibrium. This situation is illustrated in Figure 9.5, where it will be noted that the equilibrium moisture contents arising from **desorption** or loss of bound water are higher than those that occur due to **adsorption** or uptake of moisture. Such a loop is referred to as a **hysteresis loop** with the emc achieved by adsorption being from 80 to 90 per cent of that obtained by desorption. Different loops occur for different temperatures and species, but usually these differences are small and restricted to only two or three per cent moisture content at any one relative humidity.

As noted earlier, the emc is also affected by temperature, the adsorption isotherms becoming lower with increasing temperature (Skaar, 1988) (Figure 9.6). It is possible to construct a series of emc curves for different combinations of tem-

perature and relative humidity for both adsorption or desorption; a set for desorption is illustrated in Figure 9.7. From this figure it is possible to estimate the average equilibrium moisture content for wood provided the temperature and relative humidity are known.

Readers desirous of more information on the sorption of water by wood, especially on the application of various sorption theories to wood/water relationships, are referred to the texts by Simpson (1980) and Skaar (1988).

Loss in water below the fibre saturation point, that is bound water, is associated with certain dimensional changes. It is convenient to subdivide these into two groups: those that occur as a result of initial drying down from green to dry conditions – these are referred to as **shrinkage**; and those that occur as a result of daily or seasonal changes in relative humidity of the atmosphere – these are termed **movement**. Because

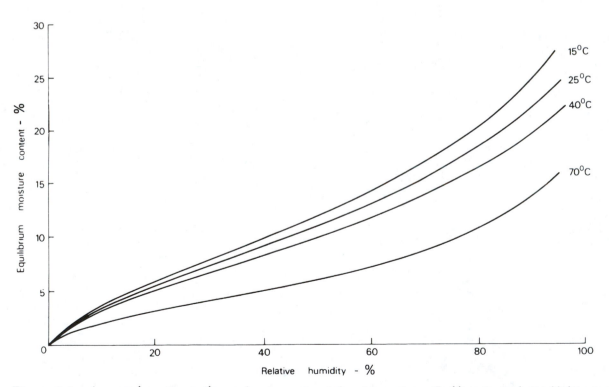

Figure 9.6 *Average desorption isotherms for six species at four temperatures (Building Research Establishment, © Crown Copyright)*

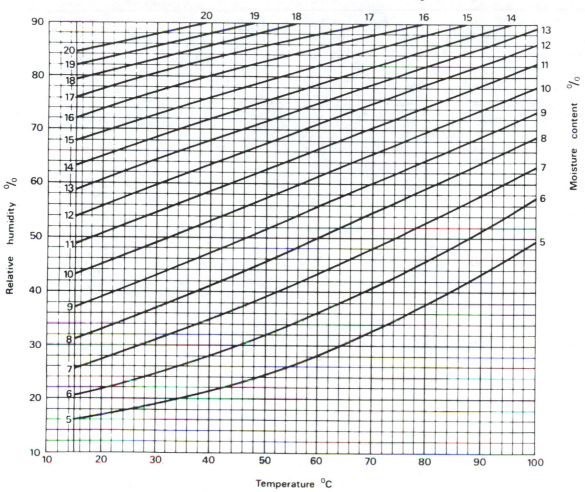

Figure 9.7 *Moisture content and relative humidity equilibrium desorption isotherms for wood obtained during drying down from 20 per cent moisture content. The isotherms represent average values for a large number of species (Building Research Establishment, © Crown Copyright)*

of the greater range in moisture contents involved, dimensional changes comprising shrinkage are considerably greater than those occurring in movement. These are discussed separately in the following sections.

9.4 Shrinkage

As explained above, water held in wood below its fibre saturation point is chemically bonded to the matrix constituents of the microfibrils comprising the cell wall. Removal of this water requires the application of a considerable amount of energy in the form of heat and the direct consequence of the reduction in moisture is to cause the wood to shrink.

The degree of shrinkage in the three principal axes is different: shrinkage in the tangential direction is on average about twice as high as that occurring in the radial direction, while shrinkage in the longitudinal direction is at least an order of magnitude lower than that in the transverse direction. Some idea of the anisotropy that occurs, as well as interspecific differences is provided in Table 9.1. It should be noted that the values given in the table represent shrinkage on drying from the 'green' state (which is defined as above the fibre saturation value of 27 per cent) to 12 per cent (and not zero) moisture content. This practice is pursued because timber at 12 per cent moisture content is in equilibrium with an atmospheric relative humidity of 65 per cent which is about average for internal conditions; consequently quoting shrinkage to 12 per cent moisture content has a greater practical significance than the values to oven-dry conditions.

While longitudinal shrinkage is usually so low that it is generally expressed as being less than 0.1 per cent, it should be appreciated that this applies to 'adult' wood, and that values up to 0.8 per cent have been recorded for longitudinal shrinkage of 'juvenile wood'. Consequently, pieces of timber containing some juvenile wood may well have a longitudinal shrinkage greater than 0.1 per cent; in the utilisation of young plantation-grown softwood, this could occur quite frequently.

From Table 9.1 it will be noted that there is about a fivefold range in shrinkage values between different species. As a fairly general rule, the shrinkage values for hardwoods are greater than those for softwoods. Inter-specific differences in shrinkage can be explained to a very large extent by corresponding differences in density. Thus, shrinkage increases with increasing density, and those factors which affect density, such as rate of growth (see Chapter 5) will also affect the level of shrinkage.

It is possible to explain the difference in shrinkage in the three principal planes in terms of the basic structure of wood. Thus the marked difference between longitudinal and horizontal shrinkage is due basically to the microfibrillar angle of the S_2 layer of the cell wall. As water is removed from the matrix surrounding the crystalline core of the microfibrils, so they move closer together and the horizontal component of this movement is several times greater than the longitudinal component (Figure

Table 9.1 *Shrinkage (per cent) on drying from 'green' to 12 per cent moisture content (from Farmer, 1972)*

Common name	Transverse		Longitudinal
	Tangential	Radial	
Hardwoods			
Afzelia	1.5	1.0	<0.1
Agba	3.0	1.5	<0.1
Poplar	5.5	2.0	<0.1
European oak	7.5	4.0	<0.1
European lime	7.5	5.0	<0.1
Beech	9.5	4.5	<0.1
Softwoods			
Western red cedar	2.5	1.5	<0.1
Yellow pine	3.5	1.5	<0.1
Norway spruce (white wood)	4.0	2.0	<0.1
Douglas fir	4.0	2.5	<0.1
Western hemlock	4.0	2.8	<0.1
Scots pine (redwood)	4.5	3.0	<0.1

9.8(i), (ii)). The explanation is supported by the evidence obtained from compression wood which, as explained in Chapter 5, is characterised by having a microfibrillar angle of the S_2 layer of up to 45°. As water is removed and the microfibrils move closer together, the components of the movements are about equal in the longitudinal and horizontal directions, in line with the known shrinkage pattern of well-developed compression wood where longitudinal and transverse shrinkage on a percentage basis can be equal (Figure 9.8(iii), (iv)). However, the relationship between shrinkage (per cent) and microfibrillar angle is not linear (Meylan, 1968).

The value for tangential shrinkage is usually about twice as high as that in the radial direction; however, in some timbers tangential shrinkage is only slightly greater than radial, while in others it can be as much as six times

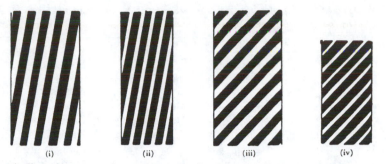

Figure 9.8   *Microfibrillar structure and its relation to shrinkage. (i) Diagrammatic representation of microfibrillar structure in green timber: the black lines represent the microfibrils and the white the films of water; (ii) the same cell when seasoned: note that the reduction in length of the cell is much less than the reduction in width; (iii) and (iv) diagrammatic representation of microfibrillar structure in green and air-dry compression wood in which the pitch of the fibrils is less than in normal wood: note that the reduction in length of the cell as the wall dries out is much greater than in normal wood, and that the percentage reduction in length is approximately equal to the percentage reduction in width.*

as great. Differences in shrinkage between tangential and radial directions have been accounted for by a number of factors (Boyd, 1974: Skaar, 1988); among the more important are: (1) the restricting effect of the ray on the radial plane: (2) the difference in degree of lignification between the radial and tangential walls; (3) small differences in microfibrillar angle between the two walls; and (4) the increased thickness of the middle lamella in the tangential direction compared with that in the radial direction. None of those variables on its own is able to account for the total extent of **differential shrinkage**; all four variables (and possibly others) probably contribute to the differences but it is not possible to estimate their relative contributions. Differential shrinkage has a direct bearing on the amount of distortion that occurs in the commercial drying of wood and this important aspect of utilisation is discussed in more detail in Chapter 13, section 13.6.1.

9.5   Movement

The second type of dimensional change caused by changes in moisture content is that known as **movement**. Movement is the dimensional change resulting from diurnal or seasonal changes in relative humidity; consequently the magnitude of the dimensional changes is very much lower than is the case of shrinkage where really quite large changes in moisture content can be involved.

For comparative and characterisation purposes, movement is quantified in terms of the changes in dimension that occur when a piece of wood that has reached equilibrium moisture content in an atmosphere of 20°C/90 per cent relative humidity is transferred to and allowed to reach equilibrium in an atmosphere of 20°C/60 per cent relative humidity.

Movement is expressed either quantitatively as a percentage value for radial and tangential directions separately, or qualitatively as belonging to one of three arbitrarily defined movement classes, where each class is defined in terms of the *sum* of the movement percentages in the radial and tangential directions. Thus, in woods with a 'small' movement the sum of radial and tangential movement is less than 3.0 per cent; in woods with 'medium' movement, the sum will lie between 3.0 and 4.5 per cent; while in those woods listed as having 'high' movement, the sum of the individual values will be greater than 4.5 per cent.

Actual movement values on percentage basis for the same hardwoods and softwoods listed in Table 9.1 are set out in Table 9.2, while in Table 9.3, movement of a larger group of woods is listed under the class system. In both Tables 9.2

Table 9.2 *Movement (per cent) on transferring timber from 90 to 60 per cent relative humidity (from Farmer, 1972)*

Common name	Transverse		Movement class
	Tangential	Radial	
Hardwoods			
Afzelia	1.0	0.5	small
Agba	1.3	0.7	small
Poplar	2.8	1.2	medium
European oak	2.5	1.5	medium
European lime	2.5	1.3	medium
Beech	3.2	1.7	large
Softwoods			
Western red cedar, imported	0.9	0.4	small
Yellow pine	1.5	0.9	small
Norway spruce (white wood)	1.5	0.7	small
Douglas fir	1.5	1.2	small
Western hemlock	1.9	0.9	small
Scots pine (redwood)	2.2	1.0	medium

Table 9.3 *Movement classes for selected commercial timbers*

Small <3.0*	Medium 3.0–4.5*	Large >4.5*
Abura	Ash (Eur.)	Ash (Jap)
Afara	Elm	Beech
Af. mahogany	Kapur	Birch
Af. walnut	Keruing	Ekki
Agba	Lime	Holly
Balsa	Maple, rock	Ramin
Guarea	Oak (Eur.), oak (Jap.)	
Iroko	Poplar	
Jelutong	Sapele	
Mahogany (Central Am.)	Sycamore	
	Utile	
Makoré	Walnut (Eur.)	
Obeche		
Teak		
Douglas fir	Parana pine	
Norway spruce	Pitch pine	
W. hemlock	Radiata pine	
Yellow pine	Scots pine	

*These values represent the *sum* of the radial and tangential movements that occur on changing the relative humidity of the atmosphere from 90 per cent to 60 per cent at 25°C.

and 9.3, longitudinal movement is not listed since the value is so small in normal wood to be of no practical significance.

It was noted in the previous section that the values for shrinkage vary considerably from species to species: the situation is similar for values of movement and in very broad terms those timbers which have high shrinkage values have corresponding large values of movement, and vice versa; this is because all dimensional changes are primarily related to the microfibrillar angle of the S_2 cell wall layer.

However, this relationship between shrinkage and movement values does not always hold good. Thus the list of timbers in Table 9.2 is set out in the same order as in Table 9.1, yet the movement values for each set of hardwoods and softwood are not in consecutive order. In addition, there are many examples where high shrinkage is combined with only 'medium' movement, for example oak, and low shrinkage with high movement, for example ramin. A full explanation of this variation is not available but it is thought that the variation among woods in their equilibrium moisture content for any one humidity is an important contributory factor.

Timber with low movement values is always in demand, particularly so for high-quality joinery work, panelling and domestic flooring. Movement of timbers with higher values results in the loosening of joints and in the development of unsightly gaps. It is not always appreciated how much variation in humidity, and consequently movement, occurs on a diurnal and seasonal basis. Maximum moisture contents are to be found in the winter months in houses not centrally heated, and minimum moisture contents in the summer months; in centrally-heated houses and offices, however, the periods of maximum and minimum moisture content are reversed. In both cases the difference in seasonal moisture content can range from 3 to 6 per cent. Diurnal changes in moisture content are much lower and are more dependent on the thickness of timber used and the quality of the finish applied; as a very rough guide, diurnal change will range from 0.1 to 1 per cent.

Movement in service can be minimised by initially selecting wood of a moisture content midway in the range to be expected. A series of established moisture contents for woodwork in different environments is illustrated in Figure 9.9 and this can be used as the basis for specifying the moisture content of the wood at the time of installation. However, it is not sufficient merely to select timber of the correct moisture content for subsequent conditions. Immediately on completion, the relative humidity of the air in new buildings is likely to be abnormally high, and timber in equilibrium with such atmospheric conditions would reach equilibrium at a moisture content of between 16 and 20 per cent, according to the season. If joinery and finishings of 10 to 12 per cent moisture content are installed in these circumstances, some swelling, which might give rise to buckling or compression set, is to be expected. On the other hand, it would not do to use timber of 16 to 20 per cent moisture content, because appreciable shrinkage would occur later. Two alternative courses are to be recommended: either temporary heating should be installed, and the building dried out before the joinery and finishings are fixed, or fixing should be delayed for three to six months to give the building a reasonable opportunity of drying out of its own accord. Sadly, due to financial pressures, neither of the alternatives is usually applied in modern construction.

The practice of baking buildings in the early days of occupation is thoroughly unsound, and may result in considerable damage to the timber, because even the most carefully seasoned and fitted joinery will shrink and distort if suddenly exposed to much drier conditions than those for

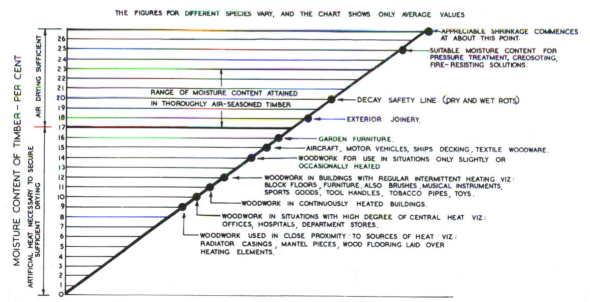

Figure 9.9 *Moisture content of timber in various environments. The figures for different species vary, and the chart shows only averages (Building Research Establishment, © Crown Copyright)*

which it was prepared. The proper course is to employ no more heat than is necessary for occupational use, and this applies both before and after the joinery is fixed. An alternative to using direct heat is to use de-humidifying equipment. In the early days of drying out a 'green' building, the results were quite dramatic, up to 1 gallon of water per hour being extracted from the structure. The intelligent use of such equipment greatly reduces shrinkage and swelling problems with second fixings and flooring, besides reducing the risk of mould growth problems if buildings have to be occupied very soon after completion.

Size, density, species, initial moisture content of the timber, and rate of air circulation and its temperature, are then the factors that influence the rate of adjustment of the moisture in wood to the relative humidity of air. It may also be mentioned that flat-sawn material will lose moisture more rapidly than quarter-sawn, and sapwood more rapidly than heartwood.

Before leaving this section on movement, mention must be made of both movement and shrinkage under restraint. In all the preceding sections we have considered the usual cases where timber is able to shrink or swell freely according to changes in moisture content. However, under certain circumstances dimensional changes may not be possible even though the moisture content of the timber has changed. A parquet floor laid tightly with well-dried blocks, if subjected to high humidity levels or flooding will want to expand; where the bonding to the floor is poor the whole floor will rise and become arched, but where the bonding is very good, each block of flooring will be restricted by its neighbour and the only means of relieving the swelling stress is for the cell walls to expand into the cell cavities. When the wood subsequently dries out it shrinks and gaps appear — in other words the wood has become permanently or irreversibly compressed, thereby increasing its density and this feature is known as **compression set**. Swelling of wooden handles with a steel collar will also lead to compression set with subsequent loosening of the handle in the collar on drying out. **Tension set** can also occur where drying out

occurs under tensile restraint. Ordinary re-wetting or further drying out of set wood will not relieve the condition, but most of the set can be removed by high-temperature treatments in a saturated atmosphere, for example steaming.

9.6 Dimensional stabilisation

Having read the sections above on shrinkage and movement, the reader will be aware that the movement of wood in service can detract from its use. At best it can result in unsightly gaps: at worst it can cause buckling of components thereby preventing their use. Various attempts have been made to reduce the effect of moisture on dimensional change, some more successfully than others.

The slicing of wood into veneers and their reconstitution into a sheet of plywood in which the grain in alternate veneers is at right angles does much to reduce movement since the tangential movement of one veneer is restricted by the almost absence of longitudinal movement in the two adjacent veneers. When compared with timber, plywood has much lower movement values both along and across the board. However, it must be appreciated that these values are higher than the longitudinal movement of timber.

Unlike plywood, the formation of 'glulam' (laminated timber) does little to reduce movement since all the lamina run in the same direction. The only advantage of glulam in this respect is that any excess of movement in individual laminates will be averaged out.

It is claimed that application of various oil-based or alkyl-based finishes will do much to slow down or even remove movement of the wood underneath: this is certainly true in the short term when daily changes in humidity will have little effect on the wood. However, where coated wood is exposed to high humidities for a long period of time, movement will still occur as the result of uptake of water vapor, albeit at a very slow rate, since no commercial synthetic finish is completely impermeable to water vapour. Paints and varnishes, therefore, slow down but certainly do not arrest movement.

Good dimensional stabilisation can be obtained by heating timber at very high temperatures (250–300°C) for short periods of time, at lower temperatures (180–200°C) in an inert gas atmosphere of 8–10 bar (Giebler, 1983), or at even lower temperatures (120–160°C) for periods of several months. Such treatment, however, will result in thermal degradation of the wood with loss in strength and toughness (Chapter 18). Indeed it is the thermal degradation of the hemicellulose which appears to reduce the propensity of wood to swell (Stamm, 1977).

A more fundamental approach to greater stability of wood in the presence of moisture is to reduce its hygroscopicity; that is the accessibility of the matrix constituents to water. This can be achieved by chemical substitution for the active hydroxyl by less polar groups; the most successful attempt has been by acetylation. In this process, acetic anhydride is used as a source of acetyl groups while pyridine is added as a catalyst (Rowell, 1984). A marked improvement in dimensional stability can be achieved in both wood and board materials with only a marginal loss in strength (Rowell, 1984; Rowell *et al.*, 1991).

Cross-linking can also lead to better stability: thus reacting wood with formaldehyde results in the formation of methylene bridges between adjacent hydroxyl groups. This process results in a marked reduction in shrinkage without bulking up the wood.

Several chemicals provide dimensional stability by holding the wood in a swollen state even after the removal of water, thereby minimising further dimensional movement. Both phenol formaldehyde and polyethylene glycol (PEG) have been used successfully, the latter chemical being widely used in the conservation of previously waterlogged items.

A further example of wall bulking is to impregnate wood with low-viscosity liquid polymers which are then polymerised *in situ* to convert them to solid polymers. The polymer impregnates the cell wall as well as coating the cell cavity, thereby almost eliminating accessibility to moisture. Wood so treated is referred to as **wood–plastic composite**, and although about four times as expensive as wood, it is being used commercially for one or two specialised purposes where dimensional stability and increased abrasion resistance are required (see Chapter 17, section 17.2.2).

References

Boyd J.D. (1974) Anisotropic shrinkage of wood: identification of the dominant determinants. *Mokuzai Gakkaishi* **20**: 473–482.

Chafe S.C. (1991) A relationship between equilibrium moisture content and specific gravity in wood. *J Inst Wood Sci* **12(3)**: 119–122.

Farmer R.H. (1972) *A handbook of hardwoods*, 2nd ed. HMSO: available only from Building Research Establishment Bookshop.

Giebler E. (1983) Dimensionsstabilisierung von Holz durch eine Fenchte/Wärme/Druck-Behandlung. *Holz als Roh- und Werkstoff* **41**: 87–94.

James W.L. (1988) Electric moisture meters for wood. *Forest Products Laboratory, Madison, General Technical Report*, FPL-GTR-6.

Meylan B.A. (1968) Cause of high longitudinal shrinkage in wood. *For Prod J* **18(4)**: 75–78.

Rowell R.M. (1984) Chemical modification of wood. *Forest Products Abstracts* **6**: 363–382.

Rowell R.M., Youngquist J.A., Rowell J.S. and Hyatt J.A. (1991) Dimensional stability of aspen fibreboard made from acetylated fiber. *Wood Sci & Technol* **23(4)**: 558–568.

Simpson W. (1980) Sorption theories applied to wood. *Wood and Fiber* **12(3)**: 183–195.

Skaar C. (1988) *Wood–Water Relations*. Springer-Verlag, Berlin.

Stamm A.J. (1977) Dimensional changes of wood and their control. In Godstein I.S. (Ed.), *Wood Technology: Chemical Aspects*. A.C.S. Symposium Series 4 American Chemical Society, Washington, D.C.

10

Other Physical Properties of Wood

The principal physical properties affecting the general performance of wood, namely density and moisture, have been discussed in detail in Chapters 8 and 9. There are however a number of other physical properties, as distinct from mechanical properties (see Chapter 11), which are of secondary importance in many general timber applications, and of particular importance in a few specialised areas. This collection of properties is covered in this chapter.

10.1 Thermal properties of wood

10.1.1 Thermal conductivity

Thermal conductivity (λ or k) of a material is a measure of its ability to transmit heat and is defined as the quantity of heat in joules which will flow through one square metre of the material one metre thick in one second where there is a one degree Celsius difference in temperature between its surfaces. It is therefore quantified in units of W/m K (W = Watts = joules/second; K = degrees Kelvin). Thus the smaller the value, the greater the resistance of the material to the passage of heat.

Dry wood is one of the poorest conductors of heat and this characteristic renders wood eminently suitable for internal wall panelling, wall sheathing in timber frame house construction, wall cladding and as handles of cooking utensils. This outstanding performance of wood is due in part to the low conductivity of the actual cell wall materials, and in part to the cellular nature of wood which in its dry state contains within the cell cavities a large volume of air – one of the poorest conductors known.

The cellular structure of wood also partly explains why heat is conducted about two to three times as rapidly along, compared with across the grain and why, in association with variation in cell-wall thickness, heat is conducted more rapidly in high-density compared with low-density woods. A range of λ values is presented in Table 10.1, illustrating the marked effects of density and grain orientation. This table also contains information on other materials in order to permit a direct comparison between timber and timber products and competing materials of construction.

The values for timber and timber products in Table 10.1 relate to oven-dry wood. Thermal conductivity increases markedly with increasing moisture content, being about twice as high at 100 per cent moisture content as it is at 10 per cent. This works to advantage in the artificial drying of green wood (kilning) where the rapid conduction of heat throughout the stack is essential.

The resistance to heat flow through a specific thickness of material (or width of cavity) of unit area is known as the thermal resistance of the material or cavity, and is expressed in units of m^2 K/W. It is equal to the thickness of the material divided by its thermal conductivity, λ, and is used directly in the calculation of the 'u' insulation values in seeking compliance with the UK Building Regulations.

10.1.2 Specific heat capacity

The specific heat capacity of a substance is the amount of heat required to raise the temperature of one gram of the substance by 1 K; it is quantified in units of J/kg K. For oven-dry wood, the specific heat capacity is about 1360 J/kg K, a value almost independent of the density of the wood. However, this value will increase with

Table 10.1   *Thermal properties*

Material	Density (kg/m³)	Thermal conductivity (W/m K) H	L	Specific heat capacity (J/kg K)	Linear thermal expansion (×10⁻⁶ per K) R	T	L
Balsa	176	0.06	—		16	24	—
Spruce	340	0.10	0.21		24	35	3.5
Douglas fir	512	0.11	—	1360	27	45	3.5
Yellow birch	660	—	—		32	39	3.6
European oak	673	0.16	0.28		—	—	—
Plywood	550						
– in plane of board		—		—			7
– through board		0.13		—			51
Chipboard	650						
– through board		0.13		—			10
Brick	2300	1.0		—			9
Concrete	2400	1.5		3350			12
Steel (mild)	7860	63		420			15
PVC (rigid)	1700	—		1000			190
Polystyrene (cellular)	—	0.03		—			—

H = mean of values in radial and tangential direction; L = along the grain.
R = in a radial direction; T = in a tangential direction.

both increasing temperature and moisture content. Specific heat capacity, like thermal conductivity, is an important parameter in determining the rate of drying of green timber in a kiln.

quoted in Table 10.1 have been derived by artificially maintaining the same moisture content at different temperatures, an occurrence unlikely to happen regularly in real-life situations.

10.1.3   Linear thermal expansion

Most materials expand on heating owing to the greater oscillation of their molecules as a result of absorbing thermal energy, and wood is no exception. However, the effect in wood is complex. At the same time as the increased thermal energy is increasing the distance between the various molecules, it is also drying out the wood and, as was noted in Chapter 9, this results in a reduction in both length and width. The net effect is that timber will actually decrease in width and length as the temperature increases. It should be appreciated therefore that the values for linear expansion of timber and timber products

10.1.4   Heat (calorific) value

Like other organic materials, wood is combustible; under suitable conditions it will burn, and its constituents undergo oxidation with the liberation of energy in the form of heat. The fuel value of a timber depends largely on the amount of wood substance in a given volume, that is, on the density, the chemical composition of the wood substance and the state of dryness of the wood. As a general rule, the denser the timber the higher its potential fuel value, but this may be modified by the presence in the wood of such substances as resin. The fuel value of resin is about twice that of wood substance and, other things being equal, resinous woods have a higher fuel value than non-resinous woods. The influence of moisture content will readily be understood: wet wood has a much lower heating value than dry wood of the same species, because much heat is lost in transforming the contained moisture into steam; the heat value of wet wood is generally of the order of 40–60 per cent of oven-dry wood, depending on the amount of water present.

As noted above, the heat value of wood from different species will be different even when expressed in terms of oven-dry mass of wood. However, the average heat value of oven-dry wood is usually taken as 18.5×10^6 J/kg. In the comparison of values between different species it should be appreciated that where wood is being burnt as a fuel to raise energy, the heat value of the wood is only one factor to be considered in assessing overall efficacy: regular combustion and low ash content are also very important, though difficult to quantify.

In comparing the heat value of wood with that of coal it is found that on an oven-dry basis, coal will yield about 1.6 times as much heat as an equal mass of wood: this assumes a 100 per cent efficiency in the burning of wood which is probably an over-estimate. Another disadvantage of wood as a fuel compared with coal is its greater volume and hence higher costs in transportation, handling and storage.

Charcoal, produced by the controlled burning of wood under reduced availability of oxygen, is an alternative fuel to ordinary wood; it has a higher fuel value than wood both on a volumetric and a weight basis. The advantages of charcoal over wood as a fuel are largely economic ones, associated with low transport and handling charges per joule of heat produced, the saving more than offsetting its cost of manufacture.

Well over half the total consumption of wood is burnt as a fuel, the largest proportion of which is consumed in a most wasteful manner in third-world countries in stoves or on open fires on the ground.

In many of the industrialised countries, wood is also burned to produce heat, but in this case the wood comprises waste from wood processing plants in the form of offcuts and trimmings. These are subsequently chipped and sometimes formed into briquettes prior to control burning in furnaces to raise steam for the production of heat and electricity required within the factory.

Within the last few years, studies have been undertaken by the UK Department of Trade and Industry to examine the technical and economic feasibility of using the vast quantities of fast growing hardwoods, especially poplar, produced by farm forestry as applied to lowland agricultural land taken out of food production and planted up with trees. These studies have indicated the feasibility of constructing a national network of farm-based mini power stations. The wood would be burned in a new style gasification reactor claimed to be 85–90 per cent efficient. Currently there is speculation that two or three very large reactors of this type, each capable of producing up to 20 million MW of electricity, could be built in some of the very large existing national forests.

10.1.5   Reaction of wood to heat

Under prolonged exposure to elevated temperatures, timber will show signs of thermal degradation. Some loss in strength will occur, but the property which is more sensitive is toughness, and this will be discussed in more detail in Chapters 11 and 18.

10.1.6   Reaction of wood to fire

When wood is heated above about 100°C, volatile gases begin to be emitted and, if heating continues, the build-up in the amount of gas released through thermal degradation of the wood is such that at surface temperatures of about 250°C, in the presence of a pilot flame, ignition of the wood occurs. This combustion raises the temperature of wood under the surface and, in consequence, the fire is kept going until all the wood is ultimately completely consumed or burned.

Timber above a certain critical thickness, which is of the order of 12 mm, cannot support self-maintained combustion, and continued burning is only possible so long as the surface receives an additional heat output from some radiant source such as the flames of a neighbouring fire. The formation of charcoal on the outside of a piece of wood probably acts as a screen against radiant or conducted heat, thereby retarding distillation of flammable gases from within. In consequence, the rate of burning is much reduced, and, when the layer of charcoal is sufficiently thick, burning may become so slow that insufficient heat is produced to continue the decomposition of wood further in, and the fire goes out. This is what happens with timbers of large dimensions, and explains why, in quite large fires, heavy timber posts and beams often survive when a building is otherwise completely gutted by fire.

The low thermal conductivity of wood has an important bearing on the way burning wood transmits a fire. Some of the heat produced is immediately radiated outwards, some is absorbed in raising the temperature of wood just inside the burning area, and the remainder is conducted to the opposite face, whence it is radiated. Further, the behaviour of the flames is important. They rise upwards, and therefore transmit more heat to wood above the source of the flames than to that below or to the sides. Hence, wood held vertically and ignited at the bottom will burn more readily than the same wood ignited at the top, or held horizontally. This explains why doors, panelling and other vertically disposed timbers constitute a greater fire hazard than beams and floors (the greater size of beams and floors is also in their favour).

Although timber is a combustible material, its behaviour in fire is predictable in terms of the rate of char formation, retention of strength in the core and absence of distortion. For most softwoods and medium-density hardwoods the rate of char formation is about 0.64 mm/minute, while for high-density hardwoods the rate is about 0.5 mm/minute. There are many facets to the behaviour of material in fire and Table 10.2 sets out the rating of timber and board materials in the various arbitrarily defined components comprising the early stages of fire development: the standard test methods used in the UK in their evaluation are also listed.

It will be noted from Table 10.2 that, although timber and board materials are rated as 'combustible', they are classified none the less as 'not easily ignitable': considerable input of heat is required before timber can be ignited.

Perhaps the most frequently quoted parameter of the performance of materials in fire is the 'surface spread of flame' rating: in this respect timber and board materials do not rate very highly, being designated class 3 where the density is above 400 kg/m^3, and class 4 where the density is below this value. The highest rating on the scale is class 1, the requirements of which are met by asbestos–cement boards and plasterboards.

Table 10.2 *The reaction of timber and board materials to the various components of fire development*

Component of reaction to fire	British Standard test method of assessment	Rating of timber and board materials
1. Non-combustibility	BS 476, Pt 4 (1970)	'Combustible'
2. Ignitability	BS 476, Pt 5 (1968)	'Not easily ignitable'
3. Surface spread of flame	BS 476, Pt 7 (1971)	Density >400 kg/m^3 – class 3 Density <400 kg/m^3 – class 4
4. Rate of heat release (fire propagation)	BS 476, Pt 6 (1968)	Does not satisfy requirements for class 0
5. Fire resistance (of the elements of a structure)	BS 476, Pt 8 (1972)	Rating depends as much on structure as on components

'Rate of heat release' is applicable to lining materials and relates to the rate at which a combustible material contributes heat to a developing fire under its influence: in this respect all wood-based materials are at a distinct disadvantage and fall far below the requirements for class 0, a rating which is demanded for materials used as linings in public buildings.

'Fire resistance' is defined as the ability of an element to continue performing its function when exposed to a fully developed fire and, as such, relates to components and not to individual materials: separate test methods are available for different components of buildings.

Timber and board materials can be effectively treated to reduce their ignitability and subsequent rate of flame spread and it is possible to achieve class 1 rating. However, it must be appreciated that these treatments have little effect on changing either their combustibility or the fire resistance rating of a component or timber structure. These treatments are no more than **flame-retardant** processes and their applicability to only one single component of fire behaviour must be clearly understood.

Flame-retardant treatments take the form of either surface applications or the impregnation of the timber with certain chemical solutions. Surface applications are used for *in situ* treatments and comprise either intumescent materials, or flame-retardant paints. Intumescent materials froth up at elevated temperatures and this ability to create a physical barrier can be used to good effect in their application to the edge of fire doors: at elevated temperatures these strips intumesce, thereby sealing the gap around the door and preventing the passage of smoke and toxic fumes. Flame-retardant paints are usually applied to the surfaces of panels and, as the name implies, are capable of reducing considerably, though certainly not arresting, the spread of flame across the surface.

Flame-retardant chemicals may also be applied to timber by impregnation, but this must be done prior to site construction. The chemical salts most commonly employed in the UK at the present time are mono- and di-ammonium phosphates, ammonium sulphate, boric acid and borax.

Most of the proprietary flame-retardants on the market at present are mixtures of these constituents, formulated to give a balance between performance and cost. Treatment of the timber involves vacuum impregnation of an aqueous solution thereby necessitating redrying of the timber following impregnation.

10.2   Electrical conductivity

Wood of zero moisture content offers practically zero conductance to an electric current and in the 1930s and 1940s wood was used for domestic plug tops and switches and for separators in lead batteries. Wood impregnated with phenolic resin was, until fairly recent times, used as an insulator in power generating plants.

Electrical conductivity increases with increasing moisture content and this relationship is put to good application in the design and use of resistance-type electric meters for the measurements of moisture content as discussed in Chapter 9, section 9.2.3. At any one moisture content, conductivity increases with increasing density of the wood: this effect is reflected in the use of different scales on the meter for different timbers.

10.3   Acoustic properties of wood

10.3.1   *Sound production*

A piece of wood, so fixed as to be allowed to vibrate freely, will emit a sound when struck, the pitch of which will depend on the natural frequency of vibration or resonance of the piece. This is a function of the elastic properties of the wood which in turn are governed by the density and the dimensions of the piece. Wood, the elasticity of which has been destroyed, as for instance by fungal decay, will give a dull sound when tapped, in contrast to the clear ring of sound wood.

The special quality imparted to notes emitted by wood is very pleasing and causes wood to be extensively used in sounding-boards and other parts of musical instruments. Uniformity of texture, which comes from extreme regularity in growth throughout the life of the tree, and freedom from defects, are the essential properties of timber for sounding-boards. Slow-grown spruce, from restricted areas in Czechoslovakia, is perhaps the most famous source of supply of high-grade piano and violin sounding-boards (see Chapter 16).

10.3.2 Sound reduction

Reduction in the level of sound can be effected by either reducing the amount of sound reflected off surfaces, or absorbing it in passing through materials.

A considerable 'deadening' effect can be achieved in a room by hanging drapes around the room or, to a lesser extent by panelling the room, including the ceiling, with wood. The cellular nature of wood markedly reduces the reflection of sound from the surface. Acoustic panels made of fibreboard with holes drilled through them every 25 mm have proven themselves to be most effective at reducing sound levels through a marked reduction in the reflection of sound waves.

The ability of a material to absorb sound is dependent on its mass and consequently it is only the high-density hardwoods which are capable of reducing the transmission of sound through them. It is in this area that certain wood-based sheet materials come into their own. For many years now, wood chipboard with a density of $680 \, kg/m^3$ has been used in the manufacture of HiFi speaker enclosures, and within the last decade cement bonded particleboard with a density of $1200 \, kg/m^3$ has been effectively used on the European continent as a sound barrier on motorways: an 18 mm thick board of this material will provide a reduction in sound level of 33 dB.

References

BS 476   *Fire tests on building materials and structures:*
   Part 4: 1970   *Non-combustibility test for materials* (reprint incorporated Amendment AMD 2483, March 1978)
   Part 5: 1979   *Method of test for ignitability*
   Part 6: 1989   *Method of test for fire propagation for products*
   Part 7: 1987   *Method for classification of the surface spread of flame of products.* BSI, London.

Strength, Elasticity and Toughness of Wood

11.1　Information

The strength of a material such as wood refers to its ability to resist applied forces that could lead to its failure, while its elasticity determines the amount of deformation that would occur under the same applied forces. These forces may be applied slowly at constant rate whereby we refer to the inherent resistance of the material as its **static strength**, or they may be applied exceptionally quickly, when we refer to the resistance of the material as its **dynamic strength**.

In the application of design to the structural use of timber it is necessary to design against both static failure and excessive deflection, the latter a manifestation of the elastic properties of wood. Either of these two parameters may be the limiting factor in a particular design. Additionally, it is frequently necessary to carry out impact testing of a prototype component in order to determine that there is sufficient resistance in the designed structure to the application of dynamic loads.

11.2　General principles

The application of a small load to a sample of wood will cause that sample to deform; the application of additional small loads will cause further deformation of the sample, and it will be found that the increments in deflection are proportional to the increments in load. This is illustrated in the lower half of the load–deformation graph illustrated in Figure 11.1 as a straight line and can be expressed as

$$\text{applied load} \propto \text{deformation}$$

or

$$\frac{\text{applied load}}{\text{deformation}} = \text{a constant}$$

The value of this constant will vary with size of the sample, hence it is necessary to express load in terms of the cross-sectional area over which it is applied, and deformation in terms of the initial length of the sample, namely:

$$\frac{\text{load (N)}}{\text{cross-sectional area (mm}^2)} = \textbf{stress } (\text{N/mm}^2)$$

$$(\text{stress is denoted by } \sigma)$$

and

$$\frac{\text{deformation (mm)}}{\text{original length (mm)}} = \textbf{strain } (\text{unitless})$$

$$(\text{strain is denoted by } \varepsilon)$$

hence

$$\frac{\text{stress } (\sigma)}{\text{strain } (\varepsilon)} = \text{a constant}$$

$$= \textbf{modulus of elasticity}$$

The **modulus of elasticity** (also known as Young's modulus) is denoted by E and expressed in units of N/mm^2.

E is a material constant characterising one piece of wood. It will be similar for other samples from the same part of the tree but, as will be described later, it will vary between different species. E is frequently referred to as the **stiffness** of wood, a popular term which conveys an appropriate image. Strictly speaking, the term stiffness is the product of the modulus and the second moment of area (I): that is, stiffness $= EI$. However, in the remainder of this chapter the popular use of the term will be used to denote E.

In the straight-line graph in Figure 11.1a wood will behave in a truly elastic fashion, and the removal of any applied load will result in zero deformation, that is loading follows the graph upwards, while unloading follows the graph back to zero, that is all the deformation is **recoverable**. In comparing different timbers, that with the highest slope will have the highest stiffness.

However, above a certain level of loading known as the **limit of proportionality**, departure from linearity occurs such that for each increment of load there is a more than proportional increment in deformation. If an applied load above the limit of proportionality is removed, the sample will not return to zero deformation, but follow a line lying parallel to the initial linear region and terminating on the horizontal axis at some finite deformation. Thus, permanent deformation has been induced in the sample which will take the form of cell crushing, if the load has been applied in longitudinal compression, or cell-wall rupture, if a longitudinally applied tensile load has been applied (Figure 11.1b).

The application of additional load will result initially in more permanent deformation and finally in failure of the sample. The stress level (load divided by cross-sectional area) at which failure occurs is deemed to be the **strength** of the wood, and the value of this will depend on the mode of stress application, for example, tension or compression.

The limit of proportionality also varies with mode of stress application. In longitudinal tensile stressing the limit occurs at about 60–65 per cent of the failure stress, while in longitudinal compressive stressing the limit is much lower at 30–50 per cent (Figure 11.2).

The linear behaviour below the limit of proportionality will be obtained only at fairly fast rates of loading. At low rates of load application there will be an effect of time under load as discussed in section 11.5.7: this will induce a measure of non-linearity into the graph.

Dynamic loading, where the rate of load application is exceptionally quick, is carried out to determine the **toughness** or impact resistance of the material. Unlike strength, which is a measure of the resistance of wood to an applied

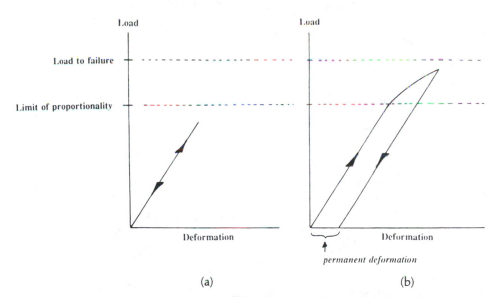

(a)   (b)

Figure 11.1 *Load–deformation curves for pieces of wood loaded to (a) below, and (b) above the limit of proportionality but below the failing load. On unloading in (a) no permanent deformation occurs; on unloading in (b) some permanent deformation will have occurred*

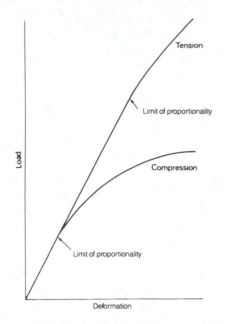

Figure 11.2 *Generalised load versus deformation diagram for wood stressed in tension and compression parallel to the grain; the limit of proportionality for each is indicated (Building Research Establishment, © Crown Copyright)*

stress, the toughness of a material is a measure of the energy required to propagate cracks in it and is quantified in units of J/mm^2. A measure of the toughness or impact resistance can be obtained by measuring the area under the load–deflection curve to maximum load – a measure of the work done or capacity to store energy before failure.

The strength of wood will vary within mode of load application; the principal modes are tension, compression (both of which can be parallel or perpendicular to the grain), bending and shear. Unlike the position with strength, modulus of elasticity in tension, compression and bending is similar and a common value for modulus of elasticity for all three modes of load application in each of the three principal planes is usually adopted.

In the case of shear, within the elastic range of wood, shear stress is proportional to shear strain, namely:

$$\text{shear stress } (\tau) \propto \text{shear strain } (\gamma)$$

or

$$\frac{\text{shear stress } (\tau)}{\text{shear strain } (\gamma)} = \text{constant} = G$$

where G is the **modulus of rigidity** and expressed in N/mm^2.

When wood is subjected to a stress in one direction it will undergo a change in dimensions at right angles to the direction of stressing. The ratio of this contraction or extension (at right angles) to the applied *strain* is known as **Poisson's ratio** and is given as:

$$\nu_{xy} = -\frac{\varepsilon_y}{\varepsilon_x}$$

where ε_x, ε_y are strains in the x, y directions resulting from an applied stress in the x direction. The minus sign indicates that, when ε_x is a tensile positive strain, ε_y is a compressive negative strain. Wood, because of its anisotropic behaviour, requires six Poisson's ratios to characterise it.

Values of strength, modulus of elasticity, modulus of rigidity and Poisson's ratios are given for selected timbers later in this chapter.

Readers desirous of understanding the theory of elasticity (applicable to isotropic materials) and its application to wood (an anisotropic material) are referred to one of the following texts: Jayne (1972), Dinwoodie (1981) or Bodig and Jayne (1982).

11.3 Determining the strength and elastic properties

11.3.1 Sample size

In the assessment of the strength and elastic properties it is possible to use two different approaches, one of which is based on small, clear specimens, while the other employs actual structural grade timbers.

Small clear test specimens
This method was for many years the basis for the derivation of working stresses for wood. However, since the mid-1970s the method has been

superseded by tests on actual structural-sized timber: the method still remains valid for characterising new timbers and for the strict academic comparison of wood from different trees or different species. The method utilises in a standardised procedure small, clear, straight-grained pieces of wood which represent the maximum quality that can be obtained. As such, the test specimens are not representative of timber actually being used without the application of a number of reducing factors. However, the method does afford the direct comparison of wood from different species, and it does permit the derivation of working stresses for either new timbers, or those timbers with only limited use, since it is a much cheaper method than full structural testing, though perhaps less representative of actual usage.

Structural-size test specimens

This approach to the derivation of working stresses for structural timber commenced in the mid-1970s and has now been applied to most of the structural softwoods and to one or two species of hardwoods used in the UK (see BS 5268: Part 2: 1991). Tests on timber of structural size more nearly reproduce actual service conditions, and they are of particular value because they allow directly for defects such as knots and splits, rather than applying a series of arbitrary reduction factors as employed in using the small clear specimens. They have the disadvantage of being costly, because of the large amount of timber required, and the length of time needed to load larger-sized test specimens to the point of failure.

11.3.2   Test methods

Using small clear tests

The procedure for tests on small, clear specimens has been standardised (BS 373). As the strength properties of wood are greatly influenced by its moisture content (see section 11.5.5), tests are made separately on 'green' materials, that is freshly-felled timber, and on material dried to a standard moisture content (usually 12 per cent), the timber being brought to this moisture content

in special conditioning chambers; alternatively, tests are made on air-dry material of known moisture content and the strength figures obtained are corrected to the standard moisture content. As it has been shown that the ultimate strength properties of a piece of wood are affected by the rate of strain, all test specimens in each test are loaded at a fixed and constant rate. For a full-scale test it has become usual to select material from five healthy trees of merchantable size, and characteristic of the average in the locality; this number is usually sufficient to give a representative measure of the variation in different pieces of wood of the same species.

The preparation of test samples and methods of test in the UK are based on BS 373: 1957 (1896) *Methods of testing small clear specimens of timber*, though some slight improvements to these methods have been made over the years and a brief account of the current methods is given below. The test samples must be conditioned to $20 \pm 3°C$ and 65 ± 2 per cent relative humidity, and tested within the same temperature and humidity ranges.

Tensile strength parallel to the grain: Owing to the very high tensile strength of timber, waisted samples are used, thereby providing sufficient wood to be gripped in the jaws to avoid crushing, yet little enough wood in the centre of the sample to allow failure to occur at loads below those which would pull out the sample from the jaws. The sample illustrated in Figure 11.3 is 300 mm in length, with a cross-section of 20×6 mm at the ends, waisted to 6×3 mm: it is usually loaded at a constant speed of 0.02 mm/s.

Tensile strength perpendicular to the grain: Although described in some textbooks, this test is rarely performed owing to difficulties in obtaining a true tension stress in the wood. Furthermore, the test has little practical significance since most of the failures that occur in practice are cleavage failures originating at one side, rather than true tensile perpendicular failures.

Compression strength parallel to the grain: One of the precautions necessary in evaluating this property is the need to ensure that the specimen does not buckle during loading, thereby subjecting it to a bending rather than a compressive stress.

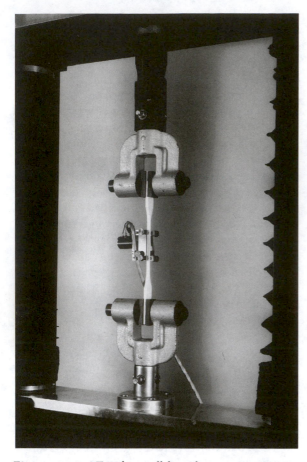

Figure 11.3 *Tensile parallel to the grain test using a waisted sample; this is fitted with an extensometer (Building Research Establishment, © Crown Copyright)*

To this end it is customary to use a special cage which ensures a uniform distribution of load over the cross-section. The sample size is $60 \times 20 \times 20$ mm and the load is applied to the piston of the cage at a rate of 0.01 mm/s.

The compression strength is obtained by dividing the load to failure by the cross-sectional area. While it is possible to obtain the modulus of elasticity for compression by recording the deformations for known loads and converting them to strain and stress respectively, this is seldom done in practice since it has been found that the modulus of elasticity in compression is similar to that in static bending which is easier to perform with accuracy.

Compression strength perpendicular to the grain: This test is seldom carried out in the UK — rather the strength is determined from the hardness of the timber since it has been established that there is a very high correlation between the two properties.

Static bending: The usual method of supporting and loading the $300 \times 20 \times 20$ mm sample is illustrated in Figure 11.4. It will be noted that the central loading head is radiused and that the ends of the sample are supported in trunnions carried on roller bearings; this provides friction-free lateral movement of the bearing points to accommodate horizontal shortenings of the beam due to deflection.

The sample is prepared so that the growth rings are parallel to one edge and the sample is tested with the growth rings parallel to direction of loading, that is it is loaded on the radial face: load is applied at a speed of 0.1 mm/s. The bending strength of wood is usually presented as a **modulus of rupture** (MOR) which is the equivalent stress in the extreme fibres of the specimen at a point of failure assuming that the simple theory of bending applies. The MOR in three-point bending is calculated from the following equation:

$$\text{MOR} = \frac{3PL}{2bd^2}$$

where MOR is the modulus of rupture, in N/mm^2

  P is the load, in N
  L is the span, in mm
  b is the width, in mm
and d is the depth, in mm.

It is customary to calculate the modulus of elasticity in bending; consequently load–deflection graphs are recorded automatically, or in the absence of this facility on the test machine, the loads corresponding to increments of deflection as recorded by a dial gauge are recorded, and the equivalent stresses and strains determined.

The modulus of elasticity (E) in three-point bending is calculated from the following equation:

$$E = \frac{P'L^3}{4\Delta' bd^3}$$

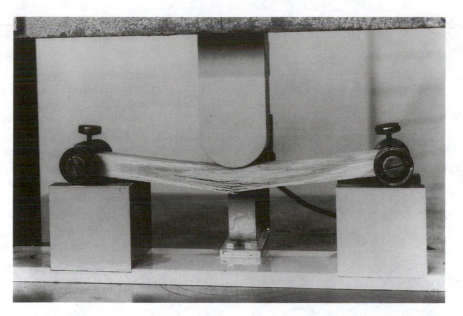

Figure 11.4 *Static bending test using 300 × 20 × 20 mm samples (Building Research Establishment, © Crown Copyright)*

where

E   is the modulus of elasticity in bending, in N/mm^2

P' is the load, in N, at the limit of proportionality

L is the span, in mm

Δ' is the deflection, in mm, at the limit of proportionality

b is the width, in mm

d is the depth, in mm.

In addition to bending strength and modulus of elasticity in bending, this test also provides values of work to maximum load and total work, both of which are a measure of toughness. Both parameters are measures of the energy absorbed and are calculated from the area under the load–deflection curve; the former parameter represents the area up to the point of inflection of the curve, while the latter parameter represents the total area under the curve: values are recorded in mm N/mm^3.

The test as described introduces shear stresses that are unimportant when comparison of the bending qualities of different timbers is made. The influence of shear stresses should, however, be eliminated when accurate figures for both modulus of rupture and modulus of elasticity are required, and this is achieved by applying the load at two points equidistant from the points of support, in what is known as a four-point bend test.

Impact bending: Several tests, some of them applied in metallurgy, have been used for this property, but the method which is presented in BS 373 and used in the UK is a modification of the Hatt–Turner impact machine, whereby samples of wood are subjected to repeated blows from a weight falling from increasing heights. The apparatus is illustrated in Figure 11.5 from which it will be noted that the 300 × 20 × 20 mm sample is supported over a span of 240 mm on supports radiused to 15 mm: spring-restrained yokes are fitted to arrest bounce. The falling weight of 1.5 kg is also radiused to 15 mm and is dropped initially from 50.8 mm, then every 25.4 mm up to 254 mm and then every 50.8 mm until complete failure occurs at which point the height is recorded in metres.

Shear strength parallel to the grain: This test is performed on a 20 mm cube of wood which is placed in a shear test rig such that when the cube

Figure 11.5 *Impact test on a 300 × 20 × 20 mm sample using a modified Hatt–Turner testing machine (Building Research Establishment, © Crown Copyright)*

is loaded at about 0.01 mm/s shear will occur along the grain. In most timbers the shear strength is similar in both the radial and tangential plane.

Cleavage: The cleavage sample some 45 mm in length and 20 × 20 mm in cross-section is cut at one end to accommodate the grips: the splitting force is applied at 0.04 mm/s and the resistance to cleavage is expressed as force per unit width. Separate samples are employed to give values for radial and tangential cleavage.

Hardness: The hardness of wood is assessed by its resistance to the impregnation of a special hardened steel tool (Janka test) rounded to a diameter of 11.3 mm which is embedded to half its diameter. Generally the tool is fitted with some type of electronic device which indicates when the correct depth has been achieved. The test is usually carried out once on each of the radial and

tangential surfaces of the static bending sample, and to avoid end-splitting the sample is surrounded by similar pieces and placed in a clamp.

Using structural-sized test specimens

The test procedures for the assessment of the strength and elastic properties of test specimens of structural-size are set out in BS 5820: 1979 (*Determination of certain physical and mechanical properties of timber in structural sizes*). Before testing, the test specimens are normally conditioned to constant mass at a temperature of $23 \pm 3°C$ and a relative humidity of 60 ± 2 per cent: these conditions are slightly different from those in BS 373 for the testing of small clear test pieces and will result in differences in moisture content between the two sets of test specimens.

Static bending: Unlike the static bend test for small clear samples where the test pieces were loaded in the middle of the sample (three-point or centre-point bending), the test for structural-sized timber employs a four-point bend test in which the distance between the two loading points is equal to the distance between one loading point and the nearer support: this particular configuration is referred to as 'third-point bending'. The distance apart of the reaction supports is 18 times the nominal depth of the specimen, though if this cannot be achieved, the geometry of the test may be modified within specified limits.

Separate tests on the same test piece are carried out for determining strength and modulus of elasticity, each with its own rate of loading (see BS 5820: 1979). In the test for modulus of elasticity which is carried out first, a record of load–deflection is made.

The modulus of rupture in four-point bending is calculated from the following equation:

$$f_m = \frac{F_{max} a}{2W}$$

where

f_m is the bending strength, in N/mm^2
F_{max} is the bending strength, in N
a is the distance between an inner load point and the nearest support, in mm
W is the section modulus determined from the actual dimensions of the section, in mm^3.

The corresponding modulus of elasticity in four-point bending is calculated from the following equation:

$$E_m = \frac{\Delta F a l_1^2}{16 I \Delta W}$$

where
E_m is modulus of elasticity in bending, in N/mm^2
ΔF is an increment of load, in N
a is as above for MOR
l_1 is the gauge length, in mm
I is the second moment of area of the section determined from its actual dimensions in mm^4
ΔW is the deflection under the increment of load ΔF, in mm.

For the calculation of the shear modulus using the single span method it is necessary to have both the true bending modulus, as previously described and the 'apparent' bending modulus as obtained from a three-point bending test, using the same test piece but supported over a short span equal to the gauge length used in the four-point bend test. The apparent modulus is calculated from the following equation:

$$E_{m\ app} = \frac{\Delta F l^3}{48 I \Delta W}$$

where
$E_{m\ app}$ is the apparent modulus of elasticity in bending, in N/mm^2
l is the span, in mm
and ΔF, I and ΔW are as above.

Tension parallel to the grain: The test employs the full cross-section of the timber specimen which must be of a length sufficient to provide a test length clear of the testing machine grips of at least nine times the nominal width.

Separate tests on the same test piece are carried out for determining tensile strength and modulus of elasticity in tension, each with its own rate of loading (see BS 5820: 1979).

Deformation is measured over a gauge length of 900 mm located not closer to the ends of the grips than twice the width of the specimen. Two extensometers are used, one each on opposite faces, and a record of load–deflection is made.

Compression parallel to the grain: The test employs the full cross-section of the test specimen

which shall be of a length six times its thickness (the smaller dimension). The end surfaces must be carefully machined to ensure that they are plane and parallel to each other and perpendicular to the length. Separate tests on the same test piece are carried out for determining compression strength and modulus of elasticity in compression, each with its own rate of loading (see BS 5820: 1979).

Deformation is measured over a central gauge length of five times the nominal width of the test specimen. Two extensometers are employed, one each on opposite faces, and a record of load–deflection is made.

Shear modulus along the grain: The single span method set out in this specification calls for the determination of both a modulus of elasticity in bending, similar to that described above, and an apparent modulus of elasticity in bending obtained in a three-point (centre-point) bend test for the same length of test specimen. Consequently, in the first test care has to be exercised that the limit of proportionality is not exceeded and the specimen damaged.

The shear modulus, (modulus of rigidity) is given by the equation:

$$G = 1.2 h^2 / l^2 \left[\frac{1}{E_{m\ app}} - \frac{1}{E_m} \right]$$

where
G = shear modulus, in N/mm^2
h = nominal depth of section, in mm
l = span in mm
$E_{m\ app}$ = apparent modulus of elasticity in centre-point loading
E_m = modulus of elasticity in third-point loading for the *same* test specimen.

11.4 Strength and elastic properties, and toughness

11.4.1 From small clear specimens

The various strength properties, toughness (impact resistance) and modulus of elasticity for a selected number of hardwoods and softwoods are presented in Table 11.1; the contents of the

Table 11.1 Values for strength and modulus of elasticity of five hardwoods and four softwoods at 12 per cent moisture content derived from small clear test pieces (extract from Lavers, 1983)

	Density (kg/m³)	Bending		Compression ∥ (N/mm²)	Compression ⊥ (N/mm²)	Tensile ∥ (N/mm²)	Tensile ⊥ (N/mm²)	Impact (m)	Hardness (N)	Shear ∥ (N/mm²)	Cleavage (radial) (N/mm width)
		Mod. of elasticity (N/mm²)	Mod. of rupture (N/mm²)								
Balsa	176	3 200	23	15.5	0.8	35	—	—	—	2.4	—
Obeche	368	5 500	54	28.2	3.91*	—	—	0.48	1 910	7.7	9.3
Agba†	497	7 600	81	43.2	5.94*	—	—	0.61	3 290	11.7	11.4
Beech†	673	10 100	108	51.8	9.41*	—	12.5	0.99	5 650	16.2	18.4
Greenheart	977	21 000	181	89.9	16.46*	—	—	1.35	10 450	20.5	17.5
Norway spruce†	417	10 200	72	36.5	4.25*	104	—	0.58	2 140	9.8	8.4
Sitka spruce†	433	10 500	74	39.2	4.25*	139	—	0.51	2 140	8.7	7.7
E. redwood	481	10 000	83	45.0	4.90*	—	3.70	0.66	2 580	11.3	9.8
Douglas fir	545	12 700	93	52.1	5.48*	138	2.90	0.86	2 980	10.8	8.9

* Calculated from hardness values (see text): compression ⊥ = 0.00147 hardness + 1.103.
† Home-grown; all other timbers imported.

Table 11.2 Values of the elastic constants for five hardwoods and four softwoods, determined on small, clear specimens

Species	Density (kg/m³)	Moisture content (%)	E_L	E_R	E_T	ν_{TR}	ν_{LR}	ν_{RT}	ν_{LT}	ν_{RL}	ν_{TL}	G_{LT}	G_{LR}	G_{TR}
Balsa	200	9	6 300	300	106	0.66	0.018	0.24	0.009	0.23	0.49	203	312	33
Af. mahogany	440	11	10 200	1 130	510	0.60	0.033	0.26	0.032	0.30	0.64	600	900	210
Birch	620	9	16 300	1 110	620	0.78	0.034	0.38	0.018	0.49	0.43	910	1 180	190
Ash	670	9	15 800	1 510	800	0.71	0.051	0.36	0.030	0.46	0.51	890	1 340	270
Beech	750	11	13 700	2 240	1 140	0.75	0.073	0.36	0.044	0.45	0.51	1 060	1 610	460
E. whitewood	390	12	10 700	710	430	0.51	0.030	0.31	0.025	0.38	0.51	620	500	23
Sitka spruce	390	12	11 600	900	500	0.43	0.029	0.25	0.020	0.37	0.47	720	750	39
E. redwood	550	10	16 300	1 100	570	0.68	0.038	0.31	0.015	0.42	0.51	680	1 160	66
Douglas fir	590	9	16 400	1 300	900	0.63	0.028	0.40	0.024	0.43	0.37	910	1 180	79

E is the modulus of elasticity in a direction indicated by the subscript (in N/mm²).

G is the modulus of rigidity in a plane indicated by the subscripts (in N/mm²).

ν_{ij} is the Poisson's ratio for an extensional stress in j direction, $= \dfrac{\text{compressive strain in } i \text{ direction}}{\text{extensional strain in } j \text{ direction}}$.

table have been extracted from *The Strength Properties of Timber* by G.M. Lavers (1983); this publication contains test data for over 400 species.

Highest strength values for any one species are obtained in tension along the grain, a direct reflection of the covalently bonded cellulose molecules comprising the core of the microfibril (Chapter 4). Lowest strength values are obtained in tension perpendicular to the grain, reflecting the presence of only secondary bonding horizontally between the cellulose molecules, the layered structure to the cell wall and the presence of a cell cavity. This is another example of the **anisotropic** nature of wood and the ratio of the longitudinal to the horizontal value of tensile strength and stiffness is about 40 : 1.

The elastic behaviour of wood can be characterised by a set of twelve elastic constants — a modulus of elasticity in each of three principal directions, a modulus of rigidity in the three principal planes and a set of six Poisson's ratios. The values of these constants for many of the species listed in Table 11.1 are set out in Table 11.2. It should be noted that the ratio of modulus of elasticity along the grain (E_L) is about 40 times greater than that horizontally ($E_R + E_T$)/2.

11.4.2   From structural-sized specimens

The strength properties and modulus of elasticity in bending for a number of structural softwoods are set out in Tables 11.3–11.6. A comparison of the strength values in these tables with those for the same timber in Table 11.1 will indicate the existence of very substantial differences, with the values obtained in the testing of structural-sized specimens being much lower, and in the case of longitudinal tension very much lower, than those obtained in the testing of small clear test pieces. The principal reason for this is that, unlike small clears, the structural-sized specimens contain defects such as knots and distorted grain which, as will be explained later, result in appreciable loss in strength, though they have only a slight effect on modulus of elasticity. This effect of size on strength is accentuated by a rate of loading

effect; thus the rate of loading is considerably slow in the case of structural size samples and this will also contribute to their lower strength values.

The effect of both these parameters is offset to some extent by the difference in the conditioning climate between the two sizes of test pieces, and in the case of bending, between the use of three- and four-point loading. The moisture content following conditioning is about 1 per cent lower for the structural-sized specimens, thereby giving an apparent increase in strength and modulus values. While the effect of the different modes of loading in bending is slight as far as strength is concerned, it does have an appreciable effect on the modulus, with values derived in four-point bending always appreciably higher than those obtained under three-point loading.

Table 11.6 sets out modulus of rupture, and ultimate tension and compression strengths for large samples matched on the basis of their modulus value in bending. This table, therefore, allows a direct comparison of bending, tension and compression strengths. Whereas in the case of small clear test pieces, the ratio of these three strength values is approximately 6 : 9 : 3, this becomes for structural-sized timbers 6 : 4 : 4; this indicates the extreme sensitivity of tensile strength to the presence of defects in structural-sized timber.

11.4.3   Reaction to stress under different modes of loading

Tensile strength parallel to the grain
Clear, straight-grained timber is at its strongest in this mode of testing and it is unfortunate that greater use is not made of this property. However, tensile strength assumes greater significance in knotty timber since these defects have a marked influence in lowering the strength of the wood. A poor relationship exists between tensile strength and density, whereas the fine structure of the cell wall, especially the microfibrillar angle of the S_2 layer, has a most marked influence on strength.

Table 11.3 *Values of bending strength and modulus of elasticity for certain softwoods derived from specimens of structural size*

Sample	No. of pieces	Average moisture content (%)	Average specific gravity	Modulus of rupture (N/mm²)		Modulus of elasticity (true) (N/mm²)	
				Mean	Standard deviation	Mean	Standard deviation
Redwood/Whitewood (Swedish)							
38 × 150	438	14.6	0.413	43.6	13.4	11 434	2 664
50 × 100	506	15.7	0.412	45.1	10.9	12 603	2 524
50 × 150	433	14.9	0.400	43.1	10.4	11 644	2 151
50 × 200	458	16.4	0.396	40.7	11.2	11 749	2 517
Redwood/Whitewood (Finnish)							
38 × 100	246	15.4	0.405	39.7	10.8	10 521	2 190
38 × 150	238	16.5	0.393	37.7	11.5	10 417	2 395
50 × 200	215	15.2	0.394	37.8	11.5	10 543	2 386
Hem-fir (Canadian)							
44 × 100	169	14.3	0.424	48.1	15.1	12 710	2 457
44 × 150	111	14.9	0.429	45.0	14.2	11 347	2 262
44 × 200	94	14.8	0.411	39.2	11.4	12 524	2 131
44 × 250	50	16.8	0.404	29.4	11.0	11 034	2 251
Sitka spruce (UK)							
50 × 100	120		0.383	41.3	8.2	10 812	2 051
47 × 97	130	15.0±1.5	0.366	32.8	10.2	9 345	2 348
47 × 97	120		0.347	31.9	11.0	8 972	2 124
Douglas fir (UK)							
38 × 100	192	15.0±1.5	0.424	35.7	12.8	10 018	2 582
47 × 200	194		0.418	34.1	11.0	10 795	2 330

Table 11.4   *Values of tensile strength and tensile modulus for Swedish redwood and whitewood derived from specimens of structural size*

Sample		No. of tests	Average moisture content (%)	Average specific gravity	Ultimate tension (N/mm²)		Modulus of elasticity E (true) (N/mm²)	
Size (mm)	Species				Mean	S	Mean	S
38 × 100	R	102	15.6	0.405	23.5	7.45	10 230	2 589
	W	94	15.6	0.405	27.7	8.56	12 130	2 637
38 × 150	R	98	16.6	0.418	22.6	9.64	9 960	2 847
	W	92	16.4	0.402	25.9	9.32	10 970	3 200
38 × 200	R	90	15.0	0.421	18.6	7.75	9 140	2 410
	W	94	16.1	0.379	21.7	7.58	9 100	1 995
Combined	R & W	570	15.9	0.405	23.3	8.87	10 260	2 828

R – Redwood (*Pinus sylvestris*); W – Whitewood (*Picea abies*).
S = Standard deviation.

When timber fails in longitudinal tension, the fracture line shows a marked tendency to run across the low-density, thin-walled early wood, but to zig-zag through the higher density, thick-walled late wood running along the grain usually within the cell wall and between the S_1 and S_2 layers: that is, failure within the late wood is basically by shear along the grain and the fracture is described as interlocked.

Compression strength parallel to the grain
High strength in longitudinal compression is required of timber used as columns, props and chair legs, though on account of the lengths of these items in relation to their cross-sectional areas, they frequently buckle at high stresses and fail in bending rather than true compression. Timber, unlike most other materials, is significantly weaker in longitudinal compression than in longitudinal tension with values as low as one-quarter. The strength of a piece of wood in compression is closely related to its density though influenced by its moisture content.

When a block of wood fails in longitudinal compression, a gross shear band occurs across the grain frequently almost horizontally on the

Table 11.5   *Values of compressive strength and compressive modulus for Polish redwood derived from specimens of structural size*

Sample	No. of tests	Average moisture content (%)	Average specific gravity	Ultimate compression stress (N/mm²)		Modulus of elasticity E (true) (N/mm²)	
				Mean	S	Mean	S
38 × 100	103	18.0	0.404	22.0	3.44	8 800	1 791
50 × 100	183	18.0	0.397	22.8	3.38	7 750	1 688

S = Standard deviation.

Table 11.6 *Comparative strengths and standard deviations (from matched samples of structural size Canadian spruce/pine/fir)*

Sample	Average specific gravity	Bending		Ultimate tension stress (N/mm^2)	Ultimate compression stress (N/mm^2)
		MOR (N/mm^2)	MOE (N/mm^2)		
2" × 4" (nominal)	0.390	43.9 (13.6)	10 814 (2 296)	27.3 (10.8)	28.2 (4.6)
2" × 8" (nominal)	0.390	38.7 (13.4)	10 422 (2 390)	26.2 (10.2)	26.7 (5.0)

(Values in brackets are the standard deviations.)

radial face and inclined at about 60° to the vertical axis on the tangential face. This shear band (Figure 11.6) comprises a zone of buckling of the cells and the occurrence within the cell walls of kinks (slip-planes) representing localised deformation of the microfibrillar structure. It should be appreciated that the gross shear band represents the final stage in a slow, progressive failure under compression: this accounts for the marked departure from linearity of the load–deflection curve at high levels of stress.

The anatomical changes associated with compression failure in wood are discussed in detail in Chapter 5 with particular reference to overstressing occurring in the living tree as a result of gale force winds or growing stresses. Suffice it here to say that dislocations of the microfibrillar structure of the cell wall (kinks) occur which spread in horizontal bands known as compression creases (Figures 5.13 and 5.15) (Dinwoodie, 1968, 1981). With increasing stress, the creases increase in size both vertically and horizontally until at the point of failure a number have coalesced to form the visible shear band.

Compression strength perpendicular to the grain

Resistance to crushing is an important property in a few selected end uses such as railway sleepers, rollers, wedges, bearing blocks and bolted timbers. Those timbers which are high in density have high compression strength across the grain.

Figure 11.6 *Block failed under longitudinal compression loading: note the prominent shear planes in which the fibres are severely buckled (Building Research Establishment, © Crown Copyright)*

The first signs of failure are the buckling of the cell walls and complete failure is associated with total collapse of the cellular structure with the elimination of the cell cavity.

Static bending and modulus of elasticity in bending

Timber is probably stressed in bending more than in any other mode; there are very many examples of where timber is used as a beam, of which the more common are floor and ceiling joists, roof truss members, table tops and chair bottoms.

Static bending is a measure of the strength of a material as a beam. In the resting position the upper half of a beam is in compression and the lower half in tension. Midway between the upper and lower surfaces is the neutral axis where both compression and tensile stresses are theoretically nil. A shearing stress operates along the neutral axis. The result of applying a load in the middle of the span is to deflect the beam out of the horizontal. This causes a shortening of the fibres on the upper, concave surface, and an elongation of those on the lower, convex surface. As the load increases compression failures develop on the upper surface, and the neutral axis moves towards the lower surface. The detailed sequence depends on the kind of wood and its physical condition. For example, in unseasoned wood the initial failure is a compression failure immediately below the point of loading, followed by either a tensile failure on the surface, or horizontal shear along the neutral axis.

The load any member can sustain is dependent on the span, that is the distance between the points of support, and on the sectional area of the member. The mathematical relationships between these three dimensions, however, are not directly proportional. For example, the effect of doubling the span is to halve the load that a beam of the same sectional area can carry. The effect of doubling the width of a beam, other factors remaining constant, is to double the load that can be sustained, but to double the depth of a beam is to increase the maximum supportable load fourfold. Because of this, beams are made rectangular, with the greater dimension in depth. There is, however, a practical limit to the magnitude of the ratio of depth to breadth in beams; a ratio greater than 4 : 1 introduces a tendency for the member to twist when loaded.

Loads applied to joists and beams involve bending stresses, but in selecting suitable sections for such members it is often necessary to allow for more than the minimum strength in bending to avoid sagging of floors and, in particular, the cracking of plaster ceilings beneath. In other words, adequate strength in bending does not necessarily ensure adequate stiffness when the permissible deflection is very small.

Shear strength parallel to the grain

This is a most important property in the structural use of timber, especially so in the region of joints. The areas surrounding the mortise or bolt hole can be subjected to very high longitudinal shear stresses. There is a fairly good correlation between shear strength and wood density: failure in longitudinal shear occurs within the cell wall, usually between the S_1 and S_2 layers.

Cleavage

This property is of lesser importance than those already described; it does give, however, an indication of how easily a timber will split, which is advantageous in the production of hand-split shingles, barrels and firewood, though a distinct disadvantage when it comes to nail or screw holding for packing case manufacture. Although density has some influence on the resistance to cleavage, the arrangement of the different tissues in the wood and the orientation of the fibres, such as interlocked grain, have a much more pronounced effect on determining cleavage than does density.

Hardness

Hardness is an important property when the timber is used for paving blocks, floors, ships decking and bearing blocks. In the case of both domestic and industrial flooring caution must be exercised, since the very hard timbers may be too slippery to provide a safe walking surface: additionally they may also be too noisy, and consequently the timbers normally used in flooring have high, but not very high levels of hardness.

Mention has already been made of the high correlation between hardness and compression perpendicular to the grain: failure in the former, like the latter property, entails initial buckling of the cell walls followed by total collapse.

Toughness or impact resistance

Toughness, or resistance to impact, is an essential requirement of timber for hammer handles, shafts and many sports goods. To resist suddenly-applied loads the timber must be tough, a property particularly associated with hickory and ash, and more recently with the South American timber pau marfin.

While these three timbers excel in their level of toughness, it should be noted that timber of all species will display higher resistance to suddenly-applied loads than to sustained loads.

Little information is available to explain the excellence of certain timbers: however, it is possible to relate the good toughness of all timber to its 'fibre-composite' structure, where the presence of weak interfaces between the wall layers absorbs energy by re-directing and arresting the primary crack (Cook and Gordon, 1964) (Figure 11.7).

Loss in toughness can result from a number of inherent and external factors (Dinwoodie, 1971). Exceptionally low density for the species of wood, the presence of very marked compression wood, or the development of brittleheart in certain low-density tropical hardwoods as described in Chapter 5, are examples of inherent factors which will induce abnormally low levels of toughness. Common examples of external variables which can also induce loss in toughness are the presence of fungal attack, thermal degradation of the wood, and the induction of dislocations (kinks) in the cell wall by high compression stressing resulting from either the effect of gale force winds on the tree, or overloading of timber in service (see Chapter 5).

Loss in toughness is usually associated with a very short brittle fracture and the terms 'brittle behaviour', or 'brashness' are frequently used as more descriptive expressions of loss in toughness.

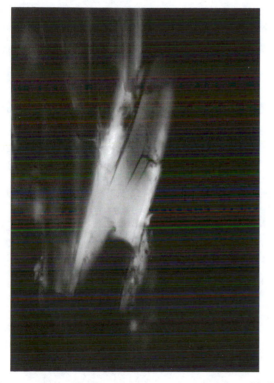

Figure 11.7 *Crack stopping in a fractured rotor blade. The primary crack running from about four o'clock has been arrested by the creation of secondary cracks in an orientation commensurate with the microfibrillar angle of the S_2 layer ($\times 750$, polarised light) (photomicrograph from the Building Research Establishment, © Crown Copyright)*

11.5 Factors influencing the strength and elasticity (stiffness) of wood

Perhaps the most important single factor influencing the strength and stiffness of timber is its density, but there are many other variables, some anatomical in origin such as knots, slope of grain and microfibrillar angle, and some environmental such as moisture content and temperature, all of which play a significant role in determining the strength and stiffness of wood.

11.5.1 Density (including ring width and ratio of early wood to late wood)

The effect of density on strength has already been discussed for each strength property in the preceding section: it will be recalled that some strength properties show a very marked correlation with density, and compression strength parallel to the grain, bending strength and hardness fall into this category (Table 11.1). The high degree of correlation obtained between density (actually specific gravity) and longitudinal compression in testing over 200 species is illustrated in Figure 11.8 (Lavers, 1983): separate regressions are presented for material of different moisture contents. A similar relationship was obtained in the plot of stiffness against density for the same test pieces; the correlation coefficient for samples at 12 per cent moisture content and 'green', were respectively 0.88 and 0.91. Evidence of this relationship is also apparent in Table 11.2. Reference to Chapter 8 will indicate how density is affected by the relative proportions of the different types of cells together with the level of wall thickness of the fibres. These variables are in turn influenced by rate of growth of the tree working through the relative proportion of early wood to late wood as previously explained in Chapter 5: the rate at which the tree grows has a very pronounced effect on many of its strength properties. Position in the tree is also significant in view of the systematic patterns of variation in

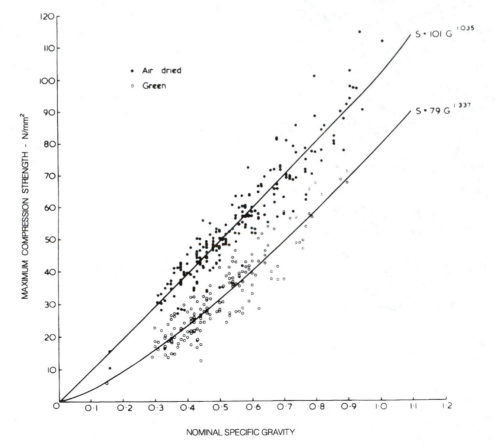

Figure 11.8   *The relation of maximum compression strength to nominal specific gravity for all species tested, green and air dried (Building Research Establishment, © Crown Copyright)*

density within the tree that were discussed in Chapter 5. Density and consequently strength will be at a minimum in the centre of the tree at its base and will increase outwards appreciably and upwards slightly.

Not all strength properties, however, are influenced by density to the same degree and, as discussed in the previous section, tensile strength, cleavage, shear and impact resistance, although being influenced to some extent by density, are determined much more either by the cellular arrangement of the wood, or by the fine structure of the cell wall.

11.5.2   Angle of the grain

In Chapter 7 the various forms of grain deviation from the vertical were described and it was shown how this could result in the creation of aesthetically pleasing figure in wood. In the structural use of timber the effect of sloping grain has a much more serious influence since it can result in a marked reduction in its strength. Wood is indeed very sensitive to the orientation of the fibres relative to the direction of applied stress. As can be seen from Figure 11.9, the degree of sensitivity varies with mode of stressing, being particularly high in the case of tensile stress along the grain. Thus, at an angle to the grain of 15° the tensile, static bending and longitudinal compressive strength are reduced to 45, 70 and 80 per cent of their respective strengths in straight-grained wood. The sensitivity of stiffness to angle of the grain appears to be similar to that for strength.

In ordinary practice the margin of safety in design and use of timber is such that an appreciable degree of tolerance in regard to sloping grain can be accepted for most purposes, excepting those where the level of stress in service is high, such as sports goods, tool handles, ladder stiles and scaffold boards. Thus for example, in BS 1129, *Portable timber ladders, steps, trestles and lightweight stagings*, it is specified that in the stiles of single section ladders, extending ladders and lightweight stagings, the combined slope of grain in the two

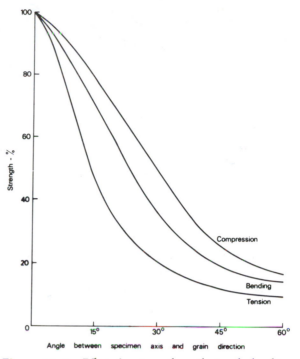

Figure 11.9   *Effect of grain angle on the tensile, bending and compression strength of timber*

longitudinal planes shall not be steeper than 1 in 10 for all the specified timbers, excluding Douglas fir where because of its superior strength the maximum combined slope of grain is specified as 1 in 8.

The directions of vessel lines, gum veins, resin ducts and seasoning checks are useful in indicating slope of grain in a piece of timber, but in the absence of these features visual inspection alone can be very misleading. Direction of the grain may be detected by raising a few fibres with the point of a penknife, or by noting the spread of an ink spot. For accurate determination of slope of grain a specially designed scribe, consisting of a cranked rod, with swivel handle, and an inclined needle at the end of the rod, should be used. Determinations should be made on both the face and edge of a piece of timber. The use of the scribe is explained and illustrated in British Standard BS 1129 (1982) (see also Figure 11.10).

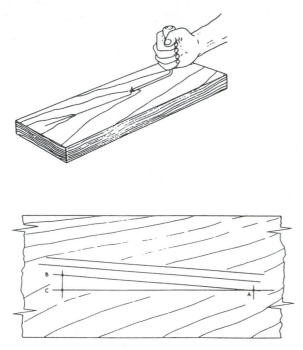

Figure 11.10 *Measurement of slope of grain is expressed as '1 in AC/BC'*

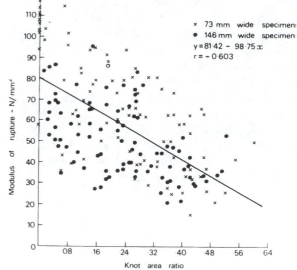

Figure 11.11 *Effect of knot area ratio on the strength of timber (Building Research Establishment, © Crown Copyright)*

11.5.3 Knots

Knots influence the strength properties and stiffness of a piece of wood to a varying degree, depending on their size, position and type. Strength properties and stiffness are adversely affected not because the wood of which the knot is composed is ordinarily inferior, but because the cells in the knot are inclined at an angle to those in the rest of the wood, and because of the irregular grain surrounding the knot (see Figures 2.4, 2.5 and 2.6).

Knots do not affect all strength properties to an equal extent: shear strength, compression strength perpendicular to the grain and modulus of elasticity are generally reduced only slightly by the presence of knots. However, compression strength parallel to the grain and bending strength suffer considerable losses with the occurrence of knots: the most marked effect, however, is on longitudinal tensile strength which is lowered appreciably by knots.

In very broad terms the effect of the size of the knot is directly proportional to its cross-sectional area. The location of the knot also plays a part in determining its significance in lowering strength. Thus in a beam for example, a knot on the compression edge will have a much less damaging effect than a similar one on the tension edge, while if it is located in the centre of the beam it will have very little effect on strength. Furthermore, any one knot will be progressively more serious in a beam the farther away it is from supports.

It is difficult to quantify the influence of knots owing to their apparent change in shape depending on how a batten of timber is cut. One method that has been used successfully is to express the shape of the knot in terms of the **knot area ratio**; this relates the sum of the cross-sectional area of all the knots at a particular cross-section to the cross-sectional area of the piece. The loss in bending strength that occurred with increasing knot area ratio in 200 home grown Douglas fir boards of two different widths is illustrated in Figure 11.11.

11.5.4   Other anatomical features

Some investigations have indicated that cell length is important in determining strength and stiffness. However, it should be appreciated that strength is not directly proportional to the length of the cells: rather it has been shown that a minimum length of cell is necessary in order to ensure sufficient overlap with the cell above and below it in order to transfer stress from one cell to the next. As strength increases owing to higher density, so the overlap between the cells must also increase in order to avoid failure due to the higher loads being transmitted: longer overlap necessitates longer cells. Thus, although the stronger woods are frequently associated with increased cell length, it should be appreciated that it is not the cell length as such which is the direct cause of the increased strength; rather the increased length is the consequence of other factors causing increased strength.

A second anatomical feature that has an influence on strength and stiffness is the microfibrillar angle of the S_2 layer. In clear, straight-grained wood, both the longitudinal tensile strength and the stiffness have been shown to be markedly affected by the microfibrillar angle: as the angle to the vertical axis increases so the tensile strength and stiffness quickly decrease. The remaining strength properties, however, appear to be only slightly affected by the angle of orientation of the microfibrils.

A third anatomical feature that influences strength and stiffness is the occurrence of compression wood and tension wood, the characteristic features of which were described in detail in Chapter 5. Compression wood tends to be stronger and stiffer in longitudinal compression, but much weaker in tension and toughness than normal wood: this has been attributed in part to the higher density and higher lignin content of the compression wood, and in part to the larger microfibrillar angle of the S_2 layer. Tension wood is stronger in tension and toughness, but weaker in compression compared with normal wood: it is much less stiff (more elastic)

than normal wood. Such changes can be accounted for in terms of the marked increase in flexibility of tension wood due to the presence of the gelatinous layer to the cell wall and its lower density.

11.5.5   Moisture content

This factor has an influence on the stiffness and almost all the strength properties of wood: the effect of moisture as far as longitudinal compression strength is concerned can be seen from Figure 11.8, where the strength values for over 200 species of a wide range of densities is considerably higher at 16 per cent moisture content than it is at about 28 per cent, namely at the fibre saturation point. A similar relationship was found for stiffness.

If for this same strength property a systematic study was made for any one timber of the change in strength as the timber dried out, it would be found that the longitudinal compressive strength of oven-dry timber was over three times that of timber at the fibre saturation point (Figure 9.2). However, the change in strength with changing moisture content is non-linear and, as Table 11.7 indicates, for Scots pine the percentage increase in strength per cent reduction in moisture content is greater at low compared with high levels of moisture content (this table is compiled from data in Lavers, 1983).

Table 11.7   *Percentage change in strength and stiffness of Scots pine timber per 1 per cent change in moisture content*

Property	Moisture range (%)		
	6–10	12–16	20–24
Modulus of elasticity (stiffness)	0.21	0.18	0.15
Modulus of rupture	4.2	3.3	2.4
Compression perpendicular	2.7	2.0	1.4
Hardness	0.058	0.053	0.045
Shear parallel	0.70	0.53	0.36

Table 11.7 also clearly shows the very marked difference among strength properties and stiffness in their sensitivity to moisture change. Modulus of rupture, like longitudinal compression strength, increases appreciably per 1 per cent reduction in moisture content, while side hardness is almost insensitive to moisture change. Only slight differences occur between different timbers.

A full explanation of the effect of moisture on strength in terms of the basic structure of the wood is still not available but it is generally accepted that the overall increase in strength with reduction in moisture is because of the shortening and consequent strengthening of the hydrogen bonds linking together the microfibrils. Longitudinal tensile strength, however, is insensitive to changes in moisture content since tensile strength is a function of the structure of the crystalline core of the microfibril which is unaffected by changing moisture content (Chapter 4).

11.5.6   Temperature

The strength and stiffness of wood are also influenced by its temperature. For most of the strength properties a good rule of thumb is that an increase in temperature of 1°C produces a 1 per cent reduction in their ultimate values. This effect of temperature is dependent on moisture content, the effect being considerably greater the higher the moisture content (Gerhards, 1982). As with the case of moisture content described in the previous section, the effect of temperature is more pronounced with some strength properties compared with others; thus the longitudinal compression strength increases almost threefold with a drop in temperature from 15 to −180°C (Figure 11.12).

It should be appreciated that long-term exposure to elevated temperatures results in a marked reduction in strength, stiffness and also toughness, the effect being greater with hardwoods than softwoods (Chapter 21). Even exposure to cyclic changes in temperature over long periods has been shown to result in thermal degradation (Moore, 1984).

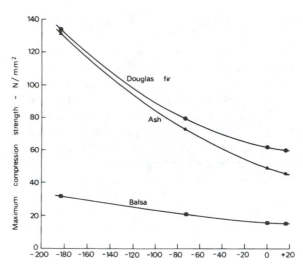

Figure 11.12 *Effect of temperature on compression strength for three timbers (Building Research Establishment, ©; Crown Copyright)*

11.5.7   Time

Rate of loading

Increase in the rate of load application in a mechanical test results in increased values of strength and stiffness. The effect is greater in testing 'green' wood where strength values can be some 50 per cent higher than those of wood at 12 per cent moisture content, a result quite contrary to that obtained when using normal rates of load application. Strain to failure appears to decrease with increased rate of load application irrespective of moisture content.

Duration of load

It is often stated that wood ages and loses strength: provided that the wood has not been attacked by fungi nor subjected to longer periods at high temperatures, then the statement is quite erroneous and there is no loss in strength of a piece of wood in an unstressed state.

The situation, however, is quite different if a load is kept on a piece of wood for a long time. We are all familiar with the sagging bookshelf. The initial load of the books is insufficient to cause the bookshelf to break, yet left on the shelf for a long period of time, the sag or deflection increases until failure generally occurs.

It is important to appreciate that failure has occurred at a much lower level of stress than would have occurred under short-term test: there is thus an 'apparent' loss in strength with time under load. It has been demonstrated by experimentation followed by mathematical projection that in order to ensure that a piece of wood will still carry its imposed load for 100 years, then that imposed load must be no greater than 50 per cent of the load the timber would fail at in the short-term test (Wood, 1951; Pearson, 1972). In all structural applications a load-endurance factor is incorporated in the design and this is usually applied as one of several contributing parts to an overall safety factor, as will be described later in Chapter 15. An example of the duration of load relationship for structural chipboard is illustrated in Figure 11.13 in which stress level is plotted against log time to failure. In order to ensure a 50 year life ($\log_{10} 7.42$) under

conditions of 20°C, 90 per cent relative humidity, the chipboard must not be loaded to more than 29.7 per cent of the stress obtained in a short-term bend test (Abbott *et al.*, 1992).

Creep

Returning to the example of the shelf loaded with books, it was recorded that deflection increases with time under load. As deflection is a manifestation of elasticity or stiffness, there is therefore an 'apparent' loss in modulus of elasticity with time under load, which parallels the 'apparent' loss in strength.

We say that the shelf has displayed **creep** behaviour where creep is defined as the additional deflection that has occurred with time after the initial elastic deflection took place when the books were first placed on it. Another everyday example of creep behaviour is to be seen on the roofs of very old buildings: the ridge frequently has a pronounced sag. An example of long-term laboratory testing of

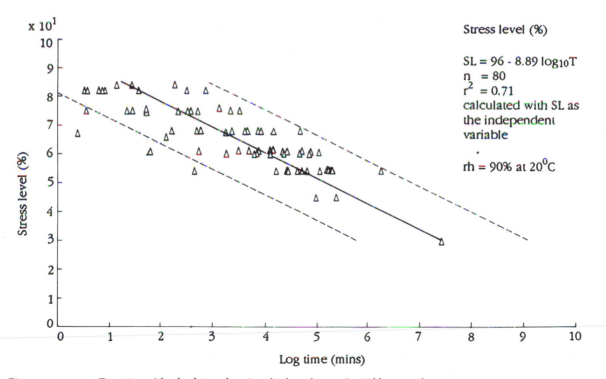

Figure 11.13 *Duration of load relationship for chipboard type C5 (Abbott et al., 1992)*

materials is illustrated in Figure 11.14 which shows the increase in deflection of a cement bonded particleboard stressed at 30 per cent of its short-term failing load in an atmosphere of 25°C/90 per cent relative humidity. Deflection increased rapidly initially and then slowed down, but it is interesting to note that, even after eight years, under-load deflection is increasing, albeit at a very slow rate.

The amount of creep that occurs is a function of the level of stress and the moisture content; creep will be greater at higher stress levels and at higher moisture contents. It will increase even faster if the moisture content changes during the time under load as illustrated in Figure 11.15. The lower graph shows creep deflection of small beech strips under a constant relative humidity of 93 per cent, while the upper graph shows the marked increase in creep under the same load when the atmosphere was cycled between 0 and 93 per cent relative humidity. After 13 cycles, the deflection was twelve times greater than had occurred at the same time under load in the sample under constant relative humidity. The sample under alternating humidity failed at 27 days, and at a load corresponding to only 37.5 per cent of its short-term strength!

It is interesting to note in passing that creep increased during the drying cycle and decreased during the wetting cycle, with the exception of the initial wetting cycle when creep increased. This type of creep behaviour under changing moisture content has been given the title 'mechano-sorptive' behaviour, but unfortunately a full explanation of the phenomenon has not yet been published. A fuller description of this type of behaviour in wood, together with additional references, is given in Dinwoodie (1989).

In the design of timber structures, therefore, the fact that deflection of timber loaded to its ordinary working stress may be several times greater under long-term loading, compared with short-term loading, must be taken into account. Hence, for many purposes a timber may be sufficiently strong in bending, but not sufficiently stiff, when required to support loads of long duration; unless allowance is made for this, deflection will, in time, become excessive. This allowance is set initially by the deflection limitation applied in the design.

A common practice to overcome excessive deflection is to camber the structure so that under dead load only, the long-term deflection will be very small.

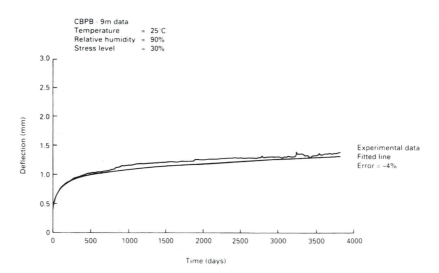

Figure 11.14 *The increase in deflection with time under load (creep) of a sample of 19 mm cement bonded particleboard. A line based on a rheological model has been fitted to the experimental data for the first 9 months and projected to the end of the experiment (Building Research Establishment, © Crown Copyright)*

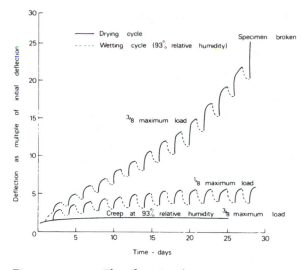

Figure 11.15 *The effect of cyclic variation in moisture content on relative creep of samples of beech timber loaded to 1/8 and 3/8 ultimate load (Building Research Establishment, © Crown Copyright)*

As noted above, long-term deflection is also affected by changes in moisture content. The deflection of loaded beams which initially are at a high moisture content and subsequently dry out under load will markedly increase. Therefore beams installed at a high moisture content should not be loaded until their moisture content is in equilibrium with service conditions: if this is not possible, then the beams should be propped up until their moisture content drops to the required level.

11.5.8 Defects

A whole range of defects can affect the strength and stiffness of wood: the most important are the presence of checks or splits that arise in seasoning, brittleheart and rot.

Checks and splits frequently arise in the seasoning of certain timbers and their influence on strength will depend not only on their frequency and depth, but also on their orientation relative to the applied load. Thus their presence would lower the bending strength of a beam if they occur in a horizontal plane, but have almost no effect if present in the vertical plane. Their significance is primarily in reducing shear resistance of the timber, and consequently the strength property most affected is shear parallel to the grain; bending strength and compression parallel to the grain are also reduced but to a lesser extent.

Brittleheart occurs in the core of many low-density hardwoods and represents a zone of wood in which longitudinal compression failure has occurred: this happens because longitudinal compressive stresses are generated in the tree during growth and these stresses are greater than the natural compressive strength of the timber. Failure takes the form of fine compression creases running across the grain: this feature is described in detail and illustrated in Chapter 5, section 5.4.11.

All strength properties of brittleheart are considerably lower than those of the same wood outside the central region of the tree. For most properties a large part of the loss can be accounted for in terms of the lower density of the core region: however, in the case of impact resistance a considerable proportion of the very extensive loss in this property can be attributed to the formation of the compression creases.

The development of rot in timber will be discussed in more detail in Chapter 19. Suffice it here to say that all strength properties and stiffness are markedly reduced with the occurrence of wood-destroying fungi.

11.6 Factors influencing the toughness of wood

Toughness, as explained earlier in this chapter in section 11.2, is a measure of the resistance of wood to the propagation of cracks, and consequently is quantified in terms of energy. It is because of this that the relationship of toughness to the set of inherent and external variables is different in some respects to that for strength and stiffness.

Wood is easily split along the grain and hence toughness is very low in this direction, unless the wood is characterised by having interlocked grain. However, across the grain wood is exceptionally tough being on a par with glass reinforced plastics. This explains why certain timbers continue to be used as handles and shafts for striking tools.

11.6.1 Density

Toughness, or impact resistance, increases proportionately with density. In the case of those ring-porous hardwoods used for striking tools, high density is associated with wide growth rings, as explained in Chapter 5. Therefore, in the case of ash, which is widely used in the UK for tool handles, it is recommended that only wood having between 4 and 16 rings per 25 mm should be used. There are many examples of serious injury arising from the use of shafts made from slowly-grown timber.

11.6.2 Angle of the grain

Toughness is even more sensitive than tensile stressing to sloping grain. It has been demonstrated that an angle to the grain of 15° will result in the timber possessing only 30 per cent of the toughness associated with straight-grained material.

11.6.3 Knots

Toughness is sensitive to the presence of knots since these tend to act as points of stress concentration.

11.6.4 Other anatomical features

One anatomical feature which has a major effect on toughness is the presence of well-developed compression wood owing to the much lower cellulose content in this abnormal tissue. Failure of compression wood is usually brittle or brash.

11.6.5 Moisture content

Toughness appears to be quite different from the strength properties in its dependence on moisture. In the majority of timbers there appears to be little effect of moisture content on level of toughness with perhaps a fairly even division in numbers between those that show a slight increase and those that display a small decrease with increasing moisture content. A few timbers, however, show an appreciable decrease in toughness with drying out: thus the toughness of green ash and teak are 10 and 30 per cent higher than wood at 16 per cent moisture content.

11.6.6 Temperature

As with strength and stiffness, there appears to be an interaction between temperature and moisture content. Thus, at high moisture contents, toughness decreases with decreasing temperature, while at low moisture contents (for example 12 per cent) toughness will increase with decreasing temperatures.

Prolonged exposure of wood to elevated temperature will result in thermal degradation resulting in extreme cases in total embrittlement, that is zero toughness (Chapter 18). Even exposure to cyclic changes in temperature over a long period will result in an appreciable loss in toughness. Thus, softwood subjected to daily temperature changes of 20–90°C over a period of three years resulted in reducing the toughness to only 44 per cent of its original value (Moore, 1984).

11.6.7 Defects

Toughness is particularly sensitive to the presence of defects. Fungal attack, even at the very early stages, will result in a marked reduction in toughness; indeed, one positive way of testing for the onset of fungal attack is to carry out a toughness or impact resistance test.

Perhaps the defect which has the most profound effect on lowering the toughness of wood is the presence of cell wall dislocations resulting from high longitudinal stressing. These can occur naturally in the tree from the effect of gale force winds (compression creases of 'thunder' shakes) or growth stresses (brittleheart) – see Chapter 5 – or artificially through overstressing of the wood in service, as discussed in an earlier section of this chapter.

11.7 Derivation of basic stresses from small clear test pieces

Nowadays, grade stresses are usually determined from the testing of structural-sized test pieces and consequently the use of the small clear system has been relegated to use for new timbers of limited volume. The system has great historical importance and for that reason alone is included here. There are two stages to the derivation of grade stresses – the first, the computation of **basic stresses**, is described here, while in Chapter 15 dealing with grading, the adjustment of these basic stresses for different qualities of wood is discussed.

In the derivation of a representative value of strength for a particular property and species, one approach would be to take the lowest recorded test value; this would lead to a very safe approach to timber use and there would be almost no risk of failure. However, by setting the representative value so low, the majority of the pieces of timber will be under-utilised as they will possess very much higher strength values. Hence, a compromise must be reached between running too high a risk of failure occurring, and yet making the best use of the potential strength available for all samples.

This compromise is determined utilising the known relationship between the standard deviation (S) and normal frequency distribution curve. In a normal distribution, 68 per cent of the results will lie in a band equivalent in width to the value of the standard deviation on each side of the mean, while 99.8 per cent of the results will lie between $+3S$ and $-3S$ of the mean. It is therefore possible to calculate a

strength value below which a certain percentage of the results will not fail; for most strength properties it is felt that the chance of getting a value lower than the estimated or characteristic value once in a hundred times is not taking too high a risk. This theoretical value above which 99 per cent of the results should fall is given by:

$$\text{Mean} - 2.33S$$

That is, by taking this value to represent or characterise the strength for this property and species, we know that 99 samples out of a hundred will be stronger than it, at least in theory. As will be explained below, various safety factors are applied in the derivation of grade stresses and these are sufficient to cover for the one sample in a hundred that is below the characteristic value.

Strength of timber is influenced by specimen size, rate of loading and duration of loading, and it is necessary to apply a safety factor to the characteristic strength value in order to accommodate the influence of these factors. Generally a value of 2.25 is used for most properties, except compression parallel to the grain where the factor is 1.4.

The **basic stress** therefore for each strength property of each species is derived from the division of the characteristic value by the safety factor:

$$\text{Basic stress} = \frac{\text{mean} - 2.33S}{2.25 \text{ (or } 1.4)}$$

The characteristic values for modulus of elasticity are not reduced to provide basic values.

It must be appreciated that while the basic stress value reflects variation in timber strength due to differences in growth rate and position in the tree, the value is based on *clear* timber used in the standard tests. Thus it was necessary to account for the effect of knots; the conversion of basic stresses to grade stresses and permissible stresses that were used in structural design is described in Chapter 15, though this chapter is concerned primarily with the derivation of grade stresses from structural size members that have been previously graded.

References

Abbott A.R., Dinwoodie J.M., Enjily V. and Page A.V. (1992) Use of structural (C5) grade chipboard. *Joint BRE/TRADA report: TRADA Design Aid DA9.*

Bodig J. and Jayne B.A. (1982) *Mechanics of Wood and Wood Composites.* Van Nostrand Reinhold, Wokingham, UK.

BS 373: 1957 (1986) *Methods of testing small clear specimens of timber.* BSI, London.

BS 1129: 1982 *Portable timber ladders, steps, trestles and lightweight stagings.* BSI, London.

BS 5268: 1991 *Structural use of timber. Part 2: Code of practice for permissible stress design, materials and workmanship.* BSI, London.

BS 5820: 1979 *Methods of test for determination of certain physical and mechanical properties of timber in structural sizes.* BSI, London.

Cook J. and Gordon J.E. (1964) A mechanism for the control of crack propagation in all brittle systems. *Proc Royal Soc A* **282**: 508.

Dinwoodie J.M. (1968) Failure in timber, Part 1. Microscopic changes in the cell wall structure associated with compression failure. *J Inst Wood Sci* **21**: 37–53.

Dinwoodie J.M. (1971) Brashness in timber and its significance. *J Inst. Wood Sci* **28(5)**: 3–11.

Dinwoodie J.M. (1981) *Timber — Its Nature and Behaviour.* Van Nostrand Reinhold, Wokingham, UK.

Dinwoodie J.M. (1989) *Wood — Nature's Cellular Polymeric Fibre-Composite.* Institute of Metals, London.

Gerhards, C.C. (1982) Effect of moisture content and temperature on the mechanical properties of wood. An analysis of immediate effects. *Wood and Fiber* **14(1)**: 4–36.

Jayne B.A. (1972) (Ed.) *Theory and Design of Wood and Fiber Composite Materials.* Syracuse Wood Sci. Series 3, Syracuse University Press, Syracuse, New York.

Lavers G.M. (1983) *The Strength Properties of Timber* (3rd edn, revised by Moore G.L.). Building Research Establishment Report. HMSO: available only from the Building Research Establishment Bookshop.

Moore G.L. (1984) The effect of long term temperature cycling on the strength of wood. *J Inst. Wood Sci* **9(6)**: 264–267.

Pearson R.G. (1972) The effect of duration of load on the bending strength of wood. *Holzforschung* **26(4)**: 153–158.

Wood L.W. (1951) Relation of strength of wood to duration of load. *Report No. 1916 Forest Products Lab. (Madison).*

Part 3

PROCESSING OF TIMBER – HOW TO CUT IT UP, DRY AND GRADE IT

12

Log Conversion

It is not the intention in this chapter to give a detailed account of sawmilling practice, and readers desirous of obtaining a comprehensive account of the subject should read one of the specialist texts on the subject. Rather it is the intention here to provide only the basic principles in an attempt to make a bridge between the properties of the tree previously described, and the utilisation of timber to be discussed in future chapters.

12.1 Restrictions on conversion

While the conversion of a log to a series of planks may appear to be a simple process, this is certainly not the case in practice. In converting a log the principal aim is to maximise output on a financial rather than a volume basis. One batten of large cross-sectional area is worth considerably more than two battens each half the cross-sectional area of the large batten; similarly one long batten is worth much more than two battens half as long. Consequently, the aim is to maximise length and cross-sectional area commensurate with the market demand for particular sizes. There is no sense in producing sizes of timber which are not required, and there are more and more applications of timber these days requiring large volumes of a few selected sizes.

This overall aim to maximise financial returns from sawn timber is tempered by a number of restrictions or limitations which frequently call for rapid decisions. A number of these are described below.

12.1.1 Shape of the log

Log taper
This is particularly important in the case of converting coniferous logs. In removing a square

cross-section from the length of the log, there will be much more 'waste' in trees with a high, compared with a low taper, and it is possible that the volume of waste could be reduced by judicious cross-cutting, thereby increasing the volume of timber obtained from the lower log.

The advent of chipper canters and associated double or quad bandsaws which reduce the log to a square cross-section has perhaps eliminated this type of decision: nonetheless, the example serves to highlight the type of question that previously had to be addressed by the sawyer.

Bowed logs
These are far from being rare and pose the same type of dilemma as to whether a greater financial return will be obtained from cutting one long length of small cross-section, or two short lengths of much larger cross-section. The answer may well be different depending on whether or not the sawyer wishes to cut a particular size (c.s.) of batten and on the degree of curvature of the log.

Large knots
This aspect is important in the production of structural softwood timber, and is particularly relevant in large-diameter hardwood logs. Decisions must be quickly made from the surface appearance as to how much the knot section in the plank will degrade its worth, and whether or not a greater financial return might be achieved by cross-cutting through the knot. The more valuable the timber, the greater the juggling between the possible conversion schemes.

12.1.2 Growth stresses

Within the living trunk of both coniferous and broad-leaved trees, internal stresses are generated: in large-diameter stems, and especially in certain genera, the magnitude of these stresses can be so high as to generate problems during conversion.

Many investigators have confirmed the presence of substantial longitudinal compressive stresses in the inner layers of a log while the outer rings show high tensile stresses. There is a progressive change in the stress across the growth rings and generally the transition from tensile to compressive stress occurs about halfway along the radius. The wood formed each year appears to be in tension such that a cumulative longitudinal compressive stress is induced upon the inner core. Tensile stresses of over $7 \, \text{N/mm}^2$ have been recorded in large-diameter European hardwoods, but Australian experience with Eucalyptus has indicated tensile stresses as high as $14 \, \text{N/mm}^2$; the corresponding compression stresses have been calculated as being of the order of $14–35 \, \text{N/mm}^2$ – see Figure 12.1.

Perhaps the best and simplest method that has been devised to demonstrate the occurrence of longitudinal growth stresses is that of Jacobs (1945). The central plank from a tree sawn through and through is cut into longitudinal strips: these will curl outwards, and when forcibly pulled back to form a composite board, it will be observed that they decrease progressively in length towards the outside of the plank.

Lateral stresses have also been recorded. The wood tissues near the periphery mature in tangential compression, thereby imposing radial tension in the wood near the pith. These lateral stresses have been found to be considerably lower than the longitudinal stresses; the transverse compressive stress rarely exceeds $3 \, \text{N/mm}^2$.

The generation of growth stresses is primarily the result of dimensional changes imposed on fibre or tracheid walls as a consequence of lateral swelling of the cell wall and axial contraction of the cell associated with the process of lignification, a view first expressed by Münch (1938), well supported over the years and recently reappraised and confirmed (Boyd, 1985).

In standing timber the high internal stresses can result in the development of **shakes**. These take the form of star-shaped cracks in the core of the tree and run up it for some considerable distance. This is a result of both the high radial tension and longitudinal compression near the pith. The first relief of stress results in a longitudinal crack through the diameter. As stress again increases, the second relief takes the form of a crack at right angles to the first. These cracks are widest in the centre and taper gradually towards the sapwood.

Longitudinal tangential cracks can also occur as a result of the high radial tension. These are most frequent in the corewood and take the form of ring shakes.

The very high compressive stresses induced in the centre of large trees of low density are usually greater than the natural compression strength of the timber: failure in compression occurs with the formation of kinks (slip-planes) and small natural compression creases. Such affected timber has been given the name **brittleheart** and was described in some detail in Chapter 5, section 5.4.11, and illustrated in Figure 5.16.

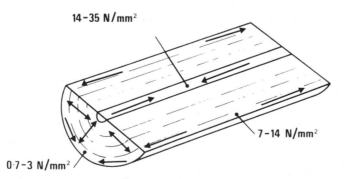

14–35 N/mm²

7–14 N/mm²

0·7–3 N/mm²

Figure 12.1   *Growth stresses recorded in a log of* Eucalyptus *spp.*

The conversion of timber containing a high level of internal stress presents two major problems. The first is simply the physical difficulty of sawing the timber, while the second is the financial loss due to timber degrade.

In cross-cutting, the longitudinal compression tends to close the kerf towards the pith thus 'pinching' or 'binding' the saw. On longitudinal–radial saw cuts there is again a tendency for the kerf to close owing to the high tangential compressive forces.

Considerable loss in timber occurs in the utilisation of trees containing shakes, or in trees which develop splits or cracks during or immediately following conversion. Where severe distortion of the sawn material occurs, usually in the form of excessive curvature, resawing may prove to be a costly necessity. The average loss of volume in beechwood due to splitting and unusable curvature has been recorded as being from 15 to 20 per cent; the financial value would be appreciably greater. The presence of shakes renders many logs unsuitable for peeling.

Difficulties are also experienced in seasoning timber containing high growth stresses. Such stresses tend to be superimposed on the normal drying stresses and result in a variety of forms of splitting and degrade, such as springing, see Chapter 13, Section 13.6.1. Generally, the practical significance is greater in hardwoods in which the stress level and consequently the degrade is highest. However, it is still present in softwoods where longitudinal stresses as low as $1 \, N/mm^2$ have been found to produce mid-span curvature in 2.5 m long boards of 3–6 mm.

Readers desirous of obtaining more information on growth stresses of a general nature are referred to the many reviews on the subject (see, for example Dinwoodie, 1966), while those interested in the mathematical modelling of the stress distribution in trees are referred to the excellent monograph by Archer (1987).

12.1.3 Possible exclusion of sapwood

As explained in Chapter 2, sapwood is that part of the tree which is physiologically active, being responsible for the conduction of mineral solutions up the tree, and for the storage of food products manufactured in the crown. In assessing the technical acceptance of sapwood it is necessary to consider its properties in comparison with those of heartwood.

Sapwood is light in colour whereas the heartwood of many timbers is coloured. In many applications the presence of two zones of different colour is not important, but in furniture manufacture this is usually unacceptable. In the manufacture of very high quality furniture, sapwood is not used in either solid wood or veneer and is removed from the log in either primary or secondary conversion: in furniture of lower quality it is common practice nowadays to stain the sapwood, thereby maximising the yield from the log.

Perhaps the most significant feature of sapwood is its lack of resistance to insect and fungal attack. In well-ventilated dry internal situations such as carcassing and joinery work, there is no objection to sapwood provided (1) the timber is thoroughly seasoned before it is installed, (2) the site conditions can reasonably be expected not to alter adversely after the timber is installed, and (3) the timber is not one prone to powder-post beetle and borer attack, or if the risk of such attack is remote. However, for outdoor uses, such as cladding, fencing, posts, gates and railway sleepers, sapwood must be excluded unless the wood is to be given adequate preservative treatment, in which case it may safely be retained.

In practice, however, especially in the case of softwoods, sapwood is not removed during conversion and it is left to the specifier to stipulate post-conversion preservative treatment of the sapwood in order to obtain a satisfactory level of durability (Chapter 22). In general, the diameters of softwood logs are too small to allow for the economic removal of sapwood: in addition, in bowed logs the 'common area' must be fully utilised in order to obtain economic return irrespective of any questions of sapwood removal.

It may be concluded that nowadays it is exceedingly rare to remove sapwood during primary conversion: in those few cases where sapwood is not acceptable it is usually removed during secondary processing.

12.1.4 *Possible exclusion of juvenile wood*

A much greater technical case can be presented for the removal of juvenile wood during primary conversion than for the removal of sapwood. Whereas sapwood can be treated after the conversion process, no enhancement of juvenile wood can be achieved.

It will be recalled that in Chapter 5 attention was drawn to the distinct patterns of variation that existed outwards from the pith in cell length, density and grain angle. In that part of the tree lying within the twelfth ring, or thereabouts, the wood is characterised by having short cells, low density and very high sloping grain, thereby severely reducing the performance of the wood in service.

Up to the early part of the century it was common practice to remove the juvenile wood in a cutting sequence known as 'boxing the heart'. Sadly, because of financial pressures and the very fast throughput of modern mills, this commendable procedure was terminated, certainly for softwood logs, leading to much justifiable criticism of the presence in service of weak, distorted timber which nearly always turns out to include juvenile wood.

12.2 Primary conversion equipment

The principal equipment required for primary conversion is a set of saws which may take the form of one or more of the following types: circular, band and frame. Many of the latest softwood mills also have a chipper canter.

The efficiency and accuracy of sawing is determined by a number of factors of which the following are the most important (see Figure 12.2):

- *The pitch of the teeth*: generally the greater the pitch, the faster the throughput, but the lower the quality of the cut surface. A greater number of teeth per saw diameter are required to cut the denser woods than those of low to medium density.

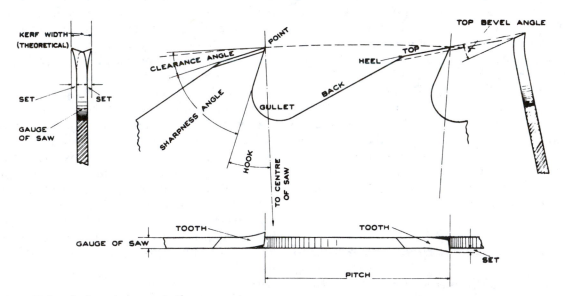

Figure 12.2 *Spring-set rip-saw teeth*

- *The speed of the saw*: quality of the cut surface and rate of throughput will increase with increasing circumferential speed of the saw.
- *The shape of the saw tooth*, as determined by the hook or rake angle and clearance angle: these factors determine the ease with which the wood is cut and the quality of surface produced.
- *The gullet space*, as determined by both the shape of the tooth and the pitch of the teeth: the gullet controls the discharge of sawdust, a most important factor in efficient saw operation.
- *The type and amount of set*: alternate teeth may be bent outwards at their tips in opposite directions (spring set) or the tip of each tooth may be splayed out on both sides (swage set). Increasing the amount of set permits higher rates of throughput, but as a result of the larger kerf formed the loss of wood due to sawdust production will increase.
- *The depth of timber being cut*: all other factors being equal, the deeper the section of wood the slower the throughput.
- *The feedspeed*: in general terms, the faster the feed speed, the poorer the quality of the cut surface.
- *The sharpness of the tooth*: rate of throughput and quality of cut will decline as the teeth lose their sharpness (Klamecki, 1979).

Many of these factors interact with one another. Thus the 'bite' of a tooth, which is perhaps one of the most important criteria in log conversion and secondary processing and refers to the depth of wood removed by each individual tooth, is a function of the combined effects of the pitch of the teeth, saw speed and feed speed. There is obviously a minimum bite below which the teeth will not penetrate the wood, while the maximum bite is limited by the poor quality of the surface produced, the inability of the gullet to discharge the sawdust produced, too high a power demand, and instability in the saw. As a very rough guide, maximum bite is usually limited to 0.75 mm in hardwoods and 1.5 mm in softwoods.

Where abrasive timbers are to be cut, saws tipped with hard alloy stelite or tungsten carbide should be used. In cutting both abrasive and very dense timbers, power demand may become the limiting factor and saw speed should be reduced.

The interdependence of density and feedspeed in determining power consumption is illustrated in Figure 12.3. Density also has a pronounced effect on the time the saw teeth take to become blunt, the time period between resharpening reducing very quickly with increasing density. The effect of density is exacerbated in certain timbers containing large quantities of gums or resins: these adhere to the saw blade, causing overheating which in turn leads to loss in tension and resulting in saw instability and a reduction in sawing accuracy.

12.2.1 Circular saws

In primary conversion, the use of a circular saw to rip the log into planks has been superseded in many mills by the bandsaw in the quest for greater output through faster cutting and reduced width of kerf. Nevertheless, some circular saws are still in use in the smaller, and sometimes less-specialised mills.

In addition to the factors described above for controlling general saw performance, the efficiency of circular saws is also determined by controlled tensioning of the blade. Thus, a tensile

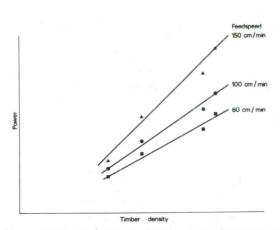

Figure 12.3 *Effect of timber density and feedspeed on the consumption of power using a circular saw to cut along the grain (rip-sawing) (Building Research Establishment, © Crown Copyright)*

stress is induced in the outer part of the saw blade by controlled hammering in the mid-region. Without tensioning, the saw would not run true as the saw temperature increases during cutting: tensioning assumes even greater significance when thin saw blades running at high rim speeds are used to maximise production through reduced sawdust production.

Where circular saws are used for ripping down a log into a series of planks, the hook angle is positive and rim speeds of 50 m/s are usually employed, though speeds up to 60 m/s are sometimes used for softwoods. Saws with different hook angles have to be used for timbers of different density classes. Generally an angle of 25° is used for rip-sawing softwoods, while an angle of 15° must be used for dense hardwoods.

Where circular saws are used for cross-cutting or for straight line edging, they are of smaller diameter with similar or slightly higher rim speeds. Saws for cross-cutting are characterised by having a negative hook angle.

12.2.2   Bandsaws

Most mills now have at least one wide bandsaw and many mills have a number being used not only for primary conversion, but also for resawing the planks into desired sizes. Generally the saw is used vertically, but examples of horizontally mounted saws, especially for the initial breakdown of the log, are known; many of these are mobile for use in the forest. The widespread adoption of bandsaws is a reflection of their superior performance to circular saws, particularly as regards the narrow width of kerf and the quality of sawn surface.

However, this improved performance of the bandsaw must be offset against the higher degree of skill required in both tensioning the blade in the machine and in setting up of the machine to ensure that the blade runs true, especially when horizontal pressure is applied to the cutting edge by the advancing timber. Blade width varies from about 150 to 300 mm and thickness from 1.0 to 2.0 mm: the teeth are always swage set and sometimes tipped. Cutting

speeds are usually of the order of 35–45 m/s and tension is applied to the blade through pneumatic pressure or the application of dead weights to the axle of one of the pulleys.

Improvements in design continue to be made in order to increase output from the saw, either through higher speeds, or smaller kerfs; the **high strain bandsaw** achieves the latter through the use of thinner saw blades and very much higher tensioning forces.

So-called 'narrower' bandsaws with 100 mm wide blades are used for some resawing, with speeds varying from 15–45 m/s; the smaller the saw, the lower is usually the speed.

12.2.3   Frame saws

These saws comprise a number of saw blades within a moving frame with reciprocating action. The frame varies in height from 1 to 3 m. When used vertically the frames carry a number of saw blades: when used horizontally there is usually only a single blade.

The frame saw was traditionally used in Scandinavia for breaking down softwood logs, and while a few remain in Europe, many have been replaced by bandsaws, primarily owing to the smaller kerf produced and the higher feed speeds possible. Frame saws had the great merit that where large quantities of battens of the same thickness were required, a frame containing up to five or six saws could produce very high outputs.

12.2.4   Chipper canters

These are now widely used as part of the primary breakdown of softwood logs, usually in conjunction with bandsaws on the same machine bed. They comprise two chipping heads facing each other which remove chips from the log, thereby producing two parallel faces to the log which is subsequently cut by bandsaw. Logs up to 60 cm in diameter can be machined at feed speeds up to 125 m/min. A variety of chip types can be produced. Alternatively the chipper can be used to produce a four-sided cant.

12.3 Conversion of hardwood logs

Well over 95 per cent of hardwoods are sawn on a 'through and through' basis to provide a series of **flat-sawn** planks of equal thickness (Figure 12.4a). The logs are held in place on the moving platform by a series of arms or 'dogs', and the log and platform are repeatedly fed into the path of the saw blade.

Where it is desired to obtain oak with a pronounced figure associated with the presence of the rays (silver figure), it is necessary for the wood to be **quarter sawn**, an expensive operation requiring much manipulation of the log in order to obtain as many radial, or near radial, faces as possible. Even then the number is limited, if parallel-sided planks are to be cut (Figure 12.4b).

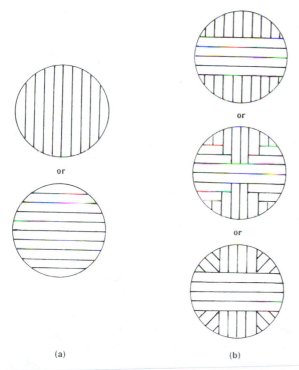

(a)   (b)

Figure 12.4 *Conversion of hardwood logs: (a) flat sawn 'through and through'; (b) three alternative 'quarter-sawn' cutting patterns to yield as many radial (or near radial) parallel-sided boards as possible*

12.4 Conversion of softwood logs

Prior to conversion, softwood logs are normally debarked in a debarking machine; this procedure extends the life of the various sawblades, yields better chips in terms of variability and cleanliness, and produces an income from the bark, much of which is used for horticulture purposes.

The last decade has seen tremendous advances in the technology associated with softwood conversion. Small sawmills have given way to large highly automated units where one operator may be in control of several saws, all by remote control. The amount and type of equipment varies considerably from one sawmill to the next, but it is possible to generalise the cutting of softwood into four different systems or methods.

The first of these relates to a small non-automated mill where the sawbench is worked by two operators — one at the saw and the second at the 'back end' of the sawbench (Figure 12.5a). The log is supported on the bench by manually applied 'dogs' and is passed through the saw to provide a flat face. The minimum width of this face from tapered logs is usually sufficient for the width of the main products being cut from the centre part of the log.

After the log carriage has returned, the log is rolled onto this face and a second slab is removed. Then the fence is positioned (set) for the cant size (width of main product) and the two-sided cant is moved out so that the second cut face is against the fence. The third cut produces a three-sided cant which is rotated so that the new cut face (third cut face) rests on the sawbench. This cant is then repeatedly fed through the saw to produce battens.

The slabs are also passed through the same saw to produce waney-edged planks: these in turn are either edged on the same saw, or passed to an edger to produce a square-edged board and then cross-cut to length.

The second method is to be found in somewhat larger mills which have an automatic saw carriage, operated by a single person positioned some distance from the sawbench (Figure 12.5b).

The softwood log is clamped on the bench by 'dogs' and passed through the saw which

(a)

Sawbench not automatic
Two operators (1 at saw : 1 at back end)

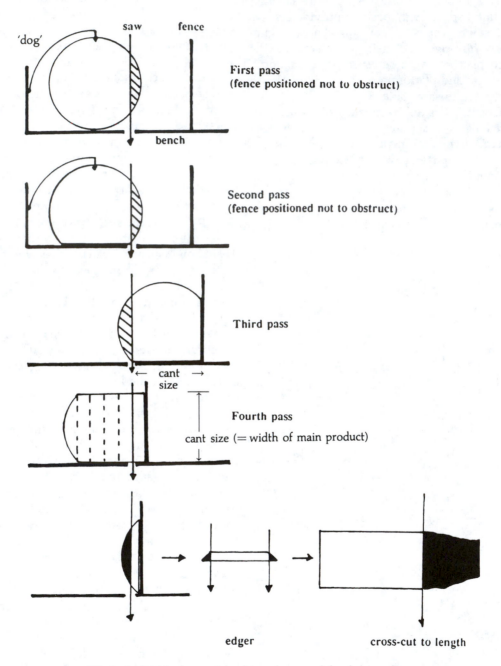

Waste (solid black areas) is chipped or used for firewood

Figure 12.5 *Four alternative methods (see also pages 139, 140 and 141) for conversion of softwood logs (J.M. Dunwoodie):* **(a)** *method 1.*

(b)
Automatic sawbench
Single operator distant from sawbench

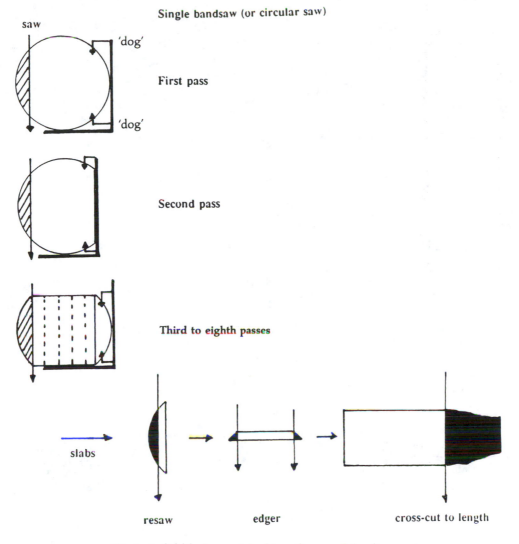

Waste (solid black areas) is chipped or used for firewood

Figure 12.5(b)   *Conversion of softwood logs: method 2*

removes one slab. The saw bench returns, the carriage knees are set at the cant width distance (width of main-product) from the saw blade, and the log is automatically spun through 180°; it is then moved and clamped against the knees before passing through the saw again to produce a two-sided 'cant'. After the removal of the third slab, the cant is turned and repeatedly fed through the saw to produce a series of square-edged battens.

The slabs are first cut parallel to the cut face on a separate saw before passing to the edger and cross-cut saw to produce square-edged trimmed boards.

(c)   **Automatic sawbench**
              Single operator distant from sawbench
              Double bandsaw

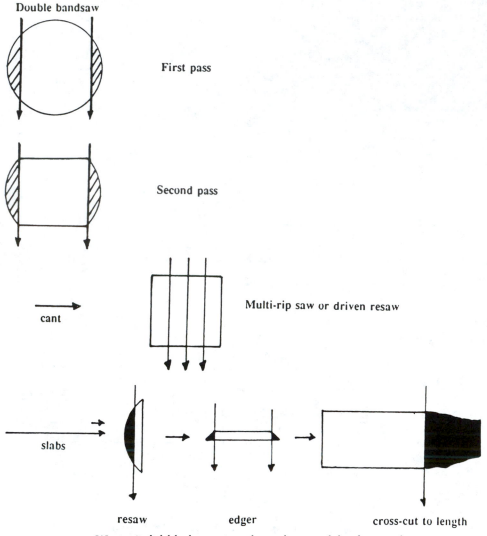

Figure 12.5(c)   *Conversion of softwood logs: method 3*

The third method relates to the automatic control of a sawbench equipped with a double bandsaw (Figure 12.5c). In the first pass, a two-sided cant is produced to the width of the main product: after the log carriage has returned, the log is automatically rotated through 90° before passing through the double bandsaw again to produce a four-sided cant. For the second cut through the double bandsaw, the width between the blades is set to allow for the thickness of battens to be subsequently cut, and the necessary saw kerfs – this stage being

(d)

**High speed automatic
Single operator and computer
Chipper canter and quad saws in single pass**

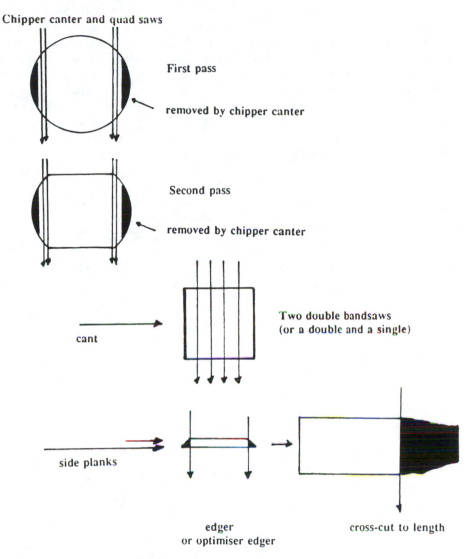

Chipper canter and quad saws

First pass

removed by chipper canter

Second pass

removed by chipper canter

cant

Two double bandsaws
(or a double and a single)

side planks

edger
or optimiser edger

cross-cut to length

Waste (solid black areas) is chipped or used for firewood

Figure 12.5(d) *Conversion of softwood logs: method 4*

reached in two fewer operations than the previous method.

The cant is then passed to a multi-rip saw or a driven resaw. The slabs are converted to boards as in the previous two methods.

The fourth method of conversion relates to the high-speed conversion of softwood logs controlled by computer and a single operator; this method is adopted in all the large softwood mills in the UK.

Data relating to the range in top diameter of the logs to be cut must first be fed into the computer which must be very accurately programmed to account for all saw kerfs. This is important because, in this highly automated method, the waste is chipped off first and the remaining part of the log must be cut up accurately to the finished thicknesses required. The log first passes through a unit containing a chipper canter and two double bandsaws (Figure 12.5d); this produces simultaneously a two-sided cant, four waney-edged planks and a load of chips. The log is automatically recirculated, rotated through 90° and fed through the unit again to produce a four-sided cant, up to four waney-edged planks and chips.

This cant then passes to two double bandsaws (or a double and a single) which reduces the cant to between two and five (depending on size of log) large-sized structural battens. The thin waney-edged battens are passed to an edger (or optimiser edger) to produce square-edged boards.

12.5   Computer-aided log conversion

Traditionally, sawmills have relied upon the experience and skill of the operators in order to convert sawlogs into the products which they require. However, the decisions taken are not always optimal, mainly because the operator cannot logically take account of all the important factors that affect the efficient conversion of a log. His task is easier and the logs converted to a consistently high yield if a computer decision-making process is used.

In recent years there has been considerable interest in and research into the extended role that computers can play in the sawmilling industry. Particular attention has been paid to improving the yield and control of sizes of sawn timber cut from logs by using computer aids.

Maximising of yield is possible through the use of optimisation computer programs, but these will inevitably vary between low and high investment sawmills. In the former, the results of background simulation runs may be used simply as an aid to operator instruction.

In higher investment sawmills, the current practice is to use a semi-automatic primary log breakdown configuration which comprises a chipper canter followed by a pair of band saws or 'quad' bandsaws. It is essential to know exactly what sizes are to be sawn from a particular log in order to set the chipper canter. The width of timber left between the two chipped faces must be an exact multiple of the sawn sizes required and saw kerf. The internal control system of this configuration includes a memory which holds pre-selected preferred cutting patterns which are linked to a sawlog top-diameter classification system. The preferred cutting pattern is accessed and the chipper canter and saws are set up correctly when the log top-diameter is entered into the control system of the machinery.

A computer optimisation program is the only way of accurately and quickly designing a log classification system and selecting the preferred cutting patterns to be put into the memory for the chipping and sawing configuration. There are two elements which should be considered in order to give an optimum conversion, within constraints, for a batch of logs:

(a) the range of top diameter in each class,
(b) the optimal conversion pattern to produce the required spread of sawn size from each class.

The use of simulation runs can lead to a finely tuned classification system. However, it is only sensible to use it if logs are being measured accurately, as in high investment sawmills which usually automatically scan log diameters in at least one plane. Log classification systems can also be used in sorting in the log yard in order that a group of logs, all of the same classification (top-diameter range), can be fed to a sawing configuration which is set up to give the optimal conversion. This system maintains very high throughput by eliminating the need to set the saws for each log and, within the constraints imposed by the equipment used, optimises the yield.

The techniques described above use optimisation programs which approximate sawlogs to truncated cones. However, these models are limited because they allow only uniform cross-sections, taper, log length, saw kerf and wane tolerance to be considered. In the UK, logs are very rarely perfectly tapered and straight, they often have irregular cross-sections and can be bent in one or more planes. Therefore, measuring systems and optimisation programs should be able to deal with true log shapes. Some improvement in optimisation of the yield from a batch of irregular shaped logs can be gained by continuously measuring the diameter in two planes at 90° to each other and linked to a common datum. This makes it possible to estimate the angular orientation of the plane of maximum bow and also the common area diameter. This can then be used for log sorting or to access a stylised conversion of the common area which has been previously derived by using optimisation programs based on truncated cones. More advanced optimisation programs allow the truncated cone to be uniformly bent in one plane or in two planes at 90° to each other. However, these programs only allow the effect of bow and irregular cross-section on yield to be estimated, and therefore do not predict the optimum yield.

The only way of accurately predicting the yield from irregular logs is to simulate their true shape, and work is currently underway writing and testing new computer software coupled with new methods of assessing log shape.

12.6 Conversion factors

The efficiency of conversion will vary widely depending on log shape, market requirements, the age of the equipment and the skill of the sawyer in the absence of computer controlled equipment. Based on the amount of solid wood produced in terms of a debarked log, the percentage out-turn will range from about 60 to 75 per cent. When the volume of usable chips is taken into account, the efficiency of conversion can rise to a maximum of 85 per cent, the remainder being sawdust.

12.7 Defects

In the conversion of a log to usable timber, a number of natural defects will manifest themselves. These will include pitch pockets, included bark, brittleheart and natural compression failures. These features are discussed fully in Chapter 5; suffice it here to say that their presence will markedly reduce the conversion ratio for wood used both aesthetically and structurally.

References

Archer R.R. (1987) *Growth Stresses and Strains in Trees.* Springer-Verlag, Berlin.

Boyd J.D. (1985) The key factor in growth stress generation in trees: lignification or crystallisation? *Bulletin* (new series) *of the International Association of Wood Anatomists* **6(2)**: 139–150.

Dinwoodie J.M. (1966) Growth stresses in timber – a review of literature. *Forestry* **39(2)**: 162–170.

Jacobs M.R. (1945) The growth stresses of woody stems. *Commonw For Bur, Aust Bull* 28.

Klamecki B.E. (1979) A review of wood cutting tool wear literature. *Holz als Roh- und Werkstoff* **37**: 265–276.

Münch E. (1938) Statik und Dynamik des schraubigen Baues der Zelland, besonders des Druck- und Zugholzes. *Flora* **N.F.32**: 357–424.

13

Seasoning of Wood

The primary aim in seasoning is to render timber as dimensionally stable as possible, thereby ensuring that once it is made up into flooring, furniture, fittings, etc., movement will be negligible or for practical purposes non-existent; simultaneously, other advantages accrue. Most wood-rotting and all sap-stain fungi can grow in timber only if the moisture content of the wood is above 22 per cent: hence, seasoning arrests the development of incipient decay in wood and removes the risk of infection of sound timber. Seasoning, however, does not confer immunity from subsequent infection should the moisture content of previously dry wood be raised above the critical minimum, as a result, for example, of prolonged exposure to damp conditions. Reduction in weight of wood accompanies loss of moisture; this is of practical importance as it reduces handling costs, and may effect economies in freight charges. Seasoning also prepares timber for various 'finishing' processes, such as painting and polishing, and it is an essential prerequisite if good penetration of wood preservatives is sought. Finally, most strength properties increase as timber dries and, although the increases may not in themselves justify the expense of seasoning, they are of more than academic significance.

Drying occurs because of differences in vapour pressure from the centre of a piece of wood outwards. As the surface layers dry, the vapour pressure in these layers falls below the vapour pressure in the wetter wood further in, and a vapour-pressure gradient is built up that results in the movement of moisture from centre to surface; further drying ('seasoning') is dependent on maintaining this vapour-pressure gradient. The steeper the gradient, the more rapidly does seasoning progress but, in practice, too steep a gradient must be avoided.

Below the fibre saturation point, drying is accompanied by shrinkage. The amount of shrinkage that will occur varies with the species and the degree of dryness attained; it will usually be greater tangentially than radially, and negligible longitudinally (Chapter 9). If the tendency of a wood to shrink on drying is high, the risk of stresses being set up in the outer layers is great with a steep moisture gradient: the outer layers want to shrink but are restrained by the wetter interior. The outer layers may set in a stretched condition, that is case-hardening occurs, or the tissues may be ruptured, that is surface checking results: these defects will be discussed more fully in section 13.6.

13.1 Principles of seasoning

The whole art of successful seasoning lies in maintaining a balance between the evaporation of water from the surface of timber and the movement of water from the interior of the wood to the surface. Three factors control water movement in wood: the humidity and temperature of the surrounding air, and the rate of circulation. Temperature has a twofold effect: by influencing the relative humidity of the air it affects the rate of evaporation of water from the surface of the wood, and within the timber it influences the rate of movement of water from the centre towards the surface.

It is important to appreciate how these three factors interact. The rate of loss of moisture from wood depends on the humidity of the air in immediate contact with the surface layers, and on the dryness of the layers themselves. The rate of movement of water outwards in a piece of wood is dependent on the vapour pressure of the outer layers being lower than the vapour pressure further in; however, the differences in vapour pressure of successive layers must not be excessive. If the outer layers are appreciably drier than the interior, greater resistance is offered to the movement of moisture outwards than when differences in vapour pressure, and consequently in moisture content, of successive layers are smaller. In extreme circumstances, resistance may be such that diffusion of moisture from the inner layers outwards is brought to a standstill, the moisture in the interior of the wood being sealed in. Resumption of moisture movements in such cases can usually be achieved only by artificial means, such as steaming in a kiln.

The relative humidity of the atmosphere, and its temperature, are all-important in the seasoning process: the lower the relative humidity of the air, the better it will be able to take up moisture from the surface of a piece of wood and, conversely, wood in contact with saturated air cannot dry at all. Even at high relative humidities it is still possible to dry wet wood provided the temperature is high enough. Thus at temperatures prevalent in the tropics, namely 27–32°C, comparatively high relative humidities, such as 70–80 per cent, still leave the air with appreciable drying powers. At these high temperatures the amount of moisture that the air requires to raise its relative humidity by 1 per cent is much greater than the amount required to raise the relative humidity by 1 per cent at, say, 16°C. This factor in the temperature–humidity relationships of the atmosphere explains why it is possible to air-dry timbers in the humid climates of the tropics to moisture contents similar to those achieved in temperate regions, and in less time. In fact, unless the site conditions of tropical storage sheds are exceptionally unfavourable, the main problem is usually to retard the rate of drying to minimise checking and distortion. At the same time, surface drying is usually not sufficiently rapid to preclude sap-stain discoloration in timbers particularly susceptible to such infestation.

When temperature is held constant, the relative humidity of the air increases as moisture is absorbed, and the affinity of the air for further moisture decreases; this, in turn, slows up the drying of the surface layers of wood exposed to such air. When air is absorbing moisture less rapidly, as a result of its relative humidity increasing, differences in moisture content of successive inner layers of wood exposed to such conditions will be less marked, the two factors thus combining to reduce seasoning stresses to a minimum. Although this mechanism is of use in drying timbers that are liable to surface splitting – in these cases air flow over the surfaces is purposely restricted – nevertheless air movement is essential to prevent excessive build-up of relative humidity which may seriously retard or even prevent any drying taking place.

Two principal methods of seasoning are in common use: **air**, sometimes called *natural*, and **kiln**, often called *artificial*. For large-sized timbers a combination of both modes of seasoning is frequently adopted.

13.2 Air seasoning

While the great majority of seasoning is carried out artificially, air seasoning is still practised in countries where the cost of kiln-drying is too high. It is also carried out in many other countries to either partially or completely dry timber of large cross-sectional areas which would be far too expensive to dry by normal kiln-drying. Thus, in timber taking more than about 4–5 weeks to kiln-dry from green, it will often be found more economical to air-dry it to about 25–30 per cent moisture content before completing drying in a kiln. In the UK, hardwood planks which have been produced by through and through conversion (Chapter 12) are usually air-dried for 18–24 months before being kiln-dried. However, large-section hardwood is frequently only air-dried because of the cost involved; beam quality oak measuring 350 × 300 mm is usually air-dried for at least 8 years prior to sale. Softwood 25 mm thick, if air stacked in the spring, should dry to about 20 per cent moisture content in 1.5–3 months, while 50 mm thick softwood will require 3–4 months. Hardwood 25 mm thick, if piled in the autumn will take 7–9 months to reach 20 per cent, while 50 mm thick hardwood will take about a year.

Air seasoning therefore, is a very slow process; it aims to make the best use of prevailing winds and the sun, while protecting timber from rain and direct sunlight. Wind, by circulating the air, prevents it from becoming saturated with moisture absorbed from seasoning timber, and the sun, by raising the temperature of the air, lowers its relative humidity. The combined effect of these two factors is to maintain the drying power of the air. Rain, on the other hand, increases the humidity of the atmosphere and, as it is accompanied by lower temperatures, reduces the drying power of the air. If at the same time the timber is actually wetted, it may pick up appreciable quantities of moisture. As a general rule, the problem is to accelerate air circulation adequately, although in the tropics, as explained earlier, with timbers prone to develop seasoning defects, it may be necessary to reduce air circulation and thus slow up the rate of drying. The extent of drying will vary with the particular environmental conditions of the locality: in the UK it is not possible to dry below about 17 per cent in the summer months and 22 per cent during the winter.

Control of the climatic factors is best achieved in properly constructed, well-ventilated sheds, but with low-quality timber such structures are impracticable on economic grounds. The most efficient shed is, moreover, only effective up to a point since the relative humidity of the air varies appreciably at different seasons of the year. Control of air circulation, whether in sheds or in the open, is effected by piling the timber in properly constructed stacks, the design of which is the most important consideration in air seasoning. Control of the movement of water in wood is more difficult. Water movement is, of course, affected indirectly by control of air circulation, but additional measures are advisable to compensate for the more rapid movement of moisture along the grain than across it. If the loss of moisture from the ends of a piece of wood is not checked, large stresses are set up that result in severe end-splitting; to minimise this problem, some form of end covering should be adopted. Three factors are therefore available for regulating air seasoning: seasoning sheds, correct stacking and end protection of the individual pieces of wood in a stack.

Seasoning sheds

In its simplest form a seasoning shed may be nothing more elaborate than a large Dutch barn with temporary roofing. On the other hand, it may be a permanent building, consisting of a roof and four walls, the walls being louvred, so that air circulation through the building can be regulated with considerable precision. In the tropics some form of shed is very necessary to protect timber against heavy rain and the very strong sun but, because of the intense heat in the middle of the day, a corrugated-iron roof should be avoided: thatch, shingles, or rough boarding are better.

Stacking

Piling technique is the most important factor in air seasoning, because such points as the position and orientation of stacks, and their method of construction, largely govern air circulation.

The important points in stack-building are the orientation, foundations, spacing and width of stacks, and the spacing and width of stickers. Two alternative methods of orienting stacks with reference to the passage ways are possible: **endwise**, that is with the timber at right angles to the passage ways, and **sidewise**, that is with the timber parallel to the passages. Endwise stacking makes for ease of inspection and tallying of the stock, but sidewise stacking ensures better air circulation from the passage ways. In endwise stacking the air is held up by the stickers and can only circulate by way of the narrow alleys between stacks. Economic considerations, and mill layout, however, usually determine the method selected, but where mechanical elevators are used for stack-building, sidewise stacking is obligatory. If several varieties of timber, requiring different seasoning periods, are dealt with in the same yard, sidewise stacking is more convenient and economical: high handling costs result when one of a series of endwise stacks is required out of turn. By far the most common failing is to crowd sheds to their maximum capacity, the excuse being made that land values are so high that the fullest use must be made of a firm's storage capacity. This argument overlooks the fact that expelling moisture from wood is inevitably an expensive matter; the quicker air-drying that is achieved by not overfilling seasoning sheds may well offset the higher rental costs per cubic metre of throughput because of the saving in fuel when such timber is finally kiln-dried just prior to use.

For the foundations of stacks, baulks of durable or preservative treated timber are commonly used, but concrete, brick, or even wooden piers, are better as they offer less resistance to the free circulation of air under a stack.

The dimensions of stacks must be kept within certain limits to secure rapid and uniform drying, and to avoid the risk of stagnant air accumulating in the centre of the pile, which can be a cause of unequal drying, leading sometimes to fungal infection. Four metres is recommended as the maximum width for stacks, and one metre as the minimum width of passage ways. Excessive height is to be avoided for similar reasons, and there is the added disadvantage that tall stacks increase handling charges.

Circulation of air through a stack is secured by separating the successive layers of timber by strips of wood known as **stickers**, the thickness of which regulates the rate of air flow. The stickers should be sound, seasoned timber and, to avoid indentation of boards in the lower part of the pile, not of a harder type than the timber in the stack. The use of softwoods for this purpose is a safeguard against the introduction of powder-post beetles through infected stickers (Chapter 20). When a stack is dismantled the stickers should receive as much consideration as is given to the seasoned timber; they should be collected, bundled and stored for further use.

The most suitable thickness for stickers depends on the thickness of the timber to be seasoned, its drying qualities and the season of stacking. For thin stock, of species not subject to serious degrade in seasoning, stickers 37.5 mm in thickness are suitable, but thick planks of species that are inclined to split or surface-check badly may require stickers as thin as 12.5 mm. Stickers should be no wider than is absolutely necessary, as the area of timber in contact with them is hindered from drying at the same rate as the remainder and, in certain timbers, such covered

portions may become stained. Stickers which are too narrow, on the other hand, cause indentation and, for this reason, the width should never be less than the thickness, and for really soft timbers it may need to be greater. A maximum of 50 mm in width should, however, suffice for the most easily bruised timber.

Stickers impede the circulation of air and, therefore, should not be unnecessarily numerous; on the other hand, an insufficiency of stickers results in the sagging of boards and planks. The distance between stickers depends on the thickness of the stock and its liability to warp; boards 12.5 mm in thickness require stickers 600–900 mm apart, but planks of 50 mm and upwards are usually sufficiently supported by stickers 1200–2400 mm apart, the spacing increasing with increase in thickness of the planks. The stickers must be in vertical rows in order to avoid unequal stresses on the lower layers of timber, which would inevitably result in a considerable amount of bowing.

As far as possible, the timbers in a stack should be of uniform length, but when this is not practicable the longest pieces should be at the bottom; projecting ends must be supported. Additionally, in any one row, all timbers must be of the same thickness, otherwise the thicker pieces carry the weight of the whole stack above them. The top layer of timber, and all projecting ends, should be covered with thin, dry boards, or, in an open yard, by a raised roof. Thin-dimensional stock should be weighted by laying heavy baulks of seasoned timber on the top of the stacks, to prevent bowing and cupping. Timber should be stacked with stickers as soon after sawing as possible.

A special form of piling, often adopted for hardwoods in the UK and in certain continental European countries and referred to as **in-stick**, is to pile each board in sequence as cut, and the log is sold as a unit. The advantages claimed for this method of piling are that the merchant is not left with narrow widths and defective boards and, with figured woods, 'matched' material is kept together.

Timbers liable to discoloration, such as sycamore, are frequently seasoned by stacking on end, thereby avoiding the use of stickers;

baulks, sleepers, squares, etc. are usually **self-piled** in various ways, the essential feature of which is that some of the pieces of timber act as stickers.

End protection

End protection is provided by coatings of various waterproof substances. Strips of wood are not recommended, because the small longitudinal shrinkage of the strip is opposed to the much greater transverse shrinkage of the timber, with the result that shrinkage is restricted between the points of attachment of the strip and stresses are set up which tend to induce end splitting. The essential qualities required of end coatings are impermeability to water vapour and the ability to adhere to a wet timber surface and retain sufficient flexibility to remain in position as the timber dries and shrinks.

Many substances fulfil this condition; wax, for example, may be used as a hot dip on the ends of small-dimensioned stock or as an emulsion applied by brush to planks and boards. Perhaps the most useful substance is a thick bituminous paint or emulsion just liquid enough to be applied by brush. This material has the added advantage that it remains effective at the higher temperatures used in a drying kiln if this treatment is subsequently found to be necessary.

13.3   Kiln seasoning

Kiln-drying is effected in a closed chamber, providing maximum control of air circulation, humidity, and temperature. In consequence, drying can be regulated so that shrinkage occurs with the minimum of degrade, and lower moisture contents can be reached than are possible with air seasoning. The great advantages of kiln seasoning are its rapidity, adaptability and precision. It also ensures a dependable supply of seasoned timber at any time of the year, and it is the only way that timber can be conditioned for interior use requiring lower equilibrium moisture contents (see Chapter 9) than those prevailing out-of-doors, or in unheated sheds.

In properly operated kilns, variation in moisture content across the section of plank and differences from plank to plank within a load can be kept low. Moreover, the drying process also sterilises the timber: the temperatures used, and the humidities maintained in the kiln, are lethal to any insect or fungus present in the timber when placed in a kiln. Such sterilisation does not, of course, protect the dried timber against fresh infestation after removal from the kiln. Further, the resins or gums in certain woods are to a large extent set or hardened in kiln-drying, so that the risk of subsequent 'bleeding' from finished surfaces is reduced. On the other hand, departure from the recommended levels of temperature and relative humidity can give rise to drying degrade and, in extreme cases, gross errors in operation can result in the load of timber being seriously damaged.

It is necessary to regulate kiln-drying to suit circumstances: different timbers and dimension of stock require drying at different rates. As a general rule, softwoods can withstand more drastic drying conditions than hardwoods, thin boards than thick planks, and partially dry stock than green timber. There are limitations to the dimensions that can be kiln seasoned economically. The time required increases rapidly with the thickness of the timber and the resulting increase in cost usually means that although it is possible to kiln dry material, from a practical point of view it is preferable first to air-dry material over, say, 75 mm thick. The occasions when large-sized timbers are required uniformly dried to low moisture contents of around 12 per cent are, however, extremely few, so that restriction to the smaller maximum dimensions suggested does not really impose limitations of practical commercial importance.

A kiln consists of some form of more or less air-tight shed, fitted with heating apparatus, a supply of water or steam sprays, and artificial means of accelerating air circulation. One of the greatest problems in kiln construction is in keeping the heat losses to a minimum. The usual method of supplying heat to kilns is by a system of steam-heated coils over which the air passes before circulating through the stacks of timber. Other methods of heating could be used, but steam is particularly suitable as it is easily regulated, and in many saw mills it is available from the burning of wood waste.

The humidity of the air can be controlled by regulating the temperature, by admitting water or water vapour, or by changing the air through removal of saturated air from the kiln and replacing it with fresh air from the outside. In practice the manipulation of temperature alone is seldom sufficient, and a system of water or steam sprays, and inlet and outlet air ducts, is installed. The circulation of air is usually by means of fans or blowers.

The rate of drying in different parts of a kiln varies because the temperature of the air and its relative humidity vary at different levels, and so arrangements have to be made to counteract this as far as possible. One method is to increase the rate of circulation so that there is less opportunity for the air to become saturated before it has passed through the stack, or the direction of circulation may be reversed by reversing the rotation of the fans after fixed time intervals.

It is important to follow closely the conditions of the air in a kiln during a run, and also to monitor the progress of drying in the timber. The first can be done by means of wet and dry bulb thermometers (**hygrometers**) suitably placed in the kiln, and illuminated, so that they can be read from outside the kiln. With readings of the two thermometers, the relative humidity of the air in a kiln is found by reference to appropriate tables or charts. All hygrometers require maintaining in good order; with the simple mercury-in-glass type, this merely involves maintaining the water level in the receptacle and the syphoning wick in perfect condition. Distilled water should be used, and the calibration of the instruments should be checked annually. Self-recording instruments are sometimes used in place of simple mercury thermometers; these incorporate inked pens, and charts (graph paper so ruled that the relative humidity is read direct). Such instruments are an essential part of a fully automatically operated kiln, and they provide a very useful record of conditions in a kiln throughout the entire run.

Progress of drying in the timber should be followed by means of **test** or **sample boards**. At the commencement of a run, the moisture content of the parcel as a whole should be determined from a sufficient number of samples by the oven-dry method described in Chapter 9. The boards or planks from which the samples are taken also serve as 'sample boards' for following the progress of drying. After the moisture-content samples have been cut from either end, the remaining length of each board should be at least 1.5 metres. The ends of the sample boards should be sealed and the boards should then be weighed and returned to the stack, but so distributed that they will provide a picture of the progress of drying in the whole consignment. The stickers above the boards are notched so that the boards can easily be withdrawn for periodical weighing, and then be replaced. In the case of the drying of hardwoods, these sample boards can be weighed three to four times during the kilning period. The mass of the boards, at the selected final moisture content, is arrived at by calculation, as explained in Chapter 9. Intermediate weighings give, by calculation, the moisture contents of the moment, and the progress of drying is, therefore, followed closely. The drying schedule can be modified according to whether it is revealed that drying is too rapid or too slow.

Control of the drying process can also be constantly monitored by the insertion of probes into the timber to provide a direct and continuous reading of changing moisture content (see Chapter 9, section 9.2.3). This signal can also be used for control drying automatically by a small computer.

In spite of reasonable care it is possible for drying stresses to be set up in the course of a kiln run, sufficiently serious to cause case-hardening or honey-combing, if not actual visible splits or checks (see Figure 13.1 and section 13.6). When such stresses are suspected, and also as part of the routine study for drying progress, the consignment should be tested for such stresses. For this purpose test pieces 12 mm thick (along the grain) are cut 22.5 cm or more from the ends of the sample boards. The centre of each

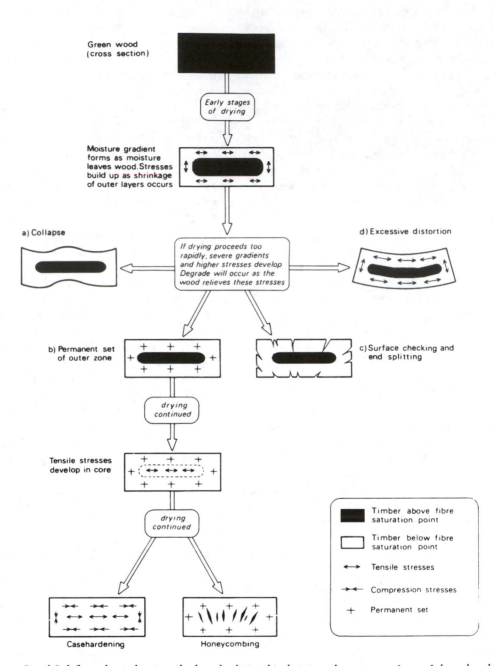

Figure 13.1 *Simplified flow chart showing the broad relationship between the common form of degrade which can occur during kiln and air drying (Building Research Establishment, © Crown Copyright) [Note 1: Combinations of the various forms of degrade can occur. Note 2: The chart is intended to outline the processes which can give rise to high levels of degrade; even careful drying cannot prevent the limited development of some categories of degrade, particularly with some timber species]*

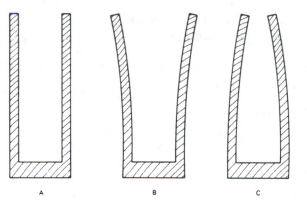

test piece is cut away to produce a 'prong' as in Figure 13.2A. A study of the behaviour of these prongs will indicate the nature and extent of the drying stresses. In the early stages of drying, tension stresses tend to be set up in the outer layers of the wood. At this stage the prongs will immediately curve outwards when cut, as in Figure 13.2B. When the samples are allowed to dry under normal room conditions for 12–24 hours, to a uniform moisture content, the slightly wetter inner faces of the prongs will shrink more than the outer faces and, if permanent stresses have occurred, the prongs will eventually bend inwards, as in Figure 13.2C. At a later stage in drying, the behaviour of the test pieces is different. By this time the stresses become reversed in the wood, and the core is in tension while the outer layers are in compression. Test pieces cut from boards in this condition will immediately curve inwards and, when allowed to dry to a uniform moisture content, the extent of curvature will be increased. When no permanent set has occurred the prongs will become parallel, as in Figure 13.2A, on reaching uniform moisture content, that is, after 12–24 hours' exposure to normal atmospheric conditions. If tests reveal that serious drying stresses have developed, the drying schedule must be modified to relieve these stresses; this aspect is discussed in the section on kiln operation.

It was necessary to stress the importance of correct stacking technique when discussing air seasoning, to which reference should be made; correct and careful stacking is of no less importance in kiln seasoning, and the trouble taken is always justified because the appearance and quality of the timber at the end of a run will be superior, compared with a poorly stacked load. With very few exceptions, stacks must be built with stickers as for air seasoning — the exceptions are certain classes of dimension stock which can be self-piled. The stickers should be selected from clean, dry timber, and finished 25 mm by 25 mm, or for thin boards 25 mm by 19 mm. Softwoods, except larch, and hardwoods not inclined to warp, of 50 mm and upwards in thickness, should have stickers spaced at 900 mm centres; for material less than 50 mm thick, the spacing should be reduced to 600 mm. Material equal to or greater than 50 mm in thickness of larch, beech, birch and other timbers that tend to warp appreciably in drying, should have stickers spaced at 600 mm centres, and for material of these species less than 50 mm in thickness the spacing should be reduced to 450 mm centres. Still more refractory timbers, such as elm, should be piled with stickers at 300 mm centres.

The same precautions must be taken regarding the alignment of stickers in vertical rows as in air seasoning, and the resulting forces must be distributed to the foundations by means of a system of cross-stringers and bearers when the stickers are spaced closer together than the distance between the main supports of a stack. Special supports must be provided for long boards that overhang the ends of a stack. Stacks of boards may be weighted to reduce distortion in the upper rows of a stack; concrete slabs are suitable, but ferrous metal weights must not be used with timbers such as oak and chestnut because the tannin in these woods may lead to serious staining when brought in contact with iron in a humid atmosphere.

In all modern kilns it is recommended that the boards in each layer of the stack should be placed edge to edge. An exception should be made when drying squares or scantlings more than about 65 mm thick, when gaps of about 10 mm should be left to allow drying from the edges.

13.3.1 Construction of kilns

Concrete floors and footings, with provision of drainage for the floor, are recommended. Walls must be constructed with materials which will provide a high degree of thermal insulation. Doors can be a serious source of heat loss: side hung doors have proved to be unsatisfactory and a centre hung door is recommended.

13.3.2 Types of kilns

There are two main types of timber drying kilns, namely **progressive** and **static loaded**.

Progressive kilns
This type of kiln is still in use in Scandinavia, though no examples are present in the UK. In progressive kilns, green timber is admitted at one end and moved gradually to the other, where it emerges dry. The air flows in the opposite direction to the movement of the timber, so that the material that has been longest in the kiln receives the hottest and driest air. In passing through the piles of timber the air absorbs moisture, which increases its relative humidity and lowers its temperature, so that at the loading end the wet timber comes in contact with relatively cool, humid air. The severest drying conditions are, therefore, at the exit end, where the timber is best able to accommodate itself to them, and the mildest at the loading end, where the timber is least able to stand up to rapid drying. After circulation through the length of the kiln, part of the air is discharged into the atmosphere, and the remainder returns below the floor of the kiln to be recirculated.

The uses of progressive kilns are limited since their successful operation depends on a steady supply of timber of the same species and dimensions. The reason for this is that the drying conditions cannot be modified as each new load is added, and while the kiln still contains partially dried loads of a particular type. In addition to lack of flexibility, progressive kilns cannot be regulated with great precision, and this renders them unsuitable for timbers that are difficult to season. The advantages of progressive kilns are held to be that, once placed in efficient operation, they require less skill to run, and the output is more or less continuous, compared with compartment kilns; they are suitable for long runs of uniform material.

Static loading kilns
These kilns, like a progressive kiln, consist of a closed chamber, but it differs in operation in that the timber remains in the same position in the kiln throughout the drying period, the temperature and humidity of the circulating air being constantly changed. Air is circulated cross-wise from top to bottom of the kiln.

Conditions at the commencement of a run are mild; that is, the relative humidity of the air in circulation is high and its temperature is low. As drying proceeds, the temperature is raised and the humidity is reduced; as a result, the drying conditions become more severe, but are kept within bounds to prevent degrade of the timber.

Fans are mounted internally and are usually positioned down one side at intervals of 1.5 metres. The size of kilns has been increasing progressively over the last decade in order to reduce unit cost of drying and new kilns now hold up to $100 \, \text{m}^3$ of timber.

The great merit of static loaded kilns is their flexibility in operation, coupled with the fact that they can be designed to give precision of control. Flexibility is desirable when the out-turn of a mill is constantly changing, both in species and dimensions of stock; and maximum control of drying conditions is essential for the successful drying of difficult timbers. In effect, compartment kilns are to be preferred to progressive kilns in all but special circumstances where mills are producing continuous supplies of a single timber of one thickness.

13.3.3 Kiln operation

Successful kiln drying is very largely dependent on the skills of the kiln operator: a poorly designed kiln will give better results in the hands of a good man than the most up-to-date and efficient kiln in unskilled hands. It is essential to confine any run to one species and one dimension of stock; it is equally important to make proper use of kiln instruments, by siting them correctly and maintaining them in good condition; and progress of drying must be followed by means of sample boards. Given these essentials, it is also advisable to use a drying schedule that provides a margin for error: typical schedules for different timbers have been evolved from experimental practice and provide such margins. Modifications may still be called for: good-quality timber will tolerate more severe drying conditions than poor-quality timber; quarter-sawn boards will dry more slowly, but with less degrade, than flat-sawn material of the same species.

Examples of typical drying schedules are given in Table 13.1; these schedules are based on the assumption that drying is carried out in a forced-draught type of kiln with an average air speed through the stack of 1–1.5 m/s. In both the illustrated schedules it will be seen that the initial temperature is lower than the final temperature, and the initial relative humidity is higher than the final relative humidity. In effect, conditions in the kiln are made progressively more severe as drying proceeds.

In general, the drying schedules included in Pratt (1986) are comparatively mild and can usually be relied upon to give satisfactory results. However, if drying is allowed to continue with the last setting in each schedule, the timber could become too dry; that is, there could be overshoot of the desired moisture content. Consequently, the schedules are followed until regular monitoring reveals that the moisture content is sufficiently low to apply a conditioning treatment: this often takes the form of steaming at 10°C above the final temperature provided that this is less than 80°C; if the kiln is already running at 80°C or above, an increase in temperature should not be necessary. The run is finished by setting equilibrium conditions for the desired moisture content.

Temperature and humidity control in the kiln are effected by manipulation of heating coil and spray valves, with occasional adjustments of air inlet and outlet dampers. Economy in operation of kilns is dependent on making the maximum use of the moisture extracted from the timber for maintaining the required humidity of the circulating air. This is effected by allowing only very slightly more moisture to escape via the ducts than is being extracted from the timber, the deficit being made good by comparatively small amounts of steam from the steam spray system.

Warming up must not be too rapid because of the time lag in heating the wood to the recorded temperature of the air. Too rapid heating of the air in a kiln may result in condensation of moisture on the surface of the timber, which takes appreciably longer to warm up. Too high temperatures in the course of a kiln run are to be avoided, as high temperatures my darken the whole consignment. Rapid cooling at the conclusion of the run is frequently possible, but there is a risk that the hot timber will heat the cool air entering the kiln, making it appreciably drier, and this may lead to a renewal of case-hardening stresses or even splitting of the wood. It is suggested that a difference of 5°C between wet and dry bulb readings should be maintained in the initial warming period, until the desired dry bulb reading is attained, and a similar difference during cooling, until the dry bulb reading has dropped to about 27°C.

The behaviour of different timbers in kiln drying varies enormously; in general softwoods are much less refractory than hardwoods. With the latter, it is common practice to partially air dry the stock first (see section 13.2), as otherwise the kiln run is too long to be economical. Commercial kiln drying schedules may occupy anything from a few days to a few weeks. Data on the drying properties of a large number of species are provided in Pratt (1986). It is possible to make a reasonable estimate of the drying time (see Pratt, 1986).

High-temperature drying

This is a fairly recently developed process now used commercially in Australia and New Zealand

Table 13.1   *Examples of standard kiln drying schedules[1]*

Schedule A[2]: (suitable for timbers which must not darken in drying and for those which have a pronounced tendency to warp but are not particularly liable to check)

		Standard schedule conditions[3] for timber thicknesses up to 38 mm			Modification to schedule for thicker timber 38–75 mm over 75 mm	
	Timber moisture content[4] (%)	Dry bulb (°C)	Wet bulb (°C)	Relative humidity (%) [approx.]	Dry bulb (°C)	Wet bulb (°C)
Stage 1	Green	35	30.5	70	31.0	32.0
2	60	35	28.5	60	29.5	30.5
3	40	40	31.0	50	32	33.0
4	30	45	32.5	40	34.0	35.0
5	20	50	35.0	35	36.5	38.0
6	15	60	40.5	30	42.5	44.5

Schedule G[2]: (suitable for timbers that dry very slowly, but which are not particularly prone to warping)

		Standard schedule conditions[3] for timber thicknesses up to 38 mm			Modification to schedule for thicker timber 38–75 mm over 75 mm	
	Timber moisture content[4] (%)	Dry bulb (°C)	Wet bulb (°C)	Relative humidity (%) [approx.]	Dry bulb (°C)	Wet bulb (°C)
Stage 1	Green	50	47.0	85	48.0	49.0
2	60	50	46.0	80	47.0	48.0
3	40	55	51.0	80	52.0	53.0
4	30	60	54.5	75	55.5	57.0
5	25	70	62.5	70	64.0	65.0
6	20	75	62.5	55	64.0	65.5
7	15	80	61.0	40	63.0	65.0

1. These schedules are reproduced from Pratt (1986) and are reproduced by courtesy of the Chief Executive, Building Research Establishment.
2. Under certain circumstances it may be necessary or advantageous to make modifications to a schedule for factors other than thickness. Guidance on schedule modifications is given in section 6.4; **particular attention is drawn to the consideration of air flow rate in 6.4.2 (Pratt, 1986)**.
3. Conditions at the air inlet.
4. Changes in kiln conditions are normally made when the wettest of the withdrawal samples has dried to the moisture content indicated for the next stage.

for the rapid drying of pine, principally *Pinius radiata*. The basic concept is to boil off the water in the wood as quickly as possible and then to recondition it to a required moisture content.

In the first phase, very fast warming up is demanded and achieved usually through the use of gas burners. Temperatures of around 120°C are reached within 2 hours and maintained for a further 12 to 15 hours; in the second phase, the stack is rapidly cooled and reconditioned to the required moisture content. The aim is to complete the drying operation within a 24-hour period.

The inherent tendency of wood to bow, twist and collapse (see section 13.6) under such extreme conditions is restrained appreciably by the use of heavy top loading during drying, together with the process of reconditioning. It is difficult to obtain direct comparison between this method and conventional drying schedules, but early indications are that degrade is similar in both regimes. High-temperature drying is certainly commercially attractive, and it is very likely that the practice will spread to the other continents. However, initial research in the UK has indicated that high-temperature drying may not be suitable for home-grown Sitka spruce.

13.4   Solar kilns

It has been recognised for very many years that air-drying of local timber is not adequate for its efficient use and that there is a need in many tropical countries for a small kiln with low running costs; further benefits could be derived if the structure was portable. Over the last decade much research has been carried out into the controlled use of solar heating for the drying of wood and this has led recently to the commercial availability of **solar kilns**. Generally these kilns are of small capacity (usually less than $10\,\text{m}^3$) and, as the rate of drying is usually very slow, they are particularly suitable for many of the difficult slow-drying tropical timbers, and for timber of different sizes. The solar kiln which is commercially available from the UK comprises a lightweight aluminium frame covered with a thin plastic glazing membrane which retains the heat built up in the kiln. Solar heat is absorbed by a false ceiling of sheets of corrugated aluminium or galvanised iron painted matt black. Some fresh air is drawn in, circulated by a fan and, together with absorbed moisture, vented to the outside. Temperature during the day rises to about 55°C and results in drying to 6–8 per cent moisture content in the dry season, and 10–12 per cent in the rainy season. Drying costs per cubic metre have been found to be only two-thirds of those using conventional kilns, while capital costs are very much lower.

13.5 Radio frequency drying of wood

It is possible to dry wood using radio frequency (RF) heating. Water molecules are polar in nature and are ordinarily randomly arranged within the timber. The application of an electric field tends to align the molecules in a direction parallel to it. Reversal of the field results in a tendency for them to re-align in the opposite direction. With rapid alternation of the direction of the field, the molecules are constantly moving resulting in the generation of heat. The heating effect takes place throughout the cross-section of the timber and if required, large quantities of heat can be developed.

There are two ways in which RF heating can be used to dry wood. The first is known as the 'boiling method' in which large quantities of RF energy are used to raise the temperature to boiling point and to drive out the moisture in the form of steam. This method is suitable for a continuous throughput of blocks of uniform size and has been used commercially to obtain very rapid drying. However, because of the need to allow the rapid escape of the generated steam, use of the method is usually restricted to permeable timbers. In less permeable timbers, internal rupture or 'bursting' is prone to occur.

For timbers of lower permeability it is usual to apply the second method known as the 'temperature gradient method'. Much lower RF power levels are employed in this method to keep the centre of the wood warm while cooler humidified air circulates over the surface, thereby setting up a temperature gradient across the wood. The temperatures employed are well below boiling point, but it is still difficult to prevent the surface layers drying out too much and the wood checking. In timbers prone to checking, energy input is lowered and longer drying times accepted. Nevertheless, it is still possible to reduce drying times to about one-quarter of those of conventional kilns.

The reduced times with RF drying must, of course, be offset by the large initial costs of RF kilns.

13.6   Seasoning defects

Next to knots, the commonest cause of degrade in timber selection and utilisation is the presence of defects which manifest themselves following seasoning (Figure 13.1). Some of these defects are present irrespective as to whether the correct seasoning technique has been adopted, since they are related to inherent limitations in the wood itself. However, faulty seasoning techniques will certainly exacerbate their presence while inducing other types to form. They conveniently fall into two broad categories – those associated with various forms of **warping** of the timber plank or batten, and those associated with **rupture** of the wood tissue.

13.6.1   *Types of warping*

Cupping applies to distortion across the width of the board and takes the form of either a convex or concave shape as seen from the end (Figure 13.3). In flat-sawn material one surface is more nearly radial than the other, and, since radial shrinkage is less than tangential as discussed in Chapter 9, that side of the board nearer the pith will shrink less than will the opposite face and the board will be cupped following seasoning. In square-section battens this type of warping will result in **diamonding** (Figure 13.3). It is possible to reduce but not eliminate cupping through correct piling and top loading the stack prior to seasoning. This defect, of course, will not develop in truly quarter-sawn material.

Bowing takes the form of concave/convex distortion along the length of a plank or batten (Figure 13.3). Bowing results from the too wide spacing of stickers which allows the timber to sag under its own mass. Its presence is due exclusively to faulty seasoning technique.

Twisting is the spiral or corkscrew distortion in a longitudinal direction of a board or plank cut through or near to the centre of the tree (Figure 13.3). In extreme cases it may render the plank useless. Twisting is due to the presence of spiral grain which was discussed in Chapters 5 and 7. It can usually be reduced, but certainly not eliminated by restrained drying executed by adding weight to the top of the stack prior to seasoning.

Spring again takes the form of concave/convex distortion along the length of the plank, but this time the distortion is in the plane of the board (Figure 13.3). Spring is not uncommon in boards cut from near the core of the log and is the result of the release of growth stresses described in Chapter 12 as the log is sawn. All timber is more or less subject to a small degree of spring, but in certain species, such as *Eucalyptus*, the degree of spring can be very high.

13.6.2   *Types of rupture of the wood tissue*

A **check** is a small separation of the wood cells that does not extend through the timber (Figure 13.1).

A **split** is a greater separation of the wood cells and extends through the timber from one face to the other. Both checks and splits can be due to the release of growth stresses, but more often than not they are indicators of too rapid drying of the wood. When this occurs the outer layers want to shrink while the interior is still saturated and stresses are set up. If the outer layers shrink their full amount, checks will result.

Honey-combing. If checks do not result in the outer layers following too rapid drying, it is possible that the outer layers may dry down and set in a stretched condition (tension set). Pieces of timber in which this occurs are said to be **case-hardened** (Figure 13.1), a condition

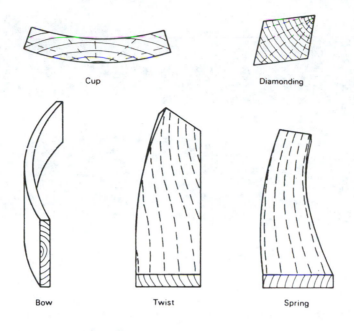

Cup

Diamonding

Bow

Twist

Spring

Figure 13.3 *Various forms of warping in timber (Building Research Establishment, © Crown Copyright)*

which can be removed quickly if noticed, by steaming in order to restore moisture to the outer layers: this is carried out at a temperature 10°C above final kiln temperature provided this is less than 80°C; if the final temperature is already above this temperature there is no need to increase the temperature. However, if case-hardening is not relieved by steaming, the outer layers set without shrinking their normal amount and, when the interior eventually dries below the fibre saturation point, it too is restrained from shrinking and interior checks results: this condition is known as honey-combing (Figure 13.1). The stresses set up are greatest tangentially, because shrinkage is greatest in this direction, and the resulting separation of the fibres is always initiated where the tissues are weakest: that is, along the rays. Honey-combing can be detected by cross-cutting the plank about 30 cm from the end and noting whether there are any internal checks along the rays on the freshly exposed end.

Collapse. Some timbers are particularly prone to this type of seasoning effect: thus it occurs fairly commonly in the seasoning of some species of *Eucalyptus*, for example *E. obliqua* and *E. regnans* (both known as Tasmanian oak). As the name implies, the cells actually collapse and become flattened as the amount of water in the cell is reduced. In the more permeable timbers, and in extreme cases, this cellular collapse manifests itself by the production on quarter-sawn faces of a corrugated surface, an effect frequently referred to as **washboarding**.

Collapse occurs during the early stages of drying and is usually associated with the too rapid drying of green wood at too high a temperature: collapse can occur with both air seasoning and kiln-drying and has been found on occasions in the drying of home-grown softwoods.

Collapse results in excessive and often irregular shrinkage, and may lead to appreciable distortion (Figure 13.4); in some extreme cases severe internal checking may occur. The defect is serious because

Figure 13.4 *Collapsed (A) and reconditioned (B) board of Tasmanian oak (Building Research Establishment, © Crown Copyright)*

of the loss of timber in trimming mis-shapen pieces of wood; collapsed air-dry wood is however, usually stronger than non-collapsed or reconditioned wood, primarily because it is more dense.

No uniform behaviour can be assigned to the effect of position in the tree in relation to the occurrence of collapse. Australian experience suggests that timber from moist or swampy places, and fast-grown immature trees, is more prone to collapse than material from other sources.

A method of **re-conditioning** has been evolved for removing collapse with complete success. It consists of heating collapsed timber in a chamber in saturated air to a temperature of about 100°C, and maintaining these conditions for some hours before allowing the timber to cool. Figure 13.4 shows the effect of re-conditioning collapsed timber for a few hours; the final moisture content was increased only by 0.5 per cent, and there was an appreciable gain in cross-sectional area. Slight checking of no commercial importance occurred on the edges; the strength properties of the re-conditioned timber were not lowered in comparison with uncollapsed material of the same moisture content. Machining qualities and working properties generally were improved. Equally successful results have been obtained with other species.

Reference

Pratt G.H. (1986) *Timber Drying Manual* (2nd edn, revised by Turner C.H.C.). Building Research Establishment Report. HMSO: available only from the Building Research Establishment Bookshop.

14

Machining of Wood and Board Materials

Following conversion from the log and subsequent seasoning, timber to be used in furniture, joinery and toy manufacture has to be further processed; this will involve some additional sawing followed by planing, spindle moulding and usually sanding.

As in Chapter 12, it is not the intention here to give a comprehensive account of the machining of wood. Rather it is the desire to give some background information on wood machining, especially those aspects influenced by the basic wood structure, thereby giving a fuller understanding of all aspects of wood utilisation. This chapter is not concerned with structural timber since this is rarely subjected to secondary processing.

Similar processing is carried out on board materials (which are described in Chapter 17) in the production of panels of differing size, shape and edge detail. Where relevant, specific points are made below.

14.1   General principles

The ultimate finish to a piece of wood or board material will generally be produced by either a planer or a spindle moulder, and efficiency in production represents a balance between the quality of finish in terms of flatness and smoothness, and maximum output. If feed speeds are too high, flatness will suffer and it may then be necessary to carry out sanding; as this represents another operation, the desire is to do as little sanding as possible.

Quality of finish is determined not only by the type of equipment used and the sharpness of the cutters (Klamecki, 1979), but also by the type of

wood being machined. Very low-density and very high-density timbers give problems, but for different reasons: low-density wood is easily compressed and frequently tears out rather than being cut cleanly, while high-density woods resist cutter insertion, are prone to burn under the cutter and cause rapid decline in the sharpness of blades.

In the majority of cases, and especially when processing abrasive woods containing minerals (Chapter 3) and board materials containing resin, saw blades, profile cutters and some planer blades are usually tungsten-carbide tipped.

Woods containing large amounts of resin (Chapter 3) will also pose problems, as the heat of the cutter causes the resin to flow; this results not only in poorer surface quality, but also in the consumption of more power to overcome the resistance generated by the resin.

The presence of tension wood (Chapter 5) has a very marked effect on quality of finish, especially in sawing and spindle moulding. This is due to the reduced density of tension wood resulting from its increased percentage of cellulose relative to lignin. The fibres tend to tear out rather than be cut cleanly and the machined items generally have to be rejected.

Moisture content has a most pronounced effect on the quality of the machined surface, especially so in the case of planing (Stewart, 1979). Thus **raised grain**, a term used to describe marked differences in height between early- and latewood on a planed surface, is the result of machining wood at either too high or too low a moisture content and then using it at a different moisture content. Raised grain, however, may also be due to very heavy 'jointing' of the planar blades.

Generally, it is the machining of wood with too high a moisture content which gives rise to raised grain. There is a tendency for the thin-walled cells to be deformed rather than cut because of their increased elasticity in the wet condition. After the cutters have passed over, these deformed areas slowly assume their previous shape, resulting in this irregular appearance to the surface which is very noticeable when the timber is dried and painted.

In order to ensure that all planer blades are cutting on a concentric circle, it is common practice to carry out the process of 'jointing' whereby the tips of the blades are lightly ground by an abrasive stick moved across them while they are rotating in the machine; this produces a small 'heel' or back-bevel on the cutter with a zero clearance angle. Should this heel exceed 0.75 mm (or less for dense timbers) the surface of the wood will be compressed and raised grain will occur as the wood recovers.

14.2 Machining equipment

14.2.1 Circular saws

The principal features of these and how the saw design and operation affect the quality of the sawn face were discussed in Chapter 12, to which reference should be made. Suffice it here to say that the sizes of the saws used in machining are usually very much smaller than they are for log conversion.

For rip-sawing timber, blades with a positive hook angle are employed, while for cross-cutting the saw blade has not only a negative hook angle, but also many more teeth. For both operations, the set on the teeth must be higher for hardwoods than for softwoods.

Sawing of board materials is usually carried out on bench saws with a saw blade diameter of 350–450 mm; portable circular saws are used where small quantities are involved. In both operations, tungsten-carbide tipped (TCT) blades should be used and for long production runs polycrystalline diamond tipped (PDT) are recommended on the basis of their improved wear resistance.

In cutting board materials, the circular saw blade should be set as low as possible to prevent chipping and scoring as the board passes the rear of the saw blade. The height of the saw blade should be positioned in order to achieve the correct hook angle relative to the board surface: the projection of the saw above the board surface has a direct influence on the cut. Breaking out of the top surface will occur if it is insufficient, and on the bottom surface if it is too great.

Since the structure of the various panel products varies considerably, the geometry of the saw required to give maximum efficiency in cutting any one board type will vary; thus, hook angle, and the various bevel and clearance angles, together with saw speed and feed speed will differ among the different materials.

14.2.2 Bandsaws (narrow)

The principal features of bandsaws were again discussed in Chapter 12: for the primary breakdown of a log, wide bandsaws (up to 300 mm) are employed, but for secondary processing blades of only 6–50 mm are used at speeds from 15 to 45 m/s: the higher speeds are employed on the larger machines. Rake angle is usually about 10° (and positive) and the clearance angle 20°: spring setting (0.25 mm) is generally used.

These 'narrow' bandsaws are used for ripping, cross-cutting and shaping.

14.2.3 Planers, routers and spindle moulders

Perhaps the single most important parameter controlling surface quality in both wood and board materials is the cutting angle of the blade. Angles of 30° or slightly above will give good surfaces with low power consumption for a wide range of timber densities, provided the material is straight-grained. However, where the grain is sloping, interlocked, or where there are knots, a 30° angle will lead to tearing (pick-up) and splitting of the surface. This can be reduced by lowering the feed speed but, if this is undesirable, blades with angles of 20°, or even 10° in very severe cases must be used, though this will result in increased power consumption and reduced blade life. Router cutters used on board materials generally have cutting angles in the range of 15-25°

Another very important variable determining quality of surface is the feed speed; since planers, routers and spindle moulders remove crescents of wood, so increasing feed speed results in an increase in the pitch of the cutter marks while the marks are of greater depth. At high speeds it is possible not only to feel these ridges lying across the machined surface, but to see them clearly on surfaces which have been sealed or polished. Thus when the wood is to be polished, the feed speed must be reduced such that the maximum pitch of the cutter marks is no greater than 1.25 mm; when wood is to be used for joinery having a varnish or painted finish, the pitch should be between 1.5 and 2.00 mm.

Generally for board materials, feed speeds are lower than they are for solid timber: too high a feed speed results in rough, fibrous cut edges. Cutters should have the largest number of cutting edges possible where high-quality edge finishing is required.

14.2.4 Sanders

Sanding after planing, moulding and routing produces a smoother surface and this is usually carried out using belt sanders incorporating either aluminium oxides or silicone-carbide-based abrasives. The latter are recommended for sanding MDF (medium density fibreboard), with belt speeds in excess of 1500 metres per minute. To remove cutter marks, 80/100 grit abrasives are used: for finishing, 120/150 grit is necessary. Moulded edges can be produced on MDF using a profiled sander.

14.3 Dust

14.3.1 Dust levels and requirements

The debit side to the machining of wood and board materials embraces not only the cost of the operation, but also the production of very large quantities of dust. This is usually in one of two different size ranges: thus fairly large particles of dust are produced in sawing, planing, spindle moulding and turning, while very fine particles of dust arise from all sanding operations. Surveys have indicated that the concentration of dust in milligrams per m^3 averaged over an 8 hour period produced from wood working machines with some form of exhaust ventilation was about 3.4 for saws, 2.3 for planers and moulders, and 8.5 for sanders (HSE, 1987).

As will be discussed in the next section on health, the inhalation of dust may induce respiratory irritation. More important, the long-term inhalation of dust causes a slowing of the dust clearance capability of the nose and an alteration to the structure of the mucous membrane lining the nasal cavity; this is accompanied by a risk of cancer in the nasal cavity and sinuses.

There are now statutory requirements for the control of wood dust in factories. As far as hardwoods are concerned, the UK Health and Safety Executive sets a control limit of 5 milligrams per m^3 for hardwoods dust for personal exposure to airborne dust average over 8 hours with effect from the 1 April 1988 (HSE, 1987). More recently, this requirement has been included in Schedule 1 of the Control of Substances Hazardous to Health Regulations (COSHH) and is now deemed to be the Maximum Exposure Limit.

The 1992 Occupational Exposure Standard (HSE, 1992) for softwood dust arising from the machining of softwood (or board materials made from softwood) is also 5 mg/m^3, expressed as an 8 hour time-weighted average.

Regulation 7 of COSHH requires that exposure to dust is 'either prevented or, if this is not realistically practicable, adequately controlled'. These limits can be met with the use of properly designed and maintained dust control equipment for most machining operations. However, for those few operations such as hand sanding where there is no method currently available to control dust, operators must wear respirators which give adequate protection, and these include either disposable ones to BS EN 149, or half-mask ones to BS 2091.

14.3.2   Health problems associated with wood processing

There are three distinct health problems among wood machinists where the incidence of disease is appreciably higher than the national average: skin irritation, respiratory irritation and nasal cancer. In the cases of irritation this is usually associated with a particular species and therefore related to a specific chemical constituent within that timber. The response to an irritant is very much dependent upon the sensitivity of the individual exposed to the wood or wood dust. Some woodworkers handling a particular timber may be unaffected, while colleagues suffer anything from mild discomfort to severe attacks that require medical treatment.

Irritants, therefore, are associated with the extractives laid down in the heartwood when it develops from sapwood (Chapter 4, section 4.5). These are largely responsible for the timber's colour, scent and level of natural durability; each species has its own characteristic group of extractives and generally the specific nature of the irritant timber can be readily explained by the chemical composition of the extractives present.

Skin irritation

This can occur either in the form of splinter wounds, or from the development of dermatitis. Fine splinters from the inner bark of ramin are well recognised as a source of irritation. However, it is **dermatitis** which is the principal skin disorder among wood workers, and it may take one of two forms – irritant dermatitis or allergic dermatitis (Orsler, 1979).

In the former type, the problem is relatively uncomplicated because the symptoms will persist so long as the sufferer remains in contact with the source of the irritation. On removing the worker from the source of irritation, the skin will slowly return to normal; re-exposure to the irritant will produce an attack of equal severity.

Allergic dermatitis is much more troublesome and is usually initiated by exposure to the fine dust of certain timbers. Usually a larger area of the body is exposed and the body sets up an allergic reaction to that dust; the worker has therefore been sensitised to that particular timber, or to timbers containing chemicals closely related to that in the irritant timber.

Mild cases of dermatitis will show as slight reddening of the exposed skin with accompanying itching; more severe cases will experience a hot burning sensation and the appearance of a rash which in many cases leads to blistering and even weeping sores.

Timbers with a well-established history of inducing dermatis include iroko, guarea, African mahogany (*Khaya anthotheca*) and mansonia. A fuller list of well-established irritant timbers, as well as those occasionally and possibly irritant, is given by Orsler (1979).

Respiratory irritation

While some timbers can cause both skin and respiratory irritation, the majority of irritant timbers cause either skin or respiratory irritation.

The inhalation of an irritant dust frequently first affects the lining of the nose and throat, the latter becoming dry and sore, while the nose and eyes run and become inflamed. Sneezing is common and nose bleeds can occur. In more extreme cases breathing difficulties can be experienced leading to asthma-like symptoms. As with dermatitis, respiratory irritation exists in both the irritant and allergenic forms.

Timbers with a well-established history of producing respiratory problems are guarea; mansonia; western red cedar, and makore. A fuller list of well established irritant timbers, as well as those occasionally and possibly irritant is given by Orsler (1979).

Nasal cancer

It was only in 1964 that a link was established between this disease and the furniture industry. Fortunately the disease is still rare, though the incidence in skilled furniture makers is at least 200 times higher in other men (Acheson, 1982).

It has been established that the cancer occurs only among those workers who have been exposed to fine wood dust: sawmill workers are unaffected. The time taken for the disease to develop and reveal itself is about 40 years, and the minimum period of exposure necessary to produce tumour development appears to be about five years. It is unclear at present as to whether the cancer-inducing mechanism derives from a single wood species or from wood in general.

The normal nose has a mechanism by which particles of dust are carried away into the back of the throat and swallowed. Studies have shown that in woodworkers in the furniture industry this mechanism is grossly impaired (Acheson, 1982). One theory is that the inhalation of wood dust over long periods slows down the clearance mechanism sufficiently to permit a carcinogen in the dust to penetrate the mucous glands and initiate carcinogenesis in the mucosal lining (Macbeth, 1965).

A study of case histories indicates that a wide range of timbers, mainly hardwoods, but embracing both temperate and tropical species, is involved (Acheson, 1982).

It is hoped that the incidence of the disease will decrease with improvements in the control of airborne fine wood dust. Nasal cancer in furniture manufacturers has been recognised by Government since 1969 as a prescribed disease under the Industrial Disease Act (Regulation 2, of Statutory Instrument 619/1969). This means that sufferers and their families may apply for increased benefits under Workers Compensation arrangements; it also means that physicians are required to inform the coroner of deaths due to this cause.

References

Acheson E.D. (1982) Nasal cancer in furniture workers: the problem. In: *Proc Seminar on the Carcinogenicity and Mutagenicity of Wood Dust*, 1 July 1981, Southampton General Hospital.

BS 2091: 1969   *Specification for respirators for protection against harmful dusts, gases and scheduled agricultural chemicals.* BSI, London.

BS EN 149: 1992   *Specification for filtering half masks to protect against particles.* BSI, London.

COSHH (1988)   *Control of Substances Hazardous to Health Regulations.*

HSE (1987)   *Control of Hardwood Dust: A Guide for Employers.* Health and Safety Executive.

HSE (1992)   *Occupational Exposure Limits.* EH40/92, Health and Safety Executive.

Klamecki B.E. (1979) A review of wood cutting tool wear literature. *Holz als Roh- und Werkstoff* **37**: 265–276.

Macbeth R.G. (1965) Malignant disease of the paranasal sinuses. *J Laryng* **79**: 592.

Orsler R.J. (1979) Health problems associated with wood processing. *Building Research Establishment Information Paper*, IP13/79.

Stewart H.A. (1979) Some surfacing defects and problems related to wood moisture content. In: *Proc Symp on Wood Moisture Content – Temperature and Humidity Relationships*, pp 70–75, Virginia Polytechnic Institute and State University, Blacksburg, Virginia. Published by USDA Forest Service, North Central Forest Experiment Station.

Timber Grading and Grade Stresses

15.1 Introduction

The quality of sawn timber varies widely depending on the species, where and how it was grown and the age at which the tree was felled. Thus, quality of wood is determined primarily by the density of wood, and strongly influenced by the size and distribution of knots and other defects which have been described in detail in Chapters 2, 5 and 11.

Grading separates the available material into groups so that marketing is rationalised and selection for a particular application is simplified. Since grading rules vary between softwoods and hardwoods, these will be treated separately.

15.2 Grading of softwoods

15.2.1 Appearance grading based on a 'visual defects' system

Until the mid-1970s, all softwood grading was based on a **defects** system, whereby the grading rules for appearance or quality set out a maximum allowable size or degree of each type of defect for each of several grades usually in terms of the dimensions of the piece of timber. Thus large defects in material of large dimensions resulted in the board being ascribed to the same grade as material of small dimensions containing small defects.

The principal defects which are usually taken into account in the application of this type of grading rules for softwood include those defects which arise from structural variability, such as knots and ring width, those which arise from conversion and drying, such as wane, splits,

twist, cup and bow, and lastly those due to fungal development taking the form of sapstain or actual rot. The derivation of grades based on these defects varies widely between countries.

Since the mid-1970s, stress grading has almost replaced appearance or quality grading in many countries, certainly as far as structural softwood is concerned. However, it is still possible to find in the market place softwood graded according to the appearance system of a specific country or group of countries. Much of this material finds its way into the so-called 'quality' market or 'joinery' grades (to distinguish it from structural softwood or 'carcassing' grades) where it is subsequently regraded according to BS 1186: the use of softwood for joinery is discussed in Chapter 16. Some examples of grading of softwood by the appearance or quality system are given below; while some of these are still applied to a small proportion of their export, others have become of only historical interest.

Softwood from Norway, Finland, Sweden, Poland and Eastern Canada

The Scandinavian countries, especially Finland, have had a unified system of grading for several decades; within the last two decades, Poland has also been exporting timber graded using the Scandinavian system.

The various lumber associations in USA and Canada have set up their own softwood grading rules; these are generally appreciably different from those used in Scandinavia. However, timber from Eastern Canada was, until the late 1970s, frequently graded using the Scandinavian system when the lumber was being exported to Europe.

The Scandinavian grading rules define six grades, namely I, II, III, IV, V and VI, in decreasing order of quality. Because of the limited avail-ability of the top grades, it is common practice to market a combination of grades I–IV under the term 'unsorted' – the proportion of each grade varies with country and even with shipper, but is usually of the order of 5 per cent of grade I, 10 per cent of II, 65 per cent of III and 20 per cent of IV – in some instances 'unsorted' contains only grades I–III.

The rules tend to be somewhat general in character, with the result that their interpretation varies among the different mills. Each mill marks every piece with its own shipping mark which identifies the quality and origin of the timber. Over the years it is possible to assess whether one particular mill is consistently below or above the average for that grade. However, heavy cut-ting in the past has necessitated the opening-up of new areas of forest, with the result that the nature of the raw material coming to the mills has changed. Shippers who have been in the habit of obtaining supplies from one locality year after year are obliged to go to different localities each year, and the graders (brackers), with only empirical experience to guide them, are depen-dent on their own judgement for maintaining continuity of quality. Moreover, when it has been customary for the total production to yield certain percentages of each quality for many years, there is a tendency to secure the same percentages, irrespective of any fall in qual-ity in the mill intake. Nowadays the produce of successive years, shipped under the same marks, may vary appreciably and, in consequence, the purchaser can no longer depend on particular shipping marks to secure the quality of timber required.

Russia

The former Soviet Union's grading rules define only five grades, grades I–III usually being exported as 'unsorted' and being approximately equivalent to 'unsorted' Scandinavian comprising their grades I–IV. The Russian grade IV is equal to, or better than Scandinavian V, while the Russian grade V is slightly better than Scandinavian VI.

West Coast, North America

While this area produces much stress-graded lumber, it still markets a considerable amount of knot-free timber which is graded according to the defects system under the grading rules of the Pacific Coast 'R' list; this provides three grades specifically for clear timber. Grade 3 clear is approximately equal to Russian I and II when these are marketed separately. Four other grades containing knots are defined and their relationship with the Scandinavian and Russian grades is presented in Table 15.1.

Brazil

Four grades of softwood are defined, but only grades 1 and 2 are exported to the UK: these are equivalent to No. 3 clear of the Pacific Coast 'R' list.

Table 15.1 *Approximate comparison of softwood grading rules based on the appearance or quality system*
(© Crown Copyright)

Norway, Sweden, Finland, Poland and Eastern Canada		I, II, III, IV unsorted	V	VI
Russia		I, II, III	IV	V
Brazil	No. 1 and No. 2			
British Columbia and Pacific Coast of North American (R list)	No. 1 clear No. 2 clear No. 3 clear	Select merchantable No. 1 merchantable	No. 2 merchantable	No. 3 common

United Kingdom

Home-grown softwood was originally graded on a defects system to one of five grades set out in BS 3819. The advent of stress-grading led to the withdrawal of this specification and timber in the UK is now graded according to BS 4978: 1988.

15.2.2   Stress grading

This method of grading softwood timber represents a more positive approach and provides the structural designer with material of assured strength; for although there is a measure of correlation between stress-graded and quality-graded timber, there is no guarantee that a piece of timber with good appearance will automatically have high strength properties.

Stress grading was originally based on the strength of small clears, this figure being reduced to account for variability and defects. It will be recalled that in Chapter 11 the effects of various factors on the strength of wood were discussed, and towards the end of the chapter a method of quantifying the variation in results of small clear timber specimens was presented. Thus it was demonstrated that the derivation of the **basic stress** for a timber took into account the variability in strength of defect-free wood that occurs within a tree and between trees of any one species (see Chapter 11, section 11.7). However, this derived basic stress does not take account of the size and distribution of knots, the slope of grain, the rate of growth, the presence of wane and splits, and the various seasoning defects such as cup, twist and bow (see Chapter 13, section 13.6).

By measuring the diameter of the knots and the size of the other defects, it is possible to describe a number of grades, each of which will possess a strength equivalent to a particular proportion of the strength of defect-free timber. The grades are therefore defined in terms of a **strength ratio**; four grades were defined, namely 40, 50, 65 and 75, representing respectively timber with strength equivalent to 40, 50, 65 and 75 per cent that of clear straight-grained timber. The **grade stresses**, therefore, are the basic stresses divided by these strength ratios.

Grade stresses for both dry (less than 18 per cent moisture content) and green timber (greater than 18 per cent) were introduced into the 1952 and 1967 versions (metricated in 1971) of the BS Code of Practice CP 112 *The structural use of timber*. Unfortunately these attempts to gain acceptance for stress-graded timber failed, primarily on account of high reject rates, and the use of very conservative stress values.

In passing, it should be noted that CP 112 contained information on the derivation of the **permissible stresses** used in structural design in timber from the product of the grade stresses and a series of **modification factors** to cover aspects such as duration of load (see Chapter 11), load-sharing between members, slenderness ratio and bearing area.

Visual grading

In an attempt to overcome the objections to the grade stresses in CP 112 (1967/71), there was published in 1973 BS 4978 *Softwood grades for structural use*. The standard was based on results obtained from structural size timber, as described in Chapter 11, sections 11.3.2 and 11.4.2. It contained a new approach to the measurement of knots that made it possible to take into account the combinations of knots which often occur in European softwoods, and is therefore less restrictive than the method used in CP 112. Knot sizes were specified in terms of **knot-area ratio**, which is the ratio of the sum of the areas of the projections of all knots at a cross-section, to the total area of the section.

Two visual stress grades were defined, namely general structural (GS) and special structural (SS): the limiting knot sizes for the lower grade (GS) have been defined so that only a very few pieces of timber would be rejected from reasonable quality parcels of fifth grade redwood and whitewood. The timber in each of these two grades must also satisfy requirements relating to slope of grain, rate of growth, wane and distortion. Stress values for those two grades were provided in *Amendment Slip No. 1* to CP 112, Part 2, published in September 1973. Although a direct comparison between the new and old grades is not strictly acceptable owing to the differences in mode of derivation of the stress values, it is generally recognised that the GS grade corresponds to between the 30 and 35 strength ratios, and the SS to between the 50 and 60 strength ratios.

Since 1973 the SS and GS grades have progressively replaced the original four grades (40, 50, 65 and 75) set out in CP 112. Full information on the visual grading of wood is provided in BS 4978: this standard also sets out the requirements for laminated timber ('glulam'). Three grades are specified, LA, LB and LC, and these grades are set primarily on the limiting knot-area ratio.

Mechanical grading

Visual grading is slow since all four sides of each piece of timber should be examined. More importantly, the system fails to take account of the experimental evidence that pieces of timber with identical defects can, and do, have appreciable differences in strength. The attractions of mechanical stress grading therefore lie in the faster and better assessment of grade; the practical implications of this are higher yields to present grade stress values with increased values of modulus of elasticity.

Mechanical stress grading originated in the late-1950s when it was discovered that for timber in structural sizes there existed significant correlations between bending strength and modulus of elasticity when measurements were made over short spans. This finding opened up the way for the development of stress grading machines in which the wood is subjected to a load as it is passed through and the deflection measured automatically. Provided the relationship between bending strength and modulus of elasticity has already been established, it is possible to ascribe grade stresses to the timber from a measurement of deflection under a standard applied load.

The prototype grading machine was developed at the former Forest Products Research Laboratory, Princes Risborough in 1962 to demonstrate the application of machine grading. Most of the machines that have subsequently been developed are either Australian or American. In 1968, that Laboratory obtained an Australian Computermatic machine and started a series of tests to determine its control settings and its effectiveness. These investigations, which indicated that machine stress grading did provide a better forecast of the bending strength of timber than was possible by visual grading, resulted in the limited introduction of machines by industry in this country. In 1969, to promote the adoption of machine grading, the Forest Products Research Laboratory recommended two grades, designated M75 and M50, which had the same grade stresses in bending as the corresponding visual grades of CP 112 (1967); these values were subsequently incorporated in *Amendment Slip No. 1* to CP 112 (1973).

The machine grade settings for one particular species or species group are determined from the experimentally derived relationship between bending strength (modulus of rupture) and modulus of elasticity. An example of this relationship, with the mean regression line and the 5th percentile exclusion line superimposed, is given in Figure 15.1.

The equation of the regression line is only one of a number of inputs to the mathematical model used to determine the machine settings for a particular timber. These settings also depend on:

(1) bandwidth (separation of the grade boundaries on the *x*-axis) which in turn is related to the grade combination being graded;

(2) the overall mean and standard deviation of the modulus of elasticity of the species;

(3) the interaction of bandwidth and the MOE parameters; and

(4) the cross-sectional size, and whether the timber piece is sawn or planed all round (PAR).

The mathematical model is only designed to select to two grades (and a reject class) in any one pass through the machine, for example:

(a) M75, M50, or REJECT,
(b) MSS, MGS, or REJECT, or
(c) SC5, SC3, or REJECT (see below).

Having established the basic relationship for each species, the grading machine can then be set up to grade the timber automatically. Depending on the type of machine, as each

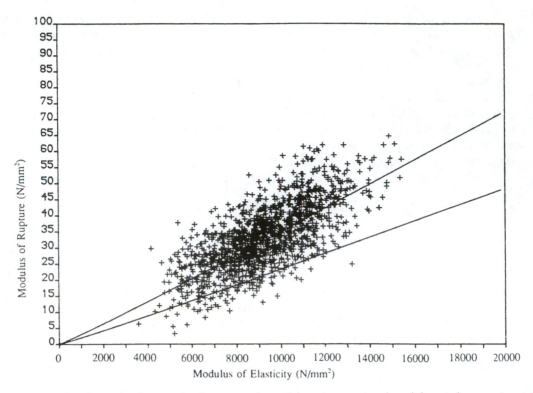

Figure 15.1 *The relationship between bending strength (modulus of rupture) and modulus of elasticity for 1348 test results, with mean regression line and 5th percentile exclusion line superimposed. The equation of the mean regression line is* $MOR = 0.002065 \ (MOE)^{1.0573}$. *The correlation coefficient* $= 0.702$ *(Building Research Establishment, © Crown Copyright)*

length of timber passes through it is either placed under a constant load and deflection is measured, or subjected to a constant deflection and load is measured. A small computer quickly assesses the appropriate stiffness indicating parameter either every 150 mm (or 100 mm) along the length of the timber. The lowest reading of load, or the highest reading of deflection, is compared with the pre-set values stored in the computer and this determines the overall grade of the piece of timber. Grade stamps are then printed on one face of the timber towards one end of the piece.

BS 4978: 1988 now defines four machine grades, M50, M75, MGS and MSS, in addition to the two visual grades GS and SS. The specification calls for the marking of all graded timber, the contents of the mark being different for visually and mechanically graded material

(Figure 15.2). Thus, for visual grades each piece of timber must carry the following marks:

(1) the grade of the piece,
(2) a mark identifying the grader or the company responsible for the grading.

For machine grades the following marks must include:

(1) the grade of the piece,
(2) the licence number of the stress grading machine,
(3) the BSI Kitemark and BS 4978,
(4) a mark indicating the species or species group.

Both visual and machine stress grading is subjected to independent third-party quality assurance by a grading agency or certification body

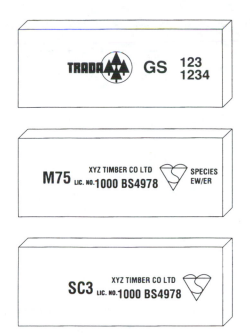

Figure 15.2 *(Top) An example of a visually graded batten of softwood showing the grade of the piece (GS), the identification mark of the grader (123), the identification mark of the company responsible for the grading (1234), together with the quality assurance mark (TRADA + symbol). Not all this information must be supplied (see section 15.2.2). (Middle) An example of a machine-graded batten of softwood graded to BS 4978 showing the grade of the piece (M75), the licence number of the stress grading machine (1000), the company name (XYZ Timber Co Ltd), the BSI quality assurance Kitemark and BS 4978, and a mark indicating the species or species group (EW/ER = European whitewood/European redwood). (Bottom) An example of a machine-graded batten of softwood graded directly to the strength class (SC3). The remainder of the mark is identical to that relating to the machine grading of wood to BS 4978 described above.*

which has been approved by the UK Timber Grading Committee: the list of Approved Certification Bodies is published by the Timber Trade Federation (1990).

These Approved Certification Bodies are also charged with the responsibility of not only overseeing the training and certification of graders and the operation of grading machines, but also of monitoring the quality of the grading carried out by the companies under their control. The

logo or mark of the certification body must form part of the grading mark applied to the timber.

Timber imported into the UK from Europe for structural use is usually already graded in accordance either with the BS 4978 grading rules, or with ECE standards. It is now rare for softwood timber to be imported ungraded for subsequent grading in the UK. Stress-graded timber imported from Canada is graded according to their National Lumber Grading Authority (NLGA) rules (1987), while that from the United States is graded according to either the National Grading Rules for Softwood Dimension Lumber (NGRDL) 1975, or the North American Export Standard for Machine Stress Rated Lumber (NAMSR) 1986.

15.2.3 Strength classes

In applying the grading rules within BS 4978, it was possible to have hundreds of grade/species combinations for structural timber. This situation was rationalised in the 1984 revision of BS 5268: Part 2 *Code of Practice for the structural uses of timber: Permissible stress design, materials and workmanship*, with the creation of a strength class system. A set of nine strength classes covers all grades and most species from the lowest grade of softwood in SC1 to the highest grade of hardwood in SC9.

The major imported structural softwoods used in the UK are to be found in SC3, SC4 and SC5. UK grown softwoods, however, are generally allocated to SC2, SC3 and SC4 when visually graded to the BS 4978 grades, and generally to SC3, SC4 and sometimes to SC5 when mechanically graded.

In order to provide the reader with some idea of the equivalence of different grades of different species, Table 15.2 sets out the grades of certain common structural softwoods that satisfy the common strength classes.

The strength class system, therefore, allows suppliers to meet a structural timber specification by supplying any combination of species and grade listed in BS 5268: Part 2 as satisfying the specified class.

Table 15.2   *Some softwood species/grade combinations which satisfy the requirements for the common strength classes when graded according to SC 4978*

Timber	Strength class			
	SC2	SC3	SC4	SC5
Imported				
Redwood		GS/MGS/M50	SS/MSS	M75
Whitewood		GS/MGS/M50	SS/MSS	M75
Douglas fir/larch		GS/MGS	SS/MSS	
British grown				
Scots pine		GS/MGS/M50	SS/MSS	M75
European spruce	M50/SS/MSS	M75		
Sitka spruce	M50/SS/MSS	M75		
Douglas fir	GS/MGS	M50/55/MSS		M75
Larch		GS/MGS	SS/MSS	

BS 4978: 1988 now makes provision for machine grading directly to the strength classes (Figure 15.2); certainly, most British grown softwood which is machine graded for structural use is now graded to these strength classes rather than to the BS 4978 grades MGS, MSS.

Current discussions in the context of European Standardisation may lead to certain modifications within the strength class system. Thus, it is proposed that the number of strength classes are increased and that there should be a new allocation of timbers to grades; this may lead to a downgrading of some UK timbers.

15.3   Grade stresses for softwoods

Grade stresses for use in structural design are included in BS 5268: Part 2: 1991 in two different formats, either of which may be used, depending solely on convenience.

First, stresses are given for the five strength classes covering softwoods, as previously described, and relate to the use of timber under dry exposure conditions (up to 85 per cent relative humidity); structural design using British grown softwood should use these values. Stress values for strength classes SC2, SC3 and SC4 are reproduced in Table 15.3 to give the reader some feel for the actual values.

Alternatively, BS 5268: Part 2: 1991 provides actual grade stresses for a range of softwood timbers graded in accordance with one of five different sets of grading rules:

(a) BS 4978: 1988, as described above, for both visual and machine graded timber – see Table 10 of BS 5268: Part 2: 1991. An example of the grade stresses for a few home-grown and imported softwoods is given in Table 15.4.

(b) NLGA (National Lumber Grading Authority – grading rules for dimension lumber, Canada, 1987): see Table 11 of BS 5268: Part 2: 1991.

(c) NGRDL (National Grading Rules for softwood Dimension Lumber, USA, 1975): see Table 12 of BS 5268: Part 2: 1991.

(d) North American export standards for machine stress rated lumber (MSR), 1986: see Table 13 of BS 5268: Part 2: 1991. And

(e) ECE sawn timber: Recommended standards for the grading of coniferous sawn timber, 1982. The S6, S8, MS6 and MS8 grades may be substituted for GS, SS, MGS and MSS respectively of BS 4978: 1988.

BS 5268: Part 2: 1991 also includes grade stresses for three grades of glue laminated members (glulam) manufactured from softwood in accordance with the requirements of BS 4169.

Table 15.3 Grade stresses and moduli of elasticity for certain strength classes for the dry exposure condition (extract from Table 9 of BS 5268: Pt 2: 1991)

Strength class	Bending parallel to grain (N/mm²)	Tension parallel to grain (N/mm²)	Compression Parallel to grain (N/mm²)	Compression Perpendicular to grain* (N/mm²)	(N/mm²)	Shear parallel to grain (N/mm²)	Modulus of elasticity Mean (N/mm²)	Modulus of elasticity Minimum (N/mm²)	Approximate density† (kg/m³)
SC2	4.1	2.5‡	5.3	2.1	1.6	0.66	8000	5000	540
SC3	5.3	3.2‡	6.8	2.2	1.7	0.67	8800	5800	540
SC4	7.5	4.5‡	7.9	2.4	1.9	0.71	9900	6600	590

*When the specification specifically prohibits wane at bearing areas, the higher values of compression perpendicular to the grain stress may be used, otherwise the lower values apply.

†Since many species may contribute to any of the strength classes, the values of density given in this table may be considered only crude approximations. When a more accurate value is required, it may be necessary to identify individual species and utilise the values given in appendix A of BS 5268: Pt 2: 1991.

‡Note the Light Framing, Stud, Structural Light Framing No. 3 and Joist and Plank No. 3 grades should not be used in tension.

Table 15.4   Grade stresses and moduli of elasticity for certain softwoods, graded to BS 4978 rules, for the dry exposure condition (extract from Table 10, BS 5268: Pt 2: 1991)

Standard name	Grade	Bending parallel to grain (N/mm²)	Tension parallel to grain* (N/mm²)	Compression		Shear parallel to grain (N/mm²)	Modulus of elasticity	
				Parallel to grain (N/mm²)	Perpendicular to grain† (N/mm²)		Mean (N/mm²)	Minimum (N/mm²)
Redwood/whitewood (imported) and Scots pine (British grown)	SS/MSS	7.5	4.5	7.9	2.1	0.82	10 500	7 000
	GS/MGS	5.3	3.2	6.8	1.8	0.82	9 000	6 000
	M75	10.0	6.0	8.7	2.4	1.32	11 000	7 000
	M50	6.6	4.0	7.3	2.1	0.82	9 000	6 000
Douglas fir-larch (Canada)	SS	7.5	4.5	7.9	2.4	0.85	11 000	7 500
	GS	5.3	3.2	6.8	2.2	0.85	10 000	6 500
Spruce-pine-fir (Canada)	SS/MSS	7.5	4.5	7.9	1.8	0.68	10 000	6 500
	GS/MGS	5.3	3.2	6.8	1.6	0.68	8 500	5 500
	M75	9.7	5.8	8.5	2.1	1.10	10 500	7 000
	M50	6.2	3.7	7.1	1.8	0.68	9 000	5 500
Sitka spruce and European spruce (British grown)	SS/MSS	5.7	3.4	6.1	1.6	0.64	8 000	5 000
	GS/MGS	4.1	2.5	5.2	1.4	0.64	6 500	4 500
	M75	6.6	4.0	6.4	1.8	1.02	9 000	6 000
	M50	4.5	2.7	5.5	1.6	0.64	7 500	5 000
Douglas fir (British grown)	SS/MSS	6.2	3.7	6.6	2.4	0.88	11 000	7 000
	GS/MGS	4.4	2.6	5.6	2.1	0.88	9 500	6 000
	M75	10.0	6.0	8.7	2.9	1.41	11 000	7 500
	M50	6.6	4.0	7.3	2.4	0.88	9 500	6 000
Larch (British grown)	SS	7.5	4.5	7.9	2.1	0.82	10 500	7 000
	GS	5.3	3.2	6.8	1.8	0.82	9 000	6 000

* Stresses applicable to timber 300 mm deep (or wide); for other section sizes see 14.6 and 16.2 of BS 5268: Pt 2: 1991.
† When the specification specifically prohibits wane at bearing areas, the SS grade compression perpendicular to the grain stress may be multiplied by 1.33 and used for all grades.

The grade stresses for both solid softwood and glulam given in BS 5268: Part 2, together with the basic loads for fasteners, apply to specific conditions and should be multiplied by the appropriate modification factors given in this standard when actual service and loading conditions are different. Some 84 modification factors are listed, covering not only softwood and glulam, but also certain grades of plywood, hardboard and chipboard; these factors, among other things, cover loading periods less than 'permanent', and wet exposure conditions.

It should be appreciated that while the design requirements of BS 5268: Part 2: 1991 are usually satisfied by calculation, using the laws of structural mechanics and the stress values and modification factors listed in that standard, they may also be satisfied by testing; provided this is carried out in accordance with section eight of the standard. In certain circumstances, testing may lead to a more economical design than by calculation, owing to the conservative nature of both stresses and modification factors.

The stress values set out in BS 5268: Part 2: 1991 are for use in permissible stress design; these grade stress values represent the stress which can safely be *permanently* sustained by softwood, glulam or certain board products of a particular strength class, or species and grade, or grade of board.

Within the next decade, BS 5268: Part 2 will be replaced by a new European standard, known as Eurocode 5. This has recently been published as a voluntary standard and will be applicable in the UK as an alternative means of designing structurally with timber. After a familiarisation period, lasting about three years, Eurocode 5 will be revised and will reappear as the definitive document for structural design throughout Europe.

Readers coming new to Eurocode 5 will find it very different from BS 5268: Part 2. Unlike the latter, the Eurocode is based on limit state design with stress values expressed in terms of **characteristic values**: these are defined as that value which has a prescribed probability of not being attained in a hypothetical unlimited test series, that is, a fractile in the distribution of the property. The characteristic value is called a lower or upper characteristic value if the prescribed value is less or greater than 0.05 respectively. In the application to solid timber and board materials, strength values are quoted as 5 percentile values, and modulus of elasticity is expressed as both 5 and 50 percentile values; in the case of board materials, swelling after cold water soaking is quoted as a 95 percentile value. These characteristic values must be reduced for the effects of duration of load and increasing severity of hazard class.

15.4   Grading of hardwoods

15.4.1   *Appearance grading based on 'visual defects' or 'cutting' systems*

While some hardwood timber is graded according to a stress-grading system, large quantities of hardwood continue to be graded according to either the **quality** or the **cutting** systems.

The quality system, analogous to that used for sawn softwood and discussed in section 15.2.1 of this chapter, appears to be adopted for grading timber of large dimensions, particularly that for a specific end use. This hardwood system takes into account not only the distribution and extent of defects, but also the size of the timber member: thus the higher grades require large pieces while only the lower grades will accept smaller planks.

The second and more frequently applied system is the cutting system which is used on planks produced from converting timber 'through and through'. The amount of timber free from defects, or with acceptable defects, is assessed in terms of rectangles measured in units of area of 12 inches^2. The grades are defined in terms of a minimum area of cuttings (that is, rectangles) acceptable, and expresses this as a fraction of the total plank area. These fractions are expressed with a denominator of 12, so that lengths in feet may be multiplied by widths in inches to give units of cutting. Additionally, the number of cuts allowed within any one grade, and the minimum size of each cutting that can be produced, depend on the size of the plank.

The method adopted, and the derivation of the grades, again varies widely from one country to the next; the most important grading systems appertaining to the UK are described below.

The National Hardwood Lumber Association of North America (NHLA)

Although there are a number of associations throughout North America, the NHLA is perhaps the most important, and its grading rules, often in modified form, are frequently used in other parts of the world. Eight grades are defined, the top five of which are exported to Europe: the grades and their associated cuttings (fractions clear) are given in Table 15.5.

Malaysia

These grading rules have evolved from the old Empire grading rules and provide for four basic grades, the lowest of which is not exported. The grades and their associated cuttings are

set out in Table 15.5. Provision is also made, under separate rules, based on a defects system, for the marketing of borer-infested timber. Much of the hardwood timber export from Malaysia is now stress graded to the Malaysian grading rules of 1968.

West Africa

There is great diversity in grading rules and their application among the different countries. In some areas hardwood is marketed using a system based on the higher grades of the NHLA rules; two groups are defined, namely FAS (1st and 2nd) and No. 1 C&S (No. 1, common and select). However, these rules are often not strictly adhered to and trade is frequently conducted on a basis of mutual understanding between buyer and seller.

United Kingdom

Originally BS 4047: 1966 set out four grades, 1, 2, 3 and 4, defined in terms of a cutting system, and four grades, A, B, C and D, based on a defects system. However, this standard was rarely applied and has now been withdrawn.

Table 15.5 *Principal hardwood grading rules*

Rules	Quality system (appearance)	Cuttings (fraction clear)
NHLA (USA)	1st	11/12
	2nd	10/12*
	Selects	11/12
	No. 1 common	8/12*
	No. 2 common	6/12*
	No. 3A common	4/12
	No. 3B common	3/12
	Below grade	
Malayan	Prime	10/12*
	Select	9/12
	Standard	8/12
	Serviceable	8/12
UK (BS 4047)	1. 10/12 (11/12 unedged)	
	2. 8/12 (10/12 unedged)	
	3. 6/12 (9/12 unedged)	
	4. 3/12 (8/12 unedged)	

* Fraction increases with decreasing size of piece.

Table 15.6 *Tropical hardwoods which satisfy the requirements for strength classes, graded to the HS grade of BS 5756 (from Table 8 of BS 5268: Part 2: 1991)*

Standard name	Strength class
Iroko Jarrah Teak	SC5
Merbau Opepe	SC6
Karri Keruing	SC7
Balau Ekki Kapur Kempas	SC8
Greenheart	SC9

Table 15.7 Grade stresses and moduli of elasticity for SC5–SC9 for dry exposure (extract from Table 9, BS 5268: Pt 2: 1991)

Strength class	Bending parallel to grain (N/mm²)	Tension parallel to grain (N/mm²)	Compression parallel to grain (N/mm²)	Compression perpendicular to grain*		Shear parallel to grain (N/mm²)	Modulus of elasticity		Approximate density† (kg/m³)
				(N/mm²)	(N/mm²)		Mean (N/mm²)	Minimum (N/mm²)	
SC5	10.0	6.0	8.7	2.8	2.4	1.00	10 700	7 100	750
SC6	12.5	7.5	12.5	3.8	2.8	1.50	14 100	11 800	840
SC7	15.0	9.0	14.5	4.4	3.3	1.75	16 200	13 600	960
SC8	17.5	10.5	16.5	5.2	3.9	2.00	18 700	15 600	1 080
SC9	20.5	12.3	19.5	6.1	4.6	2.25	21 600	18 000	1 200

* When the specification specifically prohibits wane at bearing areas, the higher values of compression perpendicular to the grain stress may be used, otherwise the lower values apply.

† Since many species may contribute to any of the strength classes, the values of density given in this table may be considered only crude approximations. When a more accurate value is required, it may be necessary to identify the individual species and utilise the values given in appendix A of the standard.

Table 15.8   Grade stresses and moduli of elasticity for a selection of structural tropical hardwoods, graded to BS 5756 rules, for the dry exposure condition (extract from Table 15 BS 5268: Pt 2: 1991)

Standard name	Grade	Bending parallel to grain* (N/mm²)	Tension parallel to grain* (N/mm²)	Compression		Shear parallel to grain (N/mm²)	Modulus of elasticity	
				Parallel to grain (N/mm²)	Perpendicular to grain† (N/mm²)		Mean (N/mm²)	Minimum (N/mm²)
Ekki	HS	25.0	15.0	24.6	5.6	3.0	18 500	15 500
Greenheart	HS	26.1	15.6	23.7	5.9	2.6	21 600	18 000
Jarrah	HS	13.8	8.2	14.2	3.1	2.0	12 400	8 700
Kapur	HS	18.1	10.9	18.0	4.1	1.9	19 200	15 800
Keruing	HS	16.2	9.7	16.0	3.6	1.7	19 300	16 100
Opepe	HS	17.0	10.2	17.6	3.8	2.1	14 500	11 300
Teak	HS	13.7	8.2	13.4	3.1	1.7	10 700	7 400

* Stress applicable to timber 300 mm deep (or wide); for other section sizes see BS 5268: Pt 2: 1991.
† When the specification specifically prohibits wane at bearing areas, the HS grade compression perpendicular to the grain stress may be multiplied by 1.33.

15.4.2   Stress grading of hardwoods

BS 5756: 1980 specifies the means of assessing the quality of tropical hardwoods for structural purposes: this standard specifies a single visual stress grade, namely Hardwood Structural (HS) grade which has been established to ensure that each piece has certain minimum strength properties. Currently there is no corresponding standard for temperate hardwoods, nor are there standards for mechanical stress grading of both tropical and temperate timbers.

As in the case of the grading of softwoods, it is a requirement that every piece of tropical hardwood timber purporting to be graded in accordance with BS 5756 shall be appropriately marked on at least one face.

15.4.3   Strength classes

In applying the grading rules within BS 5756, it is possible to have many grade/species combinations for structural tropical hardwood timber. As in the case of softwoods (section 15.2.3) the situation has been simplified with the creation of a strength class system with the structural tropical hardwoods listed under strength classes SC5 to SC9 (Table 15.6).

15.5   Grade stresses for hardwoods

Grade stresses of tropical hardwoods for use in structural design are included in BS 5268: Part 2: 1991 in two different formats, either of which may be used depending solely on convenience.

First, stresses are given for the five strength classes which cover the range of structural tropical hardwoods, and relate to the use of timber under dry exposure conditions (up to 85 per cent relative humidity). Stress values for strength classes SC5 to SC9 are reproduced in Table 15.7.

Alternatively, BS 5268: Part 2: 1991 provides actual grade stresses for a range of tropical hardwood timbers in accordance with the grading rules set out in BS 5756. In order to give the reader a feel for these values, the grade stresses for a few of these timbers are reproduced in Table 15.8.

Grade stresses for structural tropical hardwoods are used in structural design in the same way as those for structural softwoods, described in section 15.3. As noted earlier, there are no corresponding values for temperate hardwoods, but test work is currently underway to redress this omission.

References

BS CP 112: 1952, 1967, 1971.   *Code of Practice – The structural use of timber* (now withdrawn).

BS 3819: 1964   *Grading rules for sawn home-grown softwood* (now withdrawn).

BS 4047: 1966   *The grading rules for sawn home-grown hardwood* (now withdrawn).

BS 4169: 1988   *Glue-laminated timber structural members*. BSI, London.

BS 4978: 1988   *Specification for softwood grades for structural use*. BSI, London.

BS 5268: 1991   *Structural use of timber. Part 2: Code of Practice for permissible stress design, materials and workmanship*. BSI, London.

BS 5756: 1980   *Specification for tropical hardwoods graded for structural use*. BSI, London.

ECE: 1982   *Recommended standard for the stress grading of coniferous sawn timber*.

NAMSR: 1986   *North American export standard for machine stress graded lumber*.

NGRDL: 1975   *National grading rules for softwood dimension lumber*.

NLGA: 1987   *National grading rules for dimension lumber*. National Lumber Grading Authority, Canada.

Part 4

UTILISATION OF TIMBER – HOW, WHY AND WHERE IS IT USED?

16

Utilisation of Timber

The competent and efficient use of timber either in the production of manufactured items or in the wide area of timber construction depends on two important criteria: (1) the selection of the most suitable timber for the task, and (2) the choice of the most appropriate method of joining together the different pieces of timber.

This chapter therefore discusses the criteria to be considered when choosing timber for a particular end use, and then proceeds to the selection of the mode of assembly, whether by adhesives or mechanical fixings: the final section of the chapter is devoted to examples of timber in use, illustrating the combination of timber selection and mode of assembly.

16.1   Choice of timber

Basically there are two aspects to the choice of timber: the first relates to the selection of the most appropriate species for the given set of environmental and stress conditions; the second aspect is concerned with obtaining the correct grade or selecting the appropriate specification for the wood of any one species.

16.1.1   Selection of species

The selection of a timber for a particular purpose depends not only on its technical performance, but also on such factors as cost, size and availability; these four factors are closely interrelated and differ in their relative importance dependent upon the particular species and the purpose for which it is being used.

The technical performance of a timber embraces such properties as its durability, movement,

strength, stiffness and toughness, permeability and ease of processing; not all these properties will be important for any one end use and the order of their relative importance will vary with the task the timber has to perform. The selection of species of timber for a chair will depend not only on its high strength and low movement values, but also on its appearance, especially colour, as well as on its machining characteristics. Conversely, timber required for roadside fencing, or the construction of footbridges must have either high natural durability or else be permeable and therefore capable of taking preservative treatment; an ability to accept nails without splitting, as well as certain strength and stiffness characteristics, are also necessary requirements.

There are exceptional cases where it is not possible to satisfy all the necessary technical requirements and in these instances it is necessary to sacrifice performance in the secondary requirements in order to ensure compliance with the most important ones. Thus the selection of timber species for axe shafts is governed by the need for very high toughness. While this end use also needs high durability, there are no timbers which satisfy both of these requirements. Since toughness is the principal factor, timber such as ash or hickory is used for shafts even though these have very low durability ratings.

Generally, however, it is possible to satisfy all the technical requirements by a number of timbers. Final selection is then conditioned by their availability at a given time, whether they are imported in the sizes required, and finally the cost. These latter constraints reduce considerably the options on the species of timber and it is not uncommon to find that only a few timber species remain after all the constraints in selection have been exercised.

These few species will include the traditionally used timber for that particular end use, together with some potential alternatives. The timber trade in the UK has been slow to accept the use of alternative species of wood, wishing to maintain working with the 'known' timber even though substantial price rises have occurred. However, as a result of such price rises, together with difficulties in ensuring continuity of supply, the trade is slowly accepting the concept of using alternative species: such a swing away from the traditionally used timbers will certainly continue in the years to come and will probably accelerate.

The method adopted up to the present time in using alternative species is to select a timber which closest resembles *in toto* the timber to be replaced. This approach, though successful at times, has also led to disaster in certain cases owing to the failure to match up the performance of the alternative timber to the original timber for the particular end use. Thus, there has been the tendency to replace teak with a timber that resembles teak most closely, rather than by a timber whose performance matches that of teak when used for a particular end use.

This approach is to be deprecated; rather the properties that are required for a particular use must be identified, together with their relative level of importance. A timber is then selected which best satisfies these in-service requirements.

A very good example of this type of approach is given in the *Utilisation Work Book for the Institute of Wood Science Certificate Course*. The question is raised as to what timber would one use to make a wooden pencil, and the thoughts of the reader are immediately channelled into thinking in terms of the properties that are required of such a timber, namely:

(a) it must be smooth to the touch, that is the wood must machine to a fine, smooth finish;
(b) it must cut easily in order to sharpen the pencil, that is the wood must be fairly low density, of moderate stiffness, and have a fine and even texture;
(c) it must be non-toxic, since people often put pencils to their mouths;

(d) it must take surface finishes well, so it must be non-resinous;
(e) it must glue well, since pencils are made of two slats stuck together;
(f) it must be stable and with a straight grain, so it does not bend or distort in varying environmental conditions of use.

The specification which any timber must have for use as a pencil is that it must be of fine even texture, relatively low in density, moderately stiff, stable with an attractive appearance, straight-grained, non-resinous and non-toxic. One timber which satisfies all these requirements is incense cedar, and this is the timber which is currently used to manufacture pencils.

This performance type of approach to wood utilisation has been strongly advocated over the last two decades by the Building Research Establishment, which developed an end use classification system that sets out to define for each major use the properties that are of significance, together with their level of importance, and then provides a selection of the timbers available that satisfy these particular requirements (Webster, 1978; Webster *et al.*, 1984).

16.1.2 Specification of timber

In Chapter 15 the various grading systems were described and it was shown how both softwoods and hardwoods could be ascribed to a number of recognised grades according to the presence and extent of development of a range of defects. Thus timber could be ascribed to a number of 'carcassing' grades for structural use, or a number of 'appearance' grades for joinery purposes. Irrespective of the grading system, grading can be regarded as a necessary stage in the selection of timber for a particular end use. Thus, as with the structural use of timber, the timber grades are regarded as being general in nature, representing an initial sorting process in achieving a range in timber quality.

The requirements of timber for some purposes is more exacting than the conditions set out in the grading rules: such requirements are generally set out in that section of the appropriate British Standard relating to the quality of materials. In BS 1129: 1982 *Specification for portable timber ladders, steps, trestles and lightweight stagings*, for example, the species of timber and quality of timber to be used in order to achieve compliance with the British Standard are described: quality is defined in far greater detail than in the grading rules, embracing such features as slope of grain and rate of growth, in addition to knots and drying defects.

The selection of timber for scaffold boards according to BS 2482: 1981 *Specification for timber scaffold boards* again calls for a higher quality of timber than that resulting from the general grading rules. The species of timber is restricted and the selected planks, irrespective of whether they are visually or mechanically graded, must satisfy limitations in terms of the amount of wane, distortion, fungal and insect attack, and slope of grain. Higher than normal settings must be made when the boards are mechanically graded; when visual grading is employed, the board must pass strict limitations in the size and frequency of knots appearing on both the faces and the edges.

BS 3823: Part 1: 1990 *Specification for grading of wood handles for hand tools*, is concerned with ensuring among other things that the timber used for this purpose will have very high toughness, a property which is not defined in the grading rules. BS 1186: Part 1: 1991 *Timber for workmanship in joinery*, sets out to specify the quality of wood for this particular application in terms of both higher and more diverse requirements than those contained in the grading rules. Four quality classes are defined in terms of knots and knot clusters, checks, resin pockets, rate of growth, slope of grain, and presence of pin-holes.

These are only four examples of the many British Specifications relating to the use of timber for specific end uses. Where appropriate, such specifications must be used and it is necessary only to stipulate that the supply of timber must comply with the requirements set out in the particular British Standard. Sometimes it is advisable to add riders to the specification with particular reference to the moisture content of timber required, and whether preservative treatment is necessary. Unfortunately, British Standards are not in existence for all timber applications, and in these instances architects, surveyors, engineers or indeed any purchaser of timber, must compile his own timber requirement, which is frequently referred to as a 'spec' or 'specification'. Although occasionally these are well drawn up, unfortunately all too frequently they are so worded that they impose ridiculous limitations and consequently the timber is incredibly expensive or else the specification is impossible to fulfil: this state of affairs reflects ignorance on what quality of timber is required for a particular purpose, and the exact meaning of the terms employed in defining quality.

16.2 Joining together of timber parts

Where more than one piece of timber is required for the manufacture or construction of a particular item, an efficient means of joining together the separate parts must be obtained and it is most desirable that the linking agent across the joint should be as strong as the timber and able to contend with differences in the stiffness of the two members of the joint. Basically there are two systems, the first based on the application of adhesives to either flat surfaces or intricate joints, and the second concerned with mechanical means: these systems are discussed below.

16.2.1 Adhesives

The use of glues in the fabrication of timber structures and components offers a neat and efficient method of bonding together the separate pieces of wood, or of board products such as plywood, chipboard, or fibre board (see Chapter 17), which comprise the finished product. The glue bond attained must meet the strength requirements for the structure as a whole and this bond must remain unaffected by the conditions to which it will be exposed throughout its life.

The strength of a glued joint depends on many factors, the more important of which being the dimensions and physical properties of the timber used, its moisture content, the shape of the joint, (see section 16.2.2), the thickness of the glue line, and lastly the type of adhesive, which will govern the long-term performance of the joint.

Most commercial adhesives require for maximum bond strength that the two faces to be joined are brought into intimate contact. With a cellular material such as timber this ideal situation cannot be realised, and in places the glue line can be quite thick (>1 mm). Under these conditions the use of ordinary adhesives would result in considerable crazing and cracking, with associated loss in strength. For timber assembly, special glue formulations are required which will develop high bond strength in thick glue lines: these glues are known as 'gap-filling adhesives'.

Adhesive formulations exhibit a wide variation in their resistance to degradation by such agents as moisture, heat and chemicals. In general terms, the modern synthetic adhesives are more durable than the older glues based on natural materials such as fish, and animal parts together with casein: however, even among the synthetic resins, considerable variation in performance occurs.

Types of adhesives

There are two principal types of synthetic resin adhesive systems, namely thermosets and thermoplastics.

Thermosetting resins. As far as timber is concerned, these resin systems are usually based on formaldehyde; they have been available for more than 40 years, and are well tried and tested — see Table 16.1. These resins comprise a resin and a hardener, and usually the two separate components (either two liquids, or a liquid and a powder) have to be mixed prior to use; in a few cases, however, the resin is supplied as a mixture of powders to which water must be added. Epoxy resins are beginning to be used, especially in the field of structural repair.

The main advantage of the thermosetting types is that they can be used with confidence

Table 16.1 *Thermosetting adhesives for timber (extract from BRE Digest 340: 1989)*

Phenol formaldehyde	Probably the most common plywood adhesive and sometimes used as the binder in particleboards. The dark brown resin is very water-resistant when cured. Rarely used for joints between timber members
Rersorcinol formaldehyde (RF)	Rarely used on its own because of cost. Highly water resistant
Resorcinol–phenol formaldehyde (RPF or RF/PF)	These are used as assembly adhesives and can be recommended for the most severe environments. The reddish-brown resin must be mixed with a separate liquid or powder hardener.
Melamine formaldehyde (MF)	Neat MF resins are rarely used in the UK, but elsewhere they are used for plywood production. Their moisture resistance falls short of PF or RF.
Urea formaldehyde (UF)	One of the most common adhesives for assembly work and board manufacture. When cured it is almost colourless. It is usually supplied with a separate hardener. It has relatively poor resistance to long periods of wetting and drying. Recommended for low hazard situations.
Melamine–urea formaldehyde (MUF or MF/UF)	The moisture resistance of UF can be improved by adding MF. Such a mixture (MF/UF), or co-condensate (MUF) is used for the production of some particleboards.

as structural, load-bearing adhesives. Those incorporating phenol formaldehyde (PF) and resorcinol formaldehyde (RF) resins are very durable and can withstand full exposure to the weather. It should be appreciated that, although an adhesive may be moisture and weather-proof, many species of wood are less durable and in high hazard situations the choice of timber, or timber finish is just as important as the choice of adhesive. In some cases it may be necessary to use a wood preservative.

The formaldehyde resins incorporating melamine are less resistant to moisture than those with PF or RF, but can usually cope with semi-exterior situations where they are protected from direct exposure to rain or sun. They can also be used in kitchens or bathrooms where damp conditions may arise from time to time.

The urea formaldehyde resins are the least durable and are particularly susceptible to degrade in warm, moist conditions. They are usually very much cheaper than other thermosetting types and consequently are used extensively in situations where the moisture content of the wood can be maintained below about 18 per cent.

Thermoplastic adhesives. As far as timber is concerned, these resin systems are based on polyvinyl acetate (PVAC); they are available as either a one-part or a two-part system (Table 16.2).

The simple, *one-part* emulsions have long been used for furniture and indoor joinery work. The cured adhesive has a pronounced thermo-plastic behaviour which makes it unsuitable for load-bearing applications especially in warm, moist conditions. Joints made with these adhesives 'creep' if they are loaded for long periods. They are not suitable for use in thick (say <0.5 mm) glue-lines.

Despite these shortcomings, the cheapness, ease of use and light colour of one-part emulsions has resulted in them becoming one of the most widely used of all wood adhesives.

The heat and moisture resistance of these materials was first improved by adding a separate hardener. This technology has progressed over the years and some of today's *two-part* products have properties which are vastly superior to the simple one-part emulsions. Those which incorporate an isocyanate hardener produce particularly durable bonds, but it is very doubtful whether even the best of these products matches the better thermosetting types. Little information exists on the creep characteristics of

Table 16.2 *Thermoplastic PVAC adhesives for timber (extract from BRE Digest 340: 1989)*

One-part emulsions	Simple-to-use white emulsions suitable for indoor, non-load-bearing applications. They tend to have pronounced thermoplastic characteristics and relatively poor heat and moisture resistance.
One-part pre-catalysed (or self-cross-linking) emulsions	These are one-part emulsions which exhibit a certain degree of cross-linking when they cure. this imparts an improved resistance to moisture and heat.
Two-part emulsions	The most water-resistant grades of PVAC require the addition of a separate hardener or cross-linking agent. Some products are claimed to be suitable for exterior use. The hardeners may be coloured, in which case the advantage of the colourless glue-line will be lost. While these emulsions have improved heat and moisture resistance compared with both one-part emulsions and one-part pre-catalysed emulsions, they are nevertheless still thermoplastic

these improved emulsions and at the present time they should not be used for load-bearing applications.

By requiring the user to weigh out the hardener and mix it into the basic emulsion, manufacturers have detracted from the ease of use of the earlier products. However, this has been restored by manufacturers providing a one-part emulsion which is to some extent self-cross-linking. The performance of this type of cured resin falls between the simple, one-part emulsions and the two-part formulations. A disadvantage of these new products is that they have a relatively limited shelf life (because they are slowly curing from the moment they are produced), particularly at elevated temperatures. In common with the other types of PVAC, they should not be used for load-bearing applications.

Other wood adhesives. A few other adhesives have been used on timber, but their total volume is very small indeed. These include casein (produced from milk), elastomeric adhesives (low density network capable of undergoing large elastic strains and based on a solution of natural or a synthetic rubber), epoxy resins (expensive, two-part thermoset systems) and polymeric MDI (methylene di-isocyanate, a polyurethane derivative).

Specifications for adhesives until mid-1993
Prior to this date thermoset resins in the UK were specified according to BS 1204: 1979 *Synthetic resin adhesives (phenolic and aminoplastic) for wood.* Part 1 of this British Standard set out the specification for gap filling adhesives, while Part 2 contained the specification for close-contact adhesives. A gap-filling adhesive is one which, when used for load-bearing purposes, provides satisfactory bond strength in glue-lines up to 1.3 mm thick. Close-contact glue-lines are defined as being less than 0.15 mm thick.

Tests in this standard place an adhesive into one of four categories (BRE, 1989):

Type WBP — Weather and Boil Proof
Adhesives which, by standard tests and by records in service over many years, have been proved to make joints highly resistant to weather, micro-organisms, cold and boiling water, steam and dry heat. At present only a few phenol and resorcinol based adhesives satisfy this requirement.

Type BR — Boil Resistant
Joints have good resistance to weather and to the test for resistance to boiling water, but fail under prolonged exposure to the weather that WBP adhesives will withstand. The joints will withstand cold water for many years and are highly resistant to attack by micro-organisms.

In addition to phenol/resorcinol adhesives, some melamine formaldehyde (MF) and MF/UF products are able to meet these requirements.

Type MR — Moisture (and Moderately Weather) Resistant
Joints made with these adhesives will survive full exposure to the weather for only a few years. They will withstand a cold water soak for a long period, and hot water for a limited time, but fail under test for resistance to boiling water. They are resistant to attack by micro-organisms. Some of the better UF adhesives can satisfy these requirements.

Type INT — Interior
Joints are resistant to cold water but are not required to withstand attack by micro-organisms. UF adhesives will generally satisfy these requirements.

Thermoplastic adhesives used in the UK up to mid-1993 were usually specified against the German DIN standard 68 602. This standard comprised a series of tests, including water soaking and boiling, which place an adhesive into one of four categories:

Category B1 Adhesives that make joints which are stable ony at low humidities and room temperatures.

Category B2 Adhesives that make joints which are stable in fluctuating humidity and/or occasional exposure to water, such as in kitchens or bathrooms.

Category B3 Adhesives that make joints which are stable when exposed to prolonged periods of wetting with cold water. This is the category often used to specify adhesives for window and door manufacture.

Category B4 The most durable adhesives. The joints must be able to withstand several hours of exposure to boiling water and prolonged periods of soaking in cold water. Such adhesives are often assumed to be suitable for protected exterior applications.

Only a few PVAC adhesives (nearly always two-part products) satisfied the B4 requirements. The one-part self-cross-linking products usually fell into the B3 category and the simple one-part emulsions were usually classified as either B1 or B2.

The creation of a European common market has necessitated the preparation and introduction of new European standards which must be used in place of national standards by the member states. The new European specifications for timber adhesives were published in the middle of 1993 and are now the sole means of specifying adhesives in the UK (with one exception; see below); DIN 68 602 and BS 1204 (except for MR adhesives; see below) have now been withdrawn.

Specification for adhesives after mid-1993
The principal subdivision of adhesive systems is whether or not they are for structural, load-bearing use.

Adhesives (phenolic and aminoplastic) for *load-bearing* timber structures are now specified in BS EN 301: 1993. This standard classifies two types of adhesives (Class I and Class II) according to their suitability for use under load in the climate conditions given in Table 16.3.

Class I adhesives in this new European specification are therefore similar to the former WBP adhesives of BS 1204, Part 1; Class II adhesives are similar to the former BR adhesives of BS 1204, Part 1. (Both Class I and II adhesives must perform adequately on glue-lines 1.5 mm in thickness, so are effectively Part 1 adhesives.)

Class I and II adhesives must satisfy certain performance requirements as set out in BS EN 301 when tested in accordance with BS EN 302: Parts 1 to 4 using the following test methods:

(a) the tensile shear test set out in BS EN 302: Part 1 using bonded test pieces made from beech and with both thick and thin glue-lines;

(b) the delamination test set out in BS EN 302: Part 2 on adhesively bonded test pieces made out of spruce;

(c) the fibre damage test set out in BS EN 302: Part 3 on adhesively bonded test pieces made out of spruce;

(d) the shrinkage stress test set out in BS EN 302: Part 4 on adhesively bonded test pieces made out of spruce.

Non-load-bearing adhesives are specified in the new European standard BS EN 204: 1991. This standard specifies, for each of four durability classes (DI–D4, see Table 16.4), the performance of the adhesives in terms of residual adhesive strength following one or more pre-treatments, as specified in BS EN 205: 1991.

Table 16.3   *Adhesive types for use in different climatic conditions (BS EN 301)*

Temperature	Climatic conditions*	Examples	Adhesive type
>50°C	Not specified	Prolonged exposure to high temperature	I
≤50°C	>85% rh at 20°C	Full exposure to the weather	I
	≤85% rh at 20°C	Heated and ventilated building. Exterior protected from the weather. Short periods of exposure to the weather	II

* 85% rh at 20°C will result in a moisture content of approximately 20% in softwoods and most hardwoods, and a somewhat lower moisture content in wood-based panels.

Table 16.4   *Description of durability classes (BS EN 204)*

Durability class	Examples of climatic conditions and fields of application*
D1	Interior, in which the temperature only occasionally exceeds 50°C for a short time and the moisture content of the wood is 15% maximum
D2	Interior with occasional short-term exposure to running or condensed water and/or to occasional high humidity provided the moisture content of the wood does not exceed 18%
D3	Interior with frequent short-term exposure to running or condensed water and/or heavy exposure to high humidity. Exterior not exposed to weather
D4	Interior with frequent long-term exposure to running or condensed water. Exterior exposed to weather but with adequate protection by a surface coating

*If it is expected that the bonds will have to meet higher or other requirements than those shown in Table 16.3, such as for use in other climates, special agreements should be made on the type of wood and the type of adhesive, dependent on actual climatic conditions, and, if necessary, further tests should be carried out in accordance with BS EN 205.

The BS EN 204 specification is written around the performance of PVAC: indeed it is very similar to the old DIN 68 602 specification. At the present time, the old MR phenolic and aminoplastic adhesives are not covered in BS EN 204, though there are plans to revise this European standard to include them. As an interim measure, BS 1204 has been revised and published (1993) in an abridged format to cover MR adhesives specifically.

Selection of adhesives

The main criteria for selection of an adhesive are the service environment, and whether or not the joint will be load-bearing. Table 16.5 sets out, for

each of five types of exposure, recommended adhesives complying with new European specifications for load-bearing and non-load-bearing situations: the higher the hazard rating, the more resistant must be the adhesive. Such resins are usually the most expensive.

In the gluing of wood, optimum results are obtained when the timber has a moisture content of 12–15 per cent. In certain circumstances it is necessary to glue wood at higher moisture contents: for wood with a moisture content (mc) up to 22 per cent all four classes of adhesive can be used successfully, though the bond will not be quite so good as with wood at a mc of 15 per cent. At mc up to 26 per cent only RF or RF/PF resins should be used, though it is interesting to note that trials are underway on a new resin system which, it is claimed, provides very strong joints on green timber.

The effect of both fire retardants and preservative treatment on glued joints, as well as the care to be exercised in the gluing of preservative treated timber, is set out in Table 16.6.

Painting, and careful maintenance of the paint film throughout the life of a component, will usually enhance the performance of joints made with the less durable adhesives. Imperfections in the paint film will allow entry of water which could cause breakdown of the glueline. Damage to a protective coating can easily be incurred in service. The reliance on a paint film to protect a less durable adhesive in severe conditions of exposure is therefore not recommended.

16.2.2   *Joint design using adhesives*

The joiner, carpenter and furniture-maker adopt a whole range of joint designs including such well-known examples as mortise and tenon, dovetailed and lap joints. Although most of these are capable of transferring high loads from one member to the next, they are time-consuming to produce by hand, or involve the use of elaborate and expensive machines in mass production.

Most of these joint designs relate to the juxtaposition of two pieces of wood at right angles to one another. In the use of timber in

Table 16.5 *Selection of adhesives*

Exposure category	Exposure conditions	Examples	Recommended adhesive	Type	Load-bearing or non-load-bearing	Standard for compliance
Exterior High hazard	Full exposure to weather	Marine and other exterior structures. Exterior component or assemblies where glue-line is exposed to the weather	RF RF/PF or RPF PF	Cl I	L	BS EN 301
Low hazard	Exposed to weather but protected from sun and rain	Inside the roofs of open sheds and porches	RF RF/PF or RPF PF MF/UF or MUF	Cl II	L	BS EN 301
			MF/UF or MUF	MR	NL	BS 1204: 1993
			PVA	D3/D4	NL	BS EN 204
Interior High hazard	In closed buildings with warm damp conditions where a moisture content of 18% is exceeded and where the glue-line temperature can exceed 50°C	Laundries Roof spaces	RF RF/PF or RPF PF	Cl I	L	BS EN 301
			MF/UF or MUF	Cl II	L	BS EN 301
			PVA	D3/D4	NL	BS EN 204
			MF/UF or MUF	MR	NL	BS 1204: 1993
Low hazard	Heated and ventilated buildings where the moisture content of the wood will not exceed 18% moisture content and the glue-line will remain below 50°C	Inside dwelling houses, heated buildings, halls and churches	RF RF/PF or RPF PF MF/UF or MUF	Cl II	L	BS EN 301
			MF/UF or MUF* UF*	MR MR	NL NL	BS 1204: 1993 BS 1204: 1993
			PVA	D1/D2/ D3/D4	NL	BS EN 204
Special	Chemically polluted atmospheres	Structures in the neighbourhood of chemical plants. Swimming baths	RF RF/PF or RPF PF	Cl I	L	BS EN 301

*Where there is a risk of wetting and drying which would lead to degrade of an UF adhesive, it is recommended that an MUF resin be used. However, there is no longer a higher performance level for MUF adhesives in BS 1204: 1993.

Table 16.6   *(Extracted from BRE Digest 340)*

(a)   Effect of fire retardants on glued joints

There will be considerable variability in the performance of joints made with wood which has been treated with a fire retardant. Load-bearing applications should, therefore, be avoided and in non-load-bearing applications, only thermosetting adhesives should be specified. Some treatments are similar to those used for preservatives: for example, introducing the fire retardant into the wood by a vacuum/pressure process. The salts are present in high concentrations and may be slightly acid. In some cases the wood is kiln-dried after treatment but in others the wood is allowed to dry naturally. Bonding should not be attempted until the moisture content is below 18%. Fire retardants, like preservatives, may affect the accuracy of moisture content readings taken with an electrical resistance meter. They can be corrosive to ferrous metal fastenings.

The main points to be aware of are:
- The curing of PF/RF adhesives may be retarded.
- The curing of UF and MUF adhesives is generally unaffected, but the performance of the UF adhesives will be affected by fire
- The presence of some salts can impart a pronounced hygroscopic character to the wood.
- There is a risk of salt deposits on the wood surface after treatment. These should be removed by sanding or planing. Again this practice can lead to a reduced effectiveness of the treatment and care should be taken with disposal of the dust or shavings.
- The moisture content of the wood will be high after treatment and must be reduced before gluing.

(b)   Preservative treatment of glued joints

BY WATERBORNE PRESERVATIVES

This treatment involves immersion of the joint in various solutions of preservative salts, usually under pressure. If the solution is warm and the period of immersion long, most UF and PVAC resins will weaken. Some adhesives are able to recover a high proportion of their strength if they are allowed to dry but it is dangerous to rely on this. In practice, though, temperatures are not normally above ambient and the effect of commercial treatment is minimal. Fully cured PF/RF adhesives are generally able to withstand these treatments without harm.

BY ORGANIC SOLVENT-BASED PRESERVATIVES

Treatment of a joint with an organic solvent-based preservative will have little or no effect on bonding provided the adhesive is fully cured. the active ingredients have no harmful effects on the cured adhesive, and the solvent itself does not cause swelling of the wood. As the treatment is carried out at room temperature no softening of thermoplastic gluelines should occur.

(c)   Gluing preservative treated wood

WOOD TREATED WITH A WATERBORNE PRESERVATIVE

Two main problems may be encountered when attempting to bond wood which has been treated in this way:
- After treatment the wood may be wet. It is preferable to reduce the moisture content to below 18% before making a joint. If it is necessary to bond the wood before it has fully dried, use a PF/RF adhesive. Care should be taken when measuring the moisture content of treated wood with an electrical resistance meter. Some preservatives (particularly those containing copper) can seriously effect the accuracy of the reading.
  There may be deposits of salts on the wood surface; these should be removed by sanding or planing before gluing. This can reduce the effectiveness of the treatment and care must be taken in handling the dusts or shavings
- Provided the above points are covered, that is the surfaces are dry and clean, bonding should be possible with both thermosetting and thermoplastic adhesives.

WOOD TREATED WITH AN ORGANIC SOLVENT-BASED PRESERVATIVE

These products do not normally cause problems unless they contain water repellent. the main points to bear in mind are:
- Allow sufficient time for solvents to evaporate. In some cases this may take several days.
- Sand, or lightly plane the surfaces to be joined.
- In some cases, the manufacturer can supply special versions of adhesives containing solvents that counter the wetting problems introduced by water repellent.

construction it is frequently desirable to join timber end to end, though there are also instances where right-angled joints must be made: these latter usually comprise butt joints usually glued and nailed. In the case of end-jointing, butt joints would not be strong enough and for many years long sloping scarf joints were employed: these were wasteful of timber if the scarf was long, and not very strong if short.

A more recent design is the finger joint which combines the minimum wastage of timber in joint formation with the maximum surface area of overlap for adhesive bonding. The finger joint has proved to be efficient and reliable, the strength of the joint being equivalent to a piece of wood of the same dimensions containing a number of small knots. Numerous designs have been developed, two of which are illustrated in Figure 16.1. Finger jointing is also extensively used in joining together lengths of timber to be used in the manufacture of laminated beams.

Where the loading of the joint is slight, the length of the fingers can be reduced and in the extreme case these can be produced by impressing the end grain against a die, rather than using the conventional rotating cutters. The joint so formed is referred to as an 'impressed joint' (see Figure 16.1) and this has been adopted in the manufacture of window joinery.

16.2.3 *Joint design using metal connectors*

The ease with which timber can be fabricated has contributed enormously to its usefulness, but joints and fastenings have always been, until recently, the weak link in timber construction; as noted earlier in this chapter, some of the synthetic adhesives now being used result in a timber joint stronger than the timber itself. Many carpentry joints necessitate reducing the cross-section of a member, thereby reducing the strength appreciably compared with the full sectional area. Elaborate carpentry joints that reduced loss in strength of joined members to a minimum were developed by our ancestors, but for most purposes these became altogether too expensive when labour costs began their upward trend after the Industrial Revolution. Instead,

Figure 16.1 *Finger joints (centre and left) and an impressed joint (right) (Building Research Establishment, © Crown Copyright)*

bolts and sometimes only a few nails replaced good practice, frequently with loss of strength, because there was no suitable method available to replace the older empiricism. Simple bolt fastenings, although preferable to random nailing with an unspecified number of nails of indeterminate quality, reduce the strength properties of each piece of wood by the amount of timber cut away to take the bolts. Moreover, the strength of the joint is less than that of the bolt, as failure is generally induced either by shear through the timber at the bolt-hole, or through crushing of the timber bearing on the bolt itself.

The high stresses around the bolt-hole were early recognised to be a weak point of bolted joints: solutions were sought by means of bushings around the bolt, aimed at increasing the bearing area. Efforts made in this direction gave rise to rings, toothed plates and variously shaped discs, though these have been largely superseded by the use of toothed metal plates applied to the faces of butt-jointed timber members.

A US patent was granted as early as 1889 for a toothed plate for joining timber, and it is recorded in the US *Department of Commerce Bulletin* (1933) 'Modern connectors for timber construction' that even earlier than this cast-iron rings were used in US bridge construction. The years 1916 to 1922 produced urgent constructional problems that resulted in progress being made in the evolution of suitable mechanical devices for improving timber joints, which are now generally referred to as timber **connectors**. Originating in Europe, modern timber connectors reached the USA in 1930, since when rapid advances have been made, and the scope for timber construction has been greatly widened. More than sixty different types of connectors have been patented in Europe, and in several cases US patents have also been obtained. The first timber connector was the Kübler dowel evolved in Germany; appropriately enough, it was made of hardwood and was doubly conical. It fitted into recesses cut in both members to be joined, thereby increasing the bearing surface appreciably compared with a single bolt.

Figure 16.2   *A bulldog timber connector*

O. Theodorsen, of Oslo, patented the **bulldog timber connector** (Figure 16.2) which consisted of a round or square steel plate, punched at the centre to take a bolt, and with the perimeter turned alternately upwards and downwards to provide triangular teeth. The timbers to be jointed were bored to take a bolt, the connector was inserted between the two members, and the bolt slipped through. As the nut was tightened, the teeth of the connector bit into the wood. It was a comparatively simple matter to tighten the bolt when the connector was used with softwoods, but with hardwoods it was usually necessary to hammer the teeth of the connector into one of the timbers, and to use a bolt with a shank of high tensile steel that permitted the nut being tightened while the teeth were pressed into the opposite member. Once drawn together, with the teeth of the connector ring embedded, the special bolt could be replaced by an ordinary one.

A number of other connectors were also developed including: (1) Split-ring connectors which fit into pre-cut grooves of opposing members: these in turn are drawn together by a centre bolt which is independent of the ring. (2) Toothed-ring connectors, similar in conception to the above, except that the ring is toothed. (3) Claw-plate connectors, which were a development of the bulldog connector and which fitted into pre-cut recesses. In the Teco system for timber-to-timber joints the outside hub (on the

Figure 16.3   *The use of a pair of metal plate fasteners in the fabrication of a joint in a trussed rafter (Building Research Establishment, © Crown Copyright)*

face opposite the teeth) of the male plate consists of a central boss that slips into the recess of the hub of the female plate: a large bolt fits the hub snugly, the connector being flush with adjacent surfaces of the members joined; (4) Teco spike grids and Teco clamping plates, again modifications of the bulldog connector for specific applications.

The principal advantages of these connectors were the relatively high efficiency of the joint compared with a carpentry joint, and the adaptability of the system to pre-fabrication. However, on the debit side special tools were required, preferably power operated, for cutting recesses, the joints were frequently very bulky especially at the junction of three or four members and considerable timber was used in the overlap.

Few of these connectors continue to be used; in the field of trussed rafter fabricating the bulldog connectors and split rings were superseded in the 1960s by the metal plate fastener which is applied to the faces of the members across a butt joint (Figure 16.3): the reduction in bulk of the joint and in timber used is immediately apparent.

The metal plate fasteners are usually made from 18 or 20 gauge galvanised steel and may have either a pattern of holes punched in them through which nails can be driven, or they may have integral teeth. Several proprietary brands of each type are in common use in the UK; those with integral teeth are illustrated in Figure 16.4. The punched metal plate fastener with integral teeth requires a special press to form the joints and is only used when trusses are pre-fabricated in the factory. With the other type of metal plate it is only necessary to drive in the requisite number of nails at each joint; ideally this should be done in the factory, but if this is not possible it can be done on the site.

Apart from simplicity and the reduction in timber used, the metal plate fasteners have advantages over other jointing methods. All the truss members are in the same plane thus avoiding lateral eccentricity of loading. The strength of a joint can also be varied simply by changing the size of the metal plate or by varying the number of nails. In this way a proper balance can be obtained between joint strength and member strength.

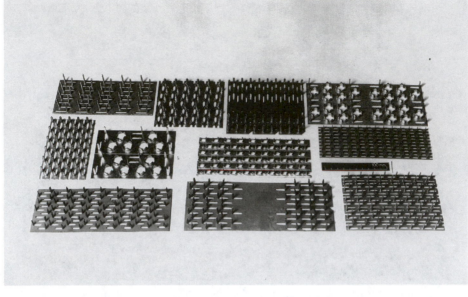

Figure 16.4   *Selection of several proprietary brands of metal plate fasteners having integral teeth (Building Research Establishment, © Crown Copyright)*

Figure 16.5   *Selection of joist hangers (Building Research Establishment, © Crown Copyright)*

A second type of timber connector which has been developed in the last three decades is the joist hanger. This is used primarily in the construction of floors or roofs to secure the joists or rafters to the horizontal beams rather than adopt elaborate joint fabrication: a selection of these galvanised supports is illustrated in Figure 16.5.

16.3   Examples of timber utilisation

16.3.1   *Timber in construction*

About 70 per cent of all sawn softwood used in the UK and 40 per cent of all sawn hardwood is to be found in construction.

Roof truss
Mention has been made above of the roof truss in association with the development and use of metal plate connectors. Perhaps the greatest single development in domestic roof construction has been the introduction of the trussed rafter in the late-1960s. This is a lightweight timber truss which is used to replace the rafters, purlins and ceiling joists of traditional roofs. It is manufactured prior to erection and so eliminates *in situ* nailing other than that required for fixing purposes. Some common trussed rafter configurations are illustrated in Figure 16.6 and of these, the fink and fan are the most widely used.

Since the application of conventional engineering principles leads to over-designing of the truss with resultant uneconomic use of timber, and since simple design methods are not applicable owing to the semi-rigid nature of the joint, the design of trussed rafters has developed on the basis of a simple analysis followed by tests on full-size trusses and subsequent modification if required.

Trusses are available to span up to 30 m. Frequently timber of different depths (though of

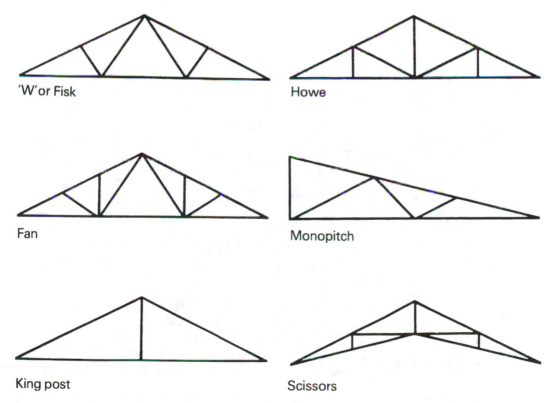

Figure 16.6   *Common trussed rafter profiles*

equal thickness) is employed within any one truss, especially in the larger spans, to best suit the different stress levels within each member (Figure 16.7).

For trussed rafter construction where not only strength but also stiffness is important, it is absolutely essential that the timber be stress graded in accordance with BS 4978: 1988 as described in Chapter 15. From the results of experimental tests and the experience gained in the use of trussed rafters over nearly a decade, it has been possible to produce tables of permissible spans for the more common types of trusses: these are set out in BS 5268: Part 3: 1985.

The trussed rafter has a number of practical advantages over the traditional type of roof which explains why most roofs are now constructed using these trusses. The trussed rafter is usually produced in the factory, thereby permitting greater control in materials and construction. The volume of timber used in a trussed rafter roof is appreciably less than in its traditional counterpart, and the erection time can be reduced to almost one-quarter. Trussed rafters are normally designed to span between the outer walls of a house, thereby avoiding the need for internal load-bearing partitions and affording the designer greater flexibility in arranging room layout.

Timber frame construction

Although timber framing, usually of oak, was widely used in house construction in Elizabethan times, the practice died out and it was not until the 1950s that timber frame housing was reintroduced. This early work by the private sector was not matched in the public sector

Figure 16.7　*The use of trussed rafters in domestic buildings: note the application of metal plate fasteners and the presence within the truss of timber members of different depths (Building Research Establishment, © Crown Copyright)*

Figure 16.8 *Timber frame construction of domestic buildings: note the softwood studding (seen through lower front window) and plywood sheathing (Building Research Establishment, © Crown Copyright)*

until the following decade when the concept was seen as the only way of achieving the large house building targets set at that time. However, it was not until the 1970s that timber frame construction really took off: especially so in the private sector. The increased use of this mode of construction reflected the high interest rates of the period, the promotion of the faster construction times of timber frames, and the shortages of skilled labour.

By the early-1980s about 25 per cent of new housing construction was timber frame. However, adverse publicity in the mid-1980s resulted in a drastic decline in timber frame construction, especially so in England and Wales. In 1992 only 8 per cent of new construction in Great Britain was timber frame, but this figure is misleading; while new starts using timber frame in England and Wales in 1992 were only 4 and 5 per cent respectively, timber frame construction in Scotland accounted for 44 per cent of new housing.

The technique of timber frame construction in the UK has developed considerably over the last thirty years, absorbing concepts and methods used both in Scandinavia and North America where housing construction is almost exclusively timber frame. Typically the UK system is based on simple factory-made panels comprising softwood studs (framing) with a wood-based panel sheathing. These panels, which have a breather membrane fixed to the outside, are assembled on site on a previously prepared concrete base (Figure 16.8). The space between the studs is infilled with thermal insulation, which is then covered with a vapour control layer prior to overlaying with plasterboard.

Externally, the walls are hung with tiles, clad with timber, or faced with bricks: in each case a cavity separates the internal and external walls. It should be appreciated that the brick skin, if used, is not load-bearing, but is present for appearance only.

Multi-storey housing is designed on the platform frame method of construction in which the shell is built one storey at a time, using the wall panels to support the intermediate floors which in turn are used as working platforms to erect the storey above.

BS 5268: Part 6: section 6.1 sets out the requirements for construction of wall panels and provides for the use of a number of proven panel products; it also gives the racking resistance values to be used in structural design, or in their absence, the test procedure to be carried out to validate a design. The most frequently used panel product has been the structural softwood plywoods described in BS 5268: Part 2. Impregnated softboard type SBS to BS 1142 (with added bracing) and high density medium board type HME again to BS 1142 have also been widely used: little chipboard has been used (types C3, C4 to BS 5669) even though the values for racking resistance are included in BS 5268: Part 6. However, especially in Scotland, OSB (with a BBA certificate) is now widely used and beginning to replace plywood as the preferred sheathing material.

Timber cladding including shingles

Irrespective of whether the internal wall of houses is built incorporating a timber frame, or more traditionally using bricks or insulating blocks, the external wall can be covered in whole or part with timber: this use is referred to as cladding and while painted softwood can, and is, used for this purpose preference should be given to the use of timber, the durability rating of which is 'moderately durable' or better, or timber of lower durability provided it can be effectively treated with preservatives.

Painted softwood cladding, though forming a pleasant contrast with brickwork, does require regular attention in the form of re-painting, and it is now usual to restrict its use to small areas; in contrast, whole surfaces of houses have been covered with preservative treated softwood. Figure 16.9 illustrates the use of softwood treated with light and dark creosote to give a pleasing finish to the cladding of a terrace of houses. In the country it is common to find garages and stables clad in timber which has been preservative treated.

Timber species whose natural durabilities are sufficiently high to impart good service as cladding include western red cedar, Douglas fir, African and American mahogany, keruing, kapur and freijo. Species frequently used for cladding following preservative treatment are redwood, whitewood and afara.

BS 1186: Part 3: 1990 lists timbers suitable for cladding and their need for treatment. Suitable preservative treatments are either pressure impregnation using waterborne copper/chrome/arsenic formulation (CCA), or double vacuum impregnation using organic solvent preservative; specifications for the preservative treatment of cladding are given in BS 5589: 1989.

The quality of wood used in cladding must be reasonably high in so far that it must be straight grained, and devoid of open knot holes

Figure 16.9 *The use of creosote preservative-treated softwood timber as cladding in terraced housing: different creosote oils have been used for the ground floor and first floor cladding (Building Research Establishment, © Crown Copyright)*

and splits. It must be capable of being machined not only to give a good surface appearance, but also to accept one of a variety of profiles which are usually applied to provide a measure of texture or relief to the clad area. There are a whole range of profiles, the standard ones being set out in BS 1186: Part 3: 1990. Cladding in the form of planking can be installed either horizontally or vertically. Care has to be exercised in the detailing of cladding around windows and this subject is well illustrated and discussed by TRADA (1993).

Plywood may be used as an alternative to solid wood provided it is at least 9 mm in thickness and manufactured for external use, that is the adhesive must now be to Class I of BS EN 301, (previously WBP) and satisfy the requirements of BS 6566: Part 8. However, most plywood, even that manufactured for external use, is manufactured from non-durable or only moderately durable species and consequently should be treated in accordance with BS 5268: Part 5. The only plywood which is made predominantly from durable species is marine grade plywood: the requirements are set out in BS 1088.

Tempered hardboard (BS 1142) has also given good performance as a cladding material, and the newer boards such as OSB and cement bonded particleboard (BS 5669: Parts 3 and 4 respectively) should also perform well, though, as with tempered hardboard, these products must be protected with appropriate paints or textured coatings.

Timber cladding may be executed using vertically hung shingles; these can also be used on pitched domestic roofing. They are also to be found as the exterior skin of many church towers. Shingles in the UK were traditionally made from cleft (split) oak heartwood, but in more recent times shingles have been imported having been manufactured from western red cedar, originally split, but more recently sawn, to produce a tapered 'tile' about 400 mm in length and from 75 to 300 mm in width. When used in the UK, shingles are normally treated with CCA to improve their long-term

durability. To special order they can also be treated with a non-leaching flame retardant. One of the attractions of using shingles is that they weather to a beautiful silver-grey colour: another attribute is that they are light in weight and require a lighter-weight frame to support them compared with concrete tiles.

Solid timber beams

Large buildings such as churches or mansion houses usually possess roofs or floors constructed with solid timber beams, frequently of substantial dimensions. Figure 16.10 indicates the very considerable volume of timber used in the

Figure 16.10 *Solid timber beams, joists and roofing boards in a parish church (Building Research Establishment, © Crown Copyright)*

Figure 16.11 *Solid timber of large dimensions used for the rafters, purlins and trusses of the roof of Kings College Chapel, Cambridge (Building Research Establishment, © Crown Copyright)*

roof of a parish church, while Figure 16.11 illustrates the magnitude of the main trusses, the horizontal purlins and the rafters in the roof of Kings College Chapel, Cambridge. Quite frequently the timber employed in these structures was either oak or pitch pine which combined the desirable properties of strength and durability. Owing to the current high cost of timber in large dimensions, few new buildings incorporate solid timber beams of any magnitude, recourse being made to the use of laminated timber beams as described in the next section.

Glue laminated beams (glulam)

The favourable strength-to-weight ratio and high flexural rigidity of wood compared with other materials render it a most attractive means of bridging large spans, whether in the form of a bridge or the supporting beams of a large roof. Since the availability of timber in long lengths is severely limited, it is customary to produce beams from shorter lengths by end jointing and lamination. One of the most attractive aspects of laminated beam usage is that it can be tailor made to fit both the load requirements and the design of the structure: thus the architect is provided with a greater degree of freedom in the use of

laminated beam construction than traditional wood or steel joists.

Softwoods are usually adopted for glulam construction in the UK, though it is possible to fabricate them from hardwoods, as is frequently done elsewhere. The timber is usually prepared in the form of 10−20 mm boards which are then generally finger jointed, though sometimes scarf jointed. The ends and faces of the boards are coated with resin, the requisite number of laminations assembled and then heat pressed: the length of the beam is increased progressively with each pressing cycle.

The selection of adhesive for glulam construction is vitally important, being dependent on the extent of hazard conditions prevailing; the various types of adhesives and their durability under various hazard levels have already been discussed in a previous section in this chapter and the comments made there on joining wood are equally applicable to the manufacture of glulam beams.

An example of glulam construction is provided in Figure 16.12, which shows the adaptability of the product. In the UK three stress grades for laminated timbers are provided in BS 5268: Part 2: 1991. Unlike the UK where glulam is used very infrequently, in both Scandinavia and

Figure 16.12 *Glued laminated beams (glulam): note the sharp radius of curvature that can be obtained (Building Research Establishment, © Crown Copyright)*

North America vast quantities are used, generally of large span, for example, to form the roof over sports stadia.

Flooring

Large quantities of timber and timber products are used for both suspended and solid floors in domestic housing. In suspended floors, joists are always of solid timber, usually softwood, while the traditional softwood tongued and grooved boarding (to BS 1297) constituting the actual floor has given way to chipboard (which should be of moisture resistance flooring grade quality C4 as set out in BS 5669: Part 2, or OSB complying with BS 5669: Part 3). These boards have the merit that not only are they laid more quickly and cheaply than tongued and grooved softwood boarding, but they do not show the distortion or ridging so characteristic of softwood boarded flooring.

In situ cast concrete sub-floors, or beam and block floors in domestic housing are frequently overlaid with a 'floating' chipboard floor, using a type C4 moisture-resistant grade to BS 5669: Part 2. The subfloor is first covered with a 35 or 50 mm layer of high density insulation over which is laid a vapour control layer, usually a polythene sheet of 500 gauge. On top of this is laid tongue and grooved chipboard panels 18 mm in thickness; the joints between adjacent rows of panels must be staggered and the tongues round one long and one short edge must be glued before assembly. The chipboard is therefore like a raft above the insulation. Sometimes the chipboard is laid across battens which rest on pads on the subfloor to which it is not attached.

Cast concrete sub-floors may also be covered by hardwood strip or parquet flooring, usually stuck to the floor with bitumen which acts as the damp proof course. Wood as a floor surface is attractive not only because of its aesthetic appeal in terms of colour, grain and figure but also because of its good wearing properties. Solid wood can be laid in strips or as parquet blocks, usually 18 mm in thickness. However, owing to rising costs these strips or blocks are now frequently composite items comprising a top lamination of solid wood on a chipboard or MDF

base. They are available in a variety of timbers, but the most common are oak, maple, sapele and pine; when used in non-domestic situations only the higher density timbers should be used. Sealers for wood floors are discussed in Chapter 23.

Board materials are also to be found in non-domestic flooring, but care must be exercised to ensure that the thickness of board used relative to the distance between supports is compatible with the loading requirement. Thus, BS 5669: Part 5 sets out minimum board thicknesses and joist spacings for both C4 grade chipboard and F2 grade OSB when used for light-duty non domestic suspended floors.

When mezzanine flooring is being installed in warehouses for additional storage or where board materials are being used for heavy-duty flooring, the thickness of board and support centres should be determined by structural analysis using the grade stresses for C5 structural grade chipboard contained in BS 5268: Part 2: 1991.

Joinery

Joinery is one of the major uses for wood in the UK; about 12 per cent of the national use of wood, equivalent to over one million cubic metres, is processed for joinery components such as windows, doors, staircases and cupboards. Even against strong competition from UPVC, some eight million doors and three million window frames are manufactured annually in the UK: a further three million doors are imported and the total annual value of doors and windows used in the UK in 1992 was over £350 million (Figure 16.13).

Ideally, timber for joinery should machine well to give a good surface for staining or painting; be suitable for manufacture to fairly close tolerances; be suitable for machine profiling; and remain stable during production and use. These requirements are completely satisfied only by fine, even-textured, slowly grown, knot-free wood. Such wood is extremely expensive nowadays, and for many joinery purposes wood of somewhat lower, though quite adequate, quality is used.

The quality of wood used in joinery is categorised in BS 1186: Part 1: 1991 by defining requirements against which the physical characteristics of the timber can be measured. Thus four grades are specified in terms of knot characteristics; splits, shakes and checks; resin pockets; straightness of grain; rate of growth and a number of other parameters.

The four classes in BS 1186 are:

Class CSH: Timber for 'clear' grades of softwood or hardwood.

Class 1:   Timber for high quality or specialised joinery.

Class 2 and 3:   Timber for general purpose joinery.

BS 1186: Part 1 specifies timber species suitable for joinery and readily available in the UK. Six softwoods and twenty-seven hardwood (including nine tropical ones) are listed together with their workability, dimensional movement and suitability for various external and internal joinery uses, with or without preservative treatment to BS 5589.

When considering timbers for joinery, there is a difference in the choice available in items produced for the domestic building market and those for prestige institutions and company buildings. The former are mass-produced items using a very limited range of timbers of which beech and European redwood are the most common. In prestige buildings a much wider range of timbers are employed for individually designed items.

In both cases, however, it is essential to consider the environmental conditions. In protected, indoor conditions there is no risk of decay and the choice of timbers is wide. However, external components such as windows and doors are used in conditions where decay can occur. Therefore, they must be manufactured from the heartwood of naturally durable timber, or of timber which if not durable, can be effectively treated with preservative according to BS 5589.

Joinery for domestic buildings is made mainly of softwood, especially European redwood which machines to a good finish and, though not naturally durable, can be easily treated with preservative for external use. European whitewood is also popular but does not machine quite so well as redwood; it has been promoted for external use, but it requires a preservative treatment which is more demanding than that used for redwood in order to obtain effective penetration. Clear western hemlock is commonly used for doors.

Hardwoods are used in joinery where a natural wood finish and better natural durability are required: however, they require more care in processing than softwoods. Many shrink more on drying and have a greater movement in service. However, many hardwoods, especially tropical hardwoods, are free of knots and so more easily meet the requirements for better quality timber for class CSH and class 1.

The properties of timbers for a variety of joinery purposes are also given by Webster (1978). This gives appropriate levels for each property for particular uses and indicates how different timbers match up to the required levels. Though many are suitable, only a few timbers are commonly used for joinery.

The conflict to produce doors of better quality from generally decreasing quality of timber has been solved in the manufacture of what are being sold as 'engineered doors'. All defects are removed from lengths of softwood, to produce relatively short lengths of clear timber. These are finger jointed and glued together to form the

Figure 16.13 *Installation of high-performance impregnated and factory-painted softwood windows (Building Establishment, © Crown Copyright)*

stiles and rails of panel doors: usually, the surface is veneered to mask the presence of the finger joints. Doors so manufactured are claimed to give far less distortion in service than doors produced by normal production methods.

Quite frequently insufficient attention is paid to the selection of adhesives in the manufacture of window frames and external doors and it is quite common to find UF and PVAC adhesives used for these products. Where fully exposed a WBP (or Class I to the new European standard) adhesive should be employed, while in partial exposure conditions a BR (or Class II to the new European standard) or better adhesive may be used.

There is strong competition from certain panel products in particular joinery applications. Thus parana pine, traditionally used for window boards (indoor sills) has been replaced in about 60 per cent of new constructions by moisture resistant MDF complying with BS 1142. The MDF is profiled and primed requiring only to be cut to length to fit the window recess.

MDF is also taking about 10 per cent of the market for skirting boards, architraves (around door frames) and dado rails; with an average 200 m run of skirting board per house this is already a large market for MDF. MDF has also been used to manufacture internal door linings (frames) as well as indoor panel and plain doors. 'Exterior grade' MDF is frequently used in the refurbishment of shop fronts.

16.3.2 Timber in marine work

Quays and groynes

Marine construction represents a particularly hazardous condition for timber and the number of timbers which combine the necessary strength, especially abrasion resistance, and durability, particularly to marine organisms, is limited.

The Handbook of Hardwoods (HMSO, 1972) lists only 21 timbers which are resistant to marine borers, of which 11 are recommended for marine work when strength is also taken into consideration. Of these 11 timbers the best

known are ekki, greenheart, okan and opepe. Generally the timbers are assembled using stainless steel or heavily galvanised bolts, though in certain applications WBP adhesives are used.

Timber still remains one of the most suitable materials for quays and marine defence work. The natural durability of certain woods combined with their abrasion resistance and ability to absorb impact loads make them most attractive and their performance is usually considerably superior to that of reinforced concrete and at a lower cost.

Boats

Inshore fishing vessels of the seine-net type have traditionally been made from timber, and continue to be so, though in reduced numbers. In the selection of timber, although durability is still important, it is customary not to use those timbers listed above for marine work primarily on account of their very high density and difficulty in machining.

A very large volume of timber is used in the construction of these vessels, not only in the superstructure and hull, but also in the keel and the ribs. It is customary to use oak in both the keel and ribs (Figure 16.14) on account of its durability, since the presence of bilge water will ensure that much of this timber remains wet.

Figure 16.14 *Solid oak ribs and keel of inshore fishing vessel (Building Research Establishment, © Crown Copyright)*

Larch is normally used for planking the hull on account of its durability, strength and availability in long lengths.

Until fairly recently, minesweepers were built of timber, primarily on the grounds that timber is non-magnetic as well as being strong and light-weight; however, from the late-1970s these vessels have been built of glass-reinforced plastic (GRP), this material having been selected partly on the basis of reduced maintenance but primarily on account of the difficulty of ensuring the supply of timber of a consistently high quality. The last Royal Naval wooden minesweeper was decommissioned in December 1993.

Some fast inshore patrol craft continue to be produced in timber though many are now built of GRP. The need for speed necessitates the use of strong, durable, lightweight timber, a series of requirements which are satisfied by African mahogany (*Khaya sp.*): this is used to provide both the ribs and the skin, the latter taking the form of plywood bonded with WBP adhesive to BS 1088.

The longevity of wooden hulled boats and yachts is most impressive. Yachts of the 19th century are frequently restored these days and recently it has been announced that a state barge dating from the 18th century (about 1750) is in use again after being restored. The barge, over 8 m in length, with carvel planking in wych elm, a traditional timber for boat building, belongs to St Michael's Mount in Cornwall.

Much mahogany and allied species continue to be used to fit out large yachts and luxury cruisers where red coloured timber with a pleasing grain is still highly prized for its aesthetic appeal.

Sailing dinghies

Although many classes of sailing dinghies are now produced in GRP, large quantities of solid wood, especially *Khaya sp.*, and marine grade plywood (BS 1088) continue to be used, particularly by those classes whose boats can be built up from a kit. About 85 000 dinghies of the Mirror class have been built over the last 30 years and with about 45 kg of timber and plywood in each, this represents a total of about 4000 tonnes for this particular class. Figure 16.15 illustrates the use of timber in the Fireball

Figure 16.15   *Fireball class racing dinghy produced in marine grade plywood (M. Dinwoodie)*

racing dinghy, the veneers for the plywood deck being selected with great care to give an attractive appearance.

16.3.3   *Timber in transportation*

Flooring

Until very recently timber was frequently used as a flooring material in both vehicles and passenger rolling stock, primarily on account of its anti-slip characteristics combined with good wearing properties; all truck floors were constructed of wood or plywood, while London Transport Underground carriages were floored in rock maple. The treads on the escalators were originally maple, but within the last decade have been replaced by rubber treads on a metal base. Currently little timber is used as flooring in transportation.

Containers

While the majority of containers are constructed in metal a small, but nonetheless significant proportion are produced using structural-grade plywood.

Railway sleepers

Although British Rail has almost completely changed over to concrete sleepers from creosote impregnated pine, wooden sleepers continue to be used in many countries throughout the world, while London Transport still use jarrah in the below-ground sections of the Underground: jarrah is selected primarily on account of its high natural durability.

Footbridges
Pedestrian footbridges are frequently constructed of timber using either a durable hardwood such as iroko, or else a softwood which has been pressure-treated by a preservative.

Fencing
Motorway and roadside fencing (Figure 16.16) continues to consume vast quantities of timber; about 11 per cent of all sawn softwood used in the UK and 10 per cent of hardwood are to be found in fencing. Fortunately, many species of timber growing in the UK are acceptable provided they are pressure-treated with a preservative to BS 5589; thus timbers such as oak, ash, beech, birch and all softwoods are used for both posts and rails as specified in BS 1722, Part 7: 1986; this British Standard also sets out limiting characteristics for the quality of timber that can be used. Additionally, this is called up in the specification for motorway fencing contained in Part 1 of the Department of Transport's *Specification for Highway Works* (1986). This latter document also

Figure 16.16 *Use of timber in motorway fencing (Building Research Establishment, © Crown Copyright)*

contains the requirements for timber quality and species used for noise barriers as well as temporary fencing.

16.3.4   *Timber in packaging*

About 14 per cent of all sawn softwood used in the UK and 9 per cent of all sawn hardwood are to be found in packaging.

Pallets
The introduction of pallets during the last decade has resulted in the use of considerable quantities of solid timber, and although the best qualities are not required, the quality selected must be good enough to ensure good performance under rough handling. Softwood, frequently maritime pine, is employed for the crossbearers and slats while home-grown elm or oak is used for the corner spacer blocks (BS 2629: Part 1: 1989).

Crates
Strong wooden crates continue to be required primarily for the export of machinery. Because of the loads to be carried, solid wood planks usually 18–25 mm in thickness are necessary and these are usually produced from imported redwood, whitewood, or maritime pine; structural grade plywood is sometimes used in equivalent thicknesses. For crates to carry less dense or lighter loads, plywood or OSB are frequently used.

Cable drums
Softwood continues to be used in the manufacture of large-diameter (1–2 m) drums to transport electricity supply cables.

16.3.5   *Timber in mining*

Although the volume of timber used annually is declining with the marked reduction in the number of coal mines, substantial quantities are still used, much of it home-grown. Specifications covering the requirements for the various categories of sawn mining timber are issued by British Coal: the bulk of the timber used is softwood, though some hardwood timber is employed.

The main categories of sawn mining timber are:

(a) *Chock timbers* which account for over half of all sawn mining timber by both volume and value: about half the consumption is hardwood.
(b) *SAS boards* (Boards for self advancing supports).
(c) *Sleepers.*
(d) *Miscellaneous uses* including crowns/baulks, pillarwood and coverboards.

16.3.6   *Timber for transmission poles*

Considerable quantities of softwood poles are still used in the UK for power and telecommunication lines. The dimensions and quality requirements for these softwood poles are set out in BS 1990: Part 1: 1984. A considerable proportion of the poles used in the UK are from home-grown softwood trees. Acceptable species, size categories, methods of preservative and method of purchase vary between the telecommunications industry and the electricity supply industry.

Scots pine has long been the preferred timber for poles and has become the yardstick for assessing the performance of other species. Corsican pine is also used by both industries having a strength similar to that of Scots pine. Larch poles are usually stronger than those of Scots pine but are more difficult to treat with preservative. Douglas fir poles are in regular use only by British Telecom while Norway spruce is used only for power transmission poles.

Long life is a prerequisite for these poles and preservative treatment is essential. Poles that comply with BS 1990 are expected to have a service life of 40 years.

16.3.7   *Timber in furniture*

Chairs, framework
After construction, the greatest volume of timber and timber products is used in furniture manufacture (Figure 16.17). In the production of chairs and the frames for tables, desks and dressing tables, solid wood is employed not only on account of its strength, but also because it can be turned or shaped as well as recessed to accommodate a joint.

The timbers employed are usually of medium density, averaging 510–900 kg/m^3, thereby ensuring good strength properties while still easy to machine. Fine texture is also important in that it produces a good finish to the article.

Figure 16.17   *Regency style chair made from longui rouge (Building Research Establishment, © Crown Copyright)*

Timbers having low movement characteristics are preferred to ensure continued tightness and evenness of joints, though the traditional use of beech is an exception in this respect. There appears to be preference for brown or red timbers, though some light coloured timbers are used, many of which are subsequently stained dark; figure is fairly low in the list of requirements for timber for chairs and frames. Advice on the selection of timbers and board materials for furniture is to be found in Webster *et al.* (1984).

Tops

The tops to tables, dressing tables and desks are usually produced from chipboard, MDF or, exceptionally, plywood. Chipboard and plywood are lipped with solid wood around the edge and veneered on the faces, the exposed face having an attractive veneer with a pronounced figure, while the 'balancing' or hidden veneer is usually plain. MDF tops are usually veneered, but the edges are frequently profiled from the board, thus utilising one of the merits of this panel product, namely its ability to machine well.

Only in the most expensive furniture is solid wood used for table tops and great care must be exercised, both during manufacture and subsequent use, to ensure that the moisture content does not decrease, thereby giving rise to splitting. The attraction of MDF, plywood and chipboard is not only their lower cost, but also their far superior dimensional stability under changing humidity conditions compared with solid wood.

Occasional furniture

Small tables, magazine racks and wooden trays are now being produced from MDF, which is sometimes embossed to indicate lines of hardwood vessels, or growth rings. As noted above, MDF lends itself to spindle moulding thereby allowing the production of moulded edges with a very high quality of surface.

Kitchen furniture

Huge quantities of chipboard are used in the manufacture of fitted kitchen units. Chipboard is generally used in preference to MDF because of its lower cost: the extra cost of MDF is not justified where a square-edged finish to the doors is required.

Frequently the chipboard is covered on faces and edges by melamine-impregnated paper: in the more expensive ranges, wood veneer and wood edging are employed. A few ranges employ solid wood for the door fronts though the carcase is usually made of chipboard: American white oak, redwood (Scots pine) and beech are the common choice of timbers. Preformed worktops are nearly always produced from Grade C3 chipboard to BS 5669.

Panelling

Timber remains one of the most attractive forms of wall covering, but because of the high cost its use is restricted to certain rooms in prestige office blocks, banks and court rooms. The selection of veneers for panelling is determined by the development of attractive figuring and colour variation (see Chapter 4): because of the very high quality and standard required in such applications, plywood is generally used as the substrate for the veneers. Where slightly lower standards are required as in shops, both MDF and chipboard are used.

Door skins can be treated under the general heading of panelling and large quantities of attractively veneered plywood are used annually in the production of doors for both domestic and public buildings.

16.3.8   *Timber in sports goods*

Rackets

Tennis, badminton and squash rackets were made of wood up to the 1970s. The requirements of a timber for so arduous a purpose were that it should have excellent toughness (impact resistance) while still being light in weight, easily machined, easily bent in production, yet fairly stiff in use. Few timbers met such demanding requirements, the species most commonly selected being European ash; some hickory and beech were also used in the frames, though the use of the former was restricted to the very expensive rackets.

In order to achieve the desired curvature the timber was cut up into thin strips. In the 1950s and early-1960s these laminations were steamed before bending, but in the 1970s thinner laminations were prepared and these were bent in the cold state and glued up using urea-formaldehyde resin.

Golf-club heads (woods)

It is not often appreciated just how many golf-club heads (woods) are manufactured annually; the production of one UK factory alone is some 250 000!

Traditionally, persimmon was used for the production of the heads on account of its hardness, durability and toughness, and while a few heads are still manufactured using this timber there appears to have been a move towards the use of laminated timber owing to the cost and difficulties in obtaining regular supplies of persimmon.

The laminated head comprises a series of plies usually about 1 mm in thickness, laid up in the same direction and bonded together by a WBP resin. The timber normally used is rock maple on account of its hardness and toughness, though beech is often used for heads in the lower price range.

Hockey sticks

The heads of hockey sticks were traditionally made of ash, primarily because of its high toughness and its good bending properties. However, with the introduction of the 'Indian' head having a tighter radius, Indian mulberry was preferred. Attempts were made to make heads with a smaller radius from ash, many of those being unsuccessful: however, one successful approach is illustrated in Figure 16.18 in which two billets of ash are first bent and subsequently glued together using urea-formaldehyde adhesive. The specimen is then shaped and notched to take the handle which is usually made of laminated cane.

Cricket bats

The only timber considered to be really suitable for cricket bats is the cricket bat willow *Salix alba var. coerulea* which is specially grown for this purpose in the eastern and south-eastern counties. Although it is not possible to define clearly what features render this one timber 'suitable', lightness of weight combined with strength, resilience and colour are perhaps the most important. The handle is usually laminated from cane imported from Indonesia.

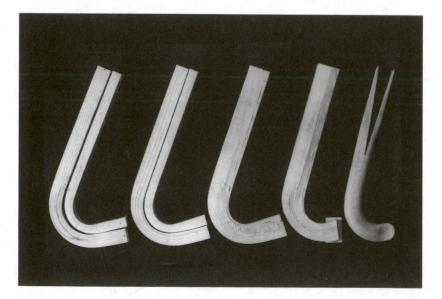

Figure 16.18 *Stages in the production of hockey sticks using European ash (Building Research Establishment, © Crown Copyright)*

Other sports items

There are many other items of sports equipment which either have used wood up to fairly recently or continue to use it and which cannot be discussed in detail in a short appreciation of the subject like this. However, they may be listed and include: cricket stumps; lacrosse and shinty sticks; polo sticks; bows and arrows; croquet mallets and balls; bowls; billiard cues; table tennis bats and tables; gymnastic equipment; and diving boards. A new application of timber in the sports area has been as a decking material for the banked velodromes for cycle racing.

16.3.9 Timber in musical instruments

Soundboards and soundboxes

Musical instruments, such as pianos, harpsichords and clavichords, incorporate a soundboard in their construction: this is positioned in close proximity to the strings, resonating when the string is struck thereby imparting a fullness to the tone of the note.

The soundboxes of the stringed instruments, though of smaller surface area, have a similar effect in enriching the tone of the notes produced on these instruments. The timber most highly prized for the production of both the soundboards of the keyed instruments and the belly of the stringed instruments is European spruce (*Picea abies*) obtained from the Carpathians in Romania and known in the trade as 'Romanian' or 'Swiss pine'. It is very carefully selected for close, even growth, preference being given to timber with at least 20 rings per inch. More recently, because of difficulties in obtaining supplies, together with high cost, carefully selected high-quality old-growth Sitka spruce from Canada is being used.

Woodwind instruments

All the species of wood which have been found suitable for woodwind instruments have very high density, reasonably fine texture and are both impermeable and stable, most important

properties in view of the amount of saliva and humid air passed into them: the timbers must work to a good finish in turning, a reflection of both the density and texture of the timber. Those timbers which satisfy these requirements are African blackwood, used for clarinets, oboes, cor anglais, bassoons and bagpipes; cocus wood for high quality flutes, and ebony, an alternative for bagpipe manufacture. As these woods become scarce and expensive, a whole range of woodwind instruments made from selected plastics has been introduced in recent years.

References

(a) National Standards

BS 1088 and BS 4079: 1966 *British specification of plywood for marine craft.* BSI, London.
BS 1129: 1982 *British Standard Specification for portable timber ladders, steps, trestles and lightweight stagings.* BSI, London.
BS 1142: 1989 *Specification for fibre building boards.* BSI, London.
BS 1186: 1991 *Timber for and workmanship in joinery. Part 1 Specification for timber.* BSI, London.
BS 1204 *Synthetic resin adhesives (phenolic and aminoplastic) for wood. Part 1: 1979 Specification for gap-filling adhesives. Part 2: 1979 Specification for close-contact adhesives.* BSI, London. [now withdrawn]
BS 1204: 1993 *Specification for type MR phenolic and aminoplastic synthetic resin and adhesives for wood.* BSI, London.
BS 1297: 1987 *Specification for tongued and grooved softwood flooring.* BSI, London.
BS 1722: Part 7: 1986 *Specification for wooden post and rail fences.* BSI, London.
BS 1990 *Wood poles for overhead power and telecommunication lines. Part 1: 1984 Specification for softwood poles.* BSI, London.
BS 2482: 1981 *Specification for timber scaffold boards.* BSI, London.
BS 2629 *Pallets for materials handling for through transit. Part 1: 1989 Specification for principal dimensions and tolerances.* BSI, London.

BS 3823: 1990   *Grading of ash and hickory wood handles for hand tools. BSI, London.*

BS 4978: 1988   *British Standard specification for softwood grades for structural use. BSI, London.*

BS 5268   *Structural use of timber.*
Part 2: 1991   *Code of Practice for permissible stress design, materials and workmanship.*
Part 3: 1985   *Code of Practice for trussed rafter roofs.*
Part 5: 1989   *Preservative, treatments for structural timber.*
Part 6   *Code of Practice for timber frame walls: Section 6.1 (1988) Dwellings not exceeding three storeys. BSI, London.*

BS 5589: 1989   *Code of Practice for the preservation of timber. BSI, London.*

BS 5669   *Particleboard.*
Part 2: 1989   *Specification for wood chipboard.*
Part 3: 1992   *Specification for oriented strand board (OSB).*
Part 4: 1989   *Specification for cement bonded particleboard.*
Part 5: 1989   *Code of Practice for the selection and application of particleboards for specific uses. BSI, London.*

BS 6566   *Plywood*
Part 8: 1985   *Specification for bond performance of veneer plywood. BSI, London.*

DIN 68 602   *Evaluation of adhesives for joining wood and derived timber products. DIN, Cologne.*

(b)   European Standards (published by BSI, London)

BS EN 204: 1991   *Classification of non-structural adhesives for joining of wood and derived timber products.*

BS EN 205: 1991   *Test methods for wood adhesives for non-structural applications – Determination of tensile shear strength of lap joints.*

BS EN 301: 1993   *Adhesives, phenolic and aminoplastic, for load-bearing timber structures: classification and performance requirements.*

BS EN 302   *Adhesives for load-bearing timber structures: test methods.*
Part 1: 1992   *Determination of bond strength in longitudinal tensile shear.*
Part 2: 1992   *Determination of resistance to delamination (Laboratory method).*
Part 3: 1992   *Determination of the effect of acid delamination to wood fibres by temperature and humidity cycling on the transverse tensile strength.*
Part 4: 1992   *Determination of the effects of wood shrinkage on the shear strength.*

(c)   Other references

BRE (1989) Choosing wood adhesives. *Building Research Establishment Digest 340*

BRE (1992) Selecting wood-based panel products. *Building Research Establishment Digest 323.*

Department of Transport (1986) *Specification for Highway Works: Part 1.* HMSO, London.

TRADA (1993) External timber cladding. *TRADA Wood Information Section 1, Sheet 20.*

Webster C (1978) *Timber Selection by Properties: the Species for the Job. Vol. 1: Windows, Doors Cladding and Flooring.* HMSO, London.

Webster C., Taylor V. and Brazier J.D. (1984) *Timber Selection by Properties: the Species for the Job. Vol. 2: Furniture.* HMSO: available only from the Building Research Establishment Bookshop.

17

Manufactured Wood Products and their Application

17.1 Introduction

In the previous chapter the utilisation of solid wood was described: with the possible exception of preservative treatment where necessary, the wood was used in its original converted and seasoned state, the various parts of the structure being joined together by adhesives, nails or connectors. In this chapter wood is treated as a raw material suitable for further processing, and attention is directed specifically at three very different areas. The first relates to the addition of resins and plastics to the timber which enhance its technical performance without basically modifying its structure: two types of product are discussed below in some detail, the first incorporating phenol formaldehyde resins which set under heat and pressure, and the second using liquid plastic monomers which are polymerised *in situ* to give a wood-plastic composite.

The second area of interest is the new generation of so-called 'engineered timber' in which pieces of wood are bonded together with an adhesive to produce 'timber' battens.

The third area relates to the cutting-up of wood and its re-formation into board materials, frequently, but not always, with the assistance of a synthetic resin. These products now constitute a third of the UK consumption of timber and timber products, and the five principal board types are discussed in considerable detail.

Wood pulp is a further manufactured wood product, very different in behaviour from both timber and the board materials and, as such, outside the scope of this book.

17.2 Enhancement with plastics

Although the two manufactured products described below embrace the chemical treatment of wood, it is not the intention to describe here the various processes of chemical modification of wood to improve its dimensional stability. Rather the reader is directed to Chapter 9, section 9.6 where this subject has been covered in considerable detail.

17.2.1 Impregnated and compressed wood

A number of 'improved wood' products were developed in the 1940s and 1950s, much of the impetus for this development occurring during the second world war as a result of either the need for specialist materials with good dielectric properties or non-magnetic characteristics, or as a straight replacement for certain metals which were in short supply.

Most of these improved woods were, and still are, produced from hardwood veneers bonded together using phenol formaldehyde resins at high temperature and pressure: this results in transverse compression of the wood and increased density. There are two principal types of product: the first uses dry but otherwise untreated veneers which are bonded together, generally using a dry film of phenol formaldehyde, to produce a densified laminate or 'super-plywood' which is sold in the UK under the trade name of 'Permawood' (see below).

The second variant of the process is to impregnate the veneers, generally under vacuum, with a solution of phenol formaldehyde prior to their assembly and pressing. Adhesive loadings as high as 50 per cent can be achieved, though generally from 20 to 30 per cent are used: a whole range of these impregnated, densified laminates have been produced in both the UK, such as 'Jigwood', 'Jablo' and 'Permali', and in the USA, such as 'Compreg'.

Although extensively used during the second world war period, especially in the production of aircraft propeller blades, demands for these improved wood products decreased markedly in the 1950s with the introduction of the synthetic polymers. However, both 'Permali' and 'Permawood' are still produced in the UK, albeit for only a few very specialised uses. Both products can be tailor made to suit necessary requirements by varying the arrangement of the wood veneers to give maximum mechanical strength in any particular direction.

Permali, the veneers of which are impregnated under vacuum before being densified under heat and pressure, has a density of about $1200 \, kg/m^3$ and very good dimensional stability in the presence of moisture: compression strengths, hardness and abrasion resistance are exceptionally high, though its bending strength is only marginally better than timber of the same density. The strength-to-weight ratio of Permali is usually considerably higher than that of steel.

It has excellent dielectric properties and its principal application nowadays is in the production of electrical insulating components which are subjected to severe mechanical forces: it is available in the form of sheets, rings and rods.

Permawood, assembled from veneers which have not been impregnated with resin, though still bonded together under pressure and heat, has a slightly lower density of $900–1200 \, kg/m^3$ depending on grade. The higher density material is sometimes called 'Hydulignum'. Dimensional stability and dielectric properties are again high and the product is used primarily in the production of certain types of large transformers.

17.2.2   Wood–plastic composite

Brief mention was made in Chapter 9, section 9.6 concerning the impregnation of permeable woods with liquid monomers which were then polymerised *in situ* to form a composite material with greatly improved dimensional stability and a marked enhancement of certain strength properties. The monomers most frequently used are monomethyl methacrylate or a mixture of styrene and acrylonitrile: polymerisation is induced by the use of either gamma irradiation or by heat in the presence of a catalyst. Swelling of the timber occurs during impregnation of the polymer, thereby indicating penetration of the cell wall, as well as coating and sometimes filling the cell cavity. The degree of swelling of treated wood in the presence of moisture is reduced to about 10 per cent that of matched untreated samples: additionally, certain strength properties, such as hardness and abrasion resistance, are from two to five times greater for the treated than the untreated wood. Many of the strength properties show little change, while the modulus of elasticity and impact resistance are usually reduced.

Following a long period of laboratory experimentation at the former Princes Risborough Laboratory, the method was taken up and developed on a small commercial scale by the Applied Radiation Chemistry Group at Harwell. The material is sold under the name of 'Curifax', though it is more frequently referred to simply as WPC (wood–plastic composite). Curifax can be produced in lengths up to 3.66 m (12 feet) but because the cost is about four times that of untreated timber, its use has been limited to the production of specialised items such as cutlery handles, brush handles and decorative flooring; there is considerable potential for its use in turnery and sports goods. An example of the wood block flooring is to be seen in the terminal building of Helsinki airport.

17.3   'Engineered structural timber'

Several products have been developed over the last decade to produce a superior grade of 'timber', often from relatively low grade wood. Although considerably more expensive than conventional timber on a volume-for-volume basis, the products can be used in smaller cross-sections because of their enhanced strength and stiffness. Therefore, when the comparison is made on the basis of structural performance, these new products are certainly competitive with conventional timber.

The enhanced performance of these products reflects their production from pieces of wood which are then partly or completely randomised before being glued together again. Consequently, defects in the original timber, in terms of juvenile wood, knots and sloping grain, are redistributed randomly in the manufactured product instead of being concentrated in particular areas of the original timber. Design stresses for structural use reflect not only this lower influence of defects, but also the much lower variability in strength and stiffness among different members of the manufactured product, that is the variability both within and between pieces of the manufactured 'timber' are much lower than occurs in normal timber.

There are four types of this product currently on the market in different parts of the world. None of these is to be confused with glulam which comprises thick laminations of wood, and which was described and discussed in Chapter 16, section 16.3.1. The new range of engineered materials are made up of much smaller units than in glulam. The four types are described below.

17.3.1 Laminated veneer lumber (LVL)

Softwood logs are rotary peeled to produce veneers about 1600 mm in length and 3 mm in thickness; these are subsequently kiln-dried and coated with a phenolic formaldehyde adhesive, laid up parallel to each other and bonded together under pressure to produce a board up to 24 m long by 1.2 m wide and up to 89 mm in thickness; this board is then cut up into required sizes of battens. Joints between veneers are staggered longitudinally by at least 100 mm and in one of the products all the veneers are scarf jointed with the exception of the middle veneer.

Two products are currently on the market — one manufactured in Finland ('Kerto') from Norway spruce, and the other in Canada ('Microlam') from a range of softwoods.

Grade stresses (on a permissible design basis) are supplied by the manufacturers: with the batten loaded in bending on its shorter face, the grade stresses for Kerto and Microlam are 13.9 and 15.6 N/mm^2 respectively while the mean modulus of elasticity parallel to the grain are 12 750 and 11 850 N/mm^2 respectively. The corresponding bending strength and MOE for whitewood and graded to BS 4978 (Chapter 15) in BS 5268: Part 2 are 7.5 N/mm^2 and 10 500 N/mm^2 for SS grade, and 5.3 N/mm^2 and 9000 N/mm^2 for GS, thereby confirming earlier comments about the superior performance of these new products compared with solid wood.

17.3.2 Parallel strand lumber (PSL)

While LVL has been on the market for over a decade, PSL is a relatively new North American product. Douglas fir or Southern pine logs are rotary peeled to produce veneer (2.5 mm in thickness) which is screened to eliminate major strength reducing defects. These sheets of veneer are cut into strands about 2400 mm in length and 3 mm in width. The strands are coated with a waterproof adhesive before being pressed together and microwave cured; lengths up to 20 m can be produced.

17.3.3 Laminated strand lumber (LSL)

This is another product which has appeared over the last decade in North America. Aspen is cut into strands some 300 mm in length and coated with an isocyanate waterproof resin before being aligned parallel to each other and bonded together in a steam injection press to produce battens of the required dimension after sanding. This product is therefore different from the two previous products in that smaller pieces of wood are used to produce small-size structural members.

17.3.4   'Scrimber'

This product was developed in Australia and, unlike the three previous products, is not produced from veneers or long strands. Instead, whole green logs are passed between heavy rollers which crush the wood to produce a fibrous net-like structure that is then coated with PF resin and pressed together to form a sheet which, after cooling, is cut up to produce battens of the required dimensions.

17.4   Board materials (wood-panel products)

The principal drawbacks of wood as a material are: (1) its variability in performance; (2) its availability in limited widths; and (3) the marked difference in property levels along and across the grain. These disadvantages are removed if the wood is reduced to smaller units and subsequently re-assembled in a particular way. Thus, the principal board materials are characterised not only by a much reduced dimensional movement in the plane of the board compared with solid timber, but also such movement is very similar both along the length and across the width of the board: unfortunately, this improvement and uniformity are achieved at the expense of greater thickness movement, though it must be admitted that this is the least important dimension as far as board utilisation is concerned.

Not only is movement similar along and across the plane of the board, but strength and stiffness also show a similar lack of variation, or at least reduced variation. Lack of variability within the board is a reflection not only of the mode of production, but also the absence of defects which characterise wood and which can be either eliminated in production, or else widely scattered throughout the board.

Perhaps the greatest attraction of boards is their large size and, although 2440×1200 mm is the commonly produced size, sheets up to $12\,500 \times 2440$ mm can be produced. This availability of large dimensions eliminates the necessity to join together planks of solid wood, as was formerly done in furniture manufacture, and also leads to the more rapid installation of domestic flooring.

There is an extensive range of board products produced from timber, but it is convenient to treat these under five categories – plywood, chipboard (particleboard), OSB (oriented strand board), fibreboard and cement bonded particleboard (CBPB). Each type of board is manufactured in a number of grades to suit a particular end use: thus, for example, although particleboard is used both in table top manufacture and as flooring in housing, the requirements of the board for each are quite different, and it is possible by changing the particle size range and distribution within the board, together with changes in the amount of resin, to produce two contrasting particleboards, each best suited for one particular end use.

Within the EEC, the total consumption of these board materials in 1989 (the last year for which complete data are available) was $30 \times 10^6\,\text{m}^3$ of which 17 per cent was plywood, 70 per cent was chipboard, OSB and CBPB, and 13 per cent was fibreboard. Production of chipboard and fibreboard almost equalled consumption, but this was certainly not the case with plywood. Consequently, the total production of board materials in the 12 member countries was only $24 \times 10^6\,\text{m}^3$ of which 7 per cent was plywood, 83 per cent chipboard, OSB and CBPB, and 10 per cent fibreboard (Figure 17.1).

In Chapter 1, it was revealed that the total consumption of wood and panel products in the UK in 1991 was $13.6 \times 10^6\,\text{m}^3$: of this, $4.6 \times 10^6\,\text{m}^3$ were board materials equivalent to 34 per cent of the total volume with a value of about £1.10×10^9. Of this $4.6 \times 10^6\,\text{m}^3$, plywood comprised 29 per cent, chipboard, OSB and CBPB 60 per cent, and fibreboard 11 per cent.

Board production in the UK in 1991 was $1.725 \times 10^6\,\text{m}^3$, equivalent to 40 per cent of the

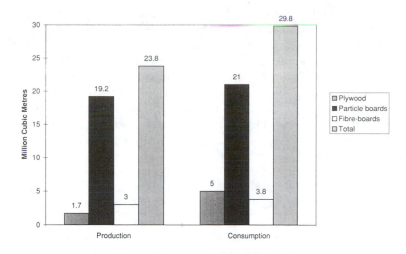

Figure 17.1   *Production and consumption of wood-based panels in the 12 EEC-member countries (1989 figures)*

consumption of boards and produced from about 1.6×10^6 tonnes of roundwood (thinnings and sneddings) and 1.4×10^6 tonnes of sawmill residues; 60 per cent of the total mass came from the publicly owned forests of the Forest Enterprise. Little plywood is produced in the UK, but chipboard, OSB and CBPB comprised 86 per cent, and fibreboard (principally MDF – medium density fibreboard) 14 per cent of the UK production. Home production, therefore, accounts for about 55 per cent of the consumption of chipboard, and about 43 per cent that of fibreboard. At the present time, the principal growth areas are in MDF and OSB.

17.4.1 Plywood

Most of the one million cubic metres of plywood consumed annually in the UK is made from softwood and is imported from the United States, Canada and Finland (Figure 17.2). Less than 1 per cent is home produced. Most temperate hardwood boards are imported from Finland, and are made either entirely from hardwood (usually birch), or are a composite of birch with softwood cores. Most tropical hardwood plywood used in the UK comes from South East Asia, mainly Indonesia and Malaysia, but the market share of South America and Africa is increasing steadily.

As noted above, solid timber has certain disadvantages by comparison with other structural materials in that its greatest strength lies in the longitudinal direction, that is along the grain, whereas it is comparatively weak across the grain. By means of cross-banding timber veneers, the high longitudinal strength

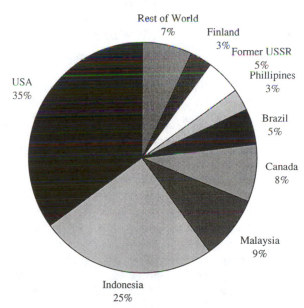

Figure 17.2   *Source of supply of plywood imported into the UK in 1991*

is imparted to two directions: however, not only tensile strength, but also shear and bending strengths, stiffness and dimensional stability are approximately equal along and across the plane of the boards.

Plywood is defined internationally as 'panels consisting of an assembly of plies bonded together with the direction of the grain in alternate plies usually at right-angles. In general, the outer and inner plies are placed symmetrically on both sides of a central ply or core'.

Plywood is subdivided internationally into two distinct groups: **veneer plywood** and **core plywood**.

Veneer plywood is defined as 'plywood in which all the plies are made of veneers up to 7 mm thick orientated with the plane parallel to the surface of the panel'. Core plywood is defined as 'plywood having a core', and this group is subdivided as follows:

(1) Wood core plywood: 'plywood having a core of solid wood or veneers'
  (a) Battenboard: 'plywood, the core of which is made of strips of solid wood more than 30 mm wide, which may or may not be glued together';
  (b) Blockboard: 'plywood, the core of which is made of strips of solid wood more than 7 mm wide but not wider than 30 mm, which may or may not be glued together';
  (c) Laminboard: 'plywood, the core of which is made of strips of solid wood or veneer not wider than 7 mm placed on edge and glued together'.
(2) Cellular plywood: 'plywood, the core of which consists of a cellular construction. There shall be at least *two* cross-banded plies on each side of the core', and
(3) Composite plywood: 'plywood, the core (or certain layers) of which are made of materials other than solid wood or veneers. Composite plywood with a core shall have at least *two* cross-banded plies on each side of the core'.

The production of plywood was a progression from the use of decorative veneers which widened the decorative scope of natural wood, and allowed valuable and attractive timbers to be used efficiently. Probably the first veneer was manufactured in ancient Egypt around 3000 BC, the probable glue used being albumin. Early in the 19th century, decorative veneers were produced by sawing, while veneer slicing developed in the middle of that century. The really major development was the production of continuous sheets of veneer with a knife by rotating round logs in a rotary lathe – an invention from 1818. The early uses of plywood were limited to furniture, joinery, etc., by the organic glues available, and constructional use of plywood only became possible by the development of synthetic resin adhesives in the 1930s.

Only certain species of solid timber are suitable for plywood production, being determined by their peeling characteristics, yield, veneer smoothness and bending properties. Manufacture often involves a 'softening' pre-treatment of selected logs by steaming or boiling; this is necessary for most of the tropical hardwoods, but the logs of certain other timbers, such as beech, birch and Douglas fir logs, may be peeled straight from the log pond. Following de-barking, the softened peeler logs are placed in a rotary lathe, which rotates the log against an adjustable knife to produce a continuous sheet of veneer of thicknesses up to 7 mm and the full width of the log. The continuous sheet is cut to remove defects such as knots, and to produce pre-determined widths. Uneven distribution of moisture content could produce stresses in the final plywood, causing twisting and warping. To eliminate this, and in order to dry the veneers, a necessary prerequisite for the application of plywood adhesives, the still wet veneers are fed through a continuous drier comprising a long heated chamber, the rate of feed and temperature being dependent on veneer thickness, adhesive type to be employed and required moisture content.

The dried veneer is then appearance-graded for face, back and core qualities. Knots may be removed and patched, and splits repaired. The narrower widths of veneer may be edge-jointed to form standard widths.

The prepared veneers are taken to the gluing area where glue is applied and the required sandwich is built up. Glue spreaders regulate the distribution and the veneers are assembled at right-angles in a balanced construction. As many as 17 plies could be assembled to produce a 25 mm plywood. A film glue, such as a phenolic resin-impregnated paper, may also be employed.

The assembled packs of veneers are finally loaded into heated multi-daylight presses for the final pressing operation. To allow even moisture distribution, the plywood so formed is stored in stacks before delivery.

The normal sheet size for plywood is 2440 × 1220 mm, but larger sizes can be produced by using either very large presses or by joining together standard sized panels using scarf- and finger-jointing techniques.

Plywood surfaces may be produced in unsanded, sanded or scraped form. Further processing may produce faces of foil, metal, plastic, phenolic film, paper and hardboard, while prime painted, texture coated and printed faces are also available. Plywood may be impregnated (ideally at the veneer stage) with preservatives and flame retardants (see Chapter 22).

Plywood bond performance depends on the type of adhesive and on the quality of the veneers. Four types of adhesive for plywood manufacture are defined in BS 1203 (1979) *Synthetic resin adhesives (phenolic and amino plastic) for plywood*, while BS 6566 Part 8 (1985) provides the specification for bond performance of veneer. The four broad categories described by adhesive type are:

WBP: *Weather and boil resistant*
These are manufactured from phenol–formaldehyde resins (PF) and are resistant to weather, micro-organisms, cold and boiling water, steam and dry heat; they are also resistant to most common solvents, wood preservatives, flame retardants and most acids. Although the adhesive will withstand severe exposure conditions without deterioration, it is important to ensure that the veneers themselves are resistant to decay. The mistake is often made of specifying 'exterior grade plywood, bonded with WBP adhesives',

without specifying the species of timber that are acceptable. If the face veneers are of birch, or other similar perishable timbers, the plywood is certainly not suitable for exterior use, notwithstanding the fact that the adhesive is weather and boil proof. However, in the case of marine grade plywood, details of which are included in BS 1088 and 4079 (1966) *Specifications for plywood for marine craft*, only those timbers the heartwood of which is sufficiently durable in its natural state to be classified as either 'moderately durable', 'durable', or 'very durable', can be used, for example African mahogany, idigbo, makoré, omu, light and dark red meranti, sapele and utile. Veneers of timbers of lower durability ratings that can be effectively treated with preservatives according to BS 4079 may be used in the core, provided the board is so marked.

CBR: *Cyclic boil resistant*
These are based on melamine–urea formaldehyde resins (MUF) and provide good resistance to weather and the cyclic boiling water test, but fail under prolonged exposure or other demanding conditions. However, they are resistant to attack by micro-organisms.

MR: *Moisture resistant*
These are urea–formaldehyde adhesives (UF) and can survive full exposure to weather for a few years. They can withstand cold water for long periods and hot water for short periods, but cannot withstand boiling water. They are resistant to attack by micro-organisms.

INT: *Interior*
These are extended urea–formaldehyde adhesives (EUF), durable in dry conditions and resistant to cold water. However, they cannot withstand boiling water and may not withstand attack by micro-organisms.

New European standards for plywood are currently being drafted and hopefully will be published in 1996/97. These will replace the current British Standards, but fortunately are not too dissimilar in concept and quality levels. The general specification will be set out in EN 636, with separate parts for the use of plywood in dry, humid and exterior conditions. The primary document supporting this specification will be EN 314 on bond quality; Part 1 contains the test methods, while Part 2 sets out the requirements.

It should be appreciated that the use of synthetic resins does not automatically guarantee good moisture resistance or durability since these resins can be extended using cheap additives, thereby lowering their performance and rendering them suitable only for internal use.

Bond performance must therefore be specified when using plywood and such a requirement is included in many national standards. In the UK, bond performance is specified in Part 8 of the Plywood specification BS 6566.

Part 7 of this same specification sets out a classification of resistance to fungal and insect attack, reference to which must be made in the use of plywood in hazardous conditions. Additional to the choice of plywood offered by type of adhesive, as quantified by its bond performance, species composition in terms of the natural durability of the veneers, and type of lay-up, plywood is also available in various surface appearance grades, defined in terms of the defects present: this is particularly important in those plywoods to be used for wall panelling, furniture and other uses where its aesthetic appeal is the dominant factor. Surface grades are specified in BS 6566: Part 6.

The method of plywood manufacture, using the grain of alternate veneers at right-angles to one another, imparts several valuable attributes to the material. Movement due to changes in relative humidity is reduced to about one-tenth that for solid timber. Taking the extreme range in equilibrium moisture contents of timber in heated buildings as being from 8 to 15 per cent as between winter and summer, this range can be expected to produce a dimensional change in timber of 2.2 per cent and only about 0.22 per cent with plywood. Another important property of plywood is its resistance to splitting, which permits nailing and screwing relatively close to the edges of the boards; this is a reflection of the removal of a line of cleavage along the grain

which is characteristic of solid timber. Impact resistance of plywood is very high and tests have shown that to initiate failure a force greater than the tensile strength of the timber species is required. Plywood is stiffer than many other materials, including mild steel sheet, when compared on a weight-for-weight basis: generally plywood has a high strength-to-weight ratio.

Plywoods tend to fall into three distinct groups. The first comprises those which are capable of being used structurally. Large quantities of softwood structural plywood are still being imported into the UK from North America, supplemented by smaller volumes from Sweden and Finland; the latter country also produced a birch/spruce structural plywood.

The use of this group of structural plywoods is controlled through the specification for bond quality (BS 6566: Part 8); grade-stresses, where required for structural design work, are included for different lay-ups, species combinations and thicknesses in BS 5268: Part 2. Requirements for plywood used as sheathing in timber frame construction are given in BS 5268: Part 6.

The second group of plywoods comprises those which are used for decorative purposes, and mention has already been made of appearance grading according to BS 6566: Part 6.

The third group of plywoods comprises those for general purpose use, which are usually of very varied performance in terms of both bond quality and strength. These are frequently used indoors for infill panels and certain type of furniture.

Guidance on the principal types of plywood available in the UK and selection of plywood for specific applications in construction is given in BRE Digests 323 (1992a) and 394 (1994a), while information on specific plywoods is available from those organisations listed in Appendix III.

17.4.2 *Chipboard (particleboard)*

The term particleboard is used to characterise boards produced from pressing particles of cellulosic material bonded together with some type of resin. While inevitably most particleboard is produced from wood particles, wood is certainly not used exclusively and commercial particleboard is produced from flax shives in Belgium and from various husks in India.

Interest here is centred on wood particleboard which in the UK is usually called **chipboard**. The value of chipboard used in the UK is about two-thirds that of plywood, but on a volume basis it greatly exceeds plywood consumption; current consumption of chipboard is in excess of $3 \times 10^6 \, \text{m}^3$ compared with about $1 \times 10^6 \, \text{m}^3$ of plywood. Commercial production in the UK started in the late-1940s and after a slow start demand began to increase almost exponentially. Over the years this demand has been met partly by UK production and partly by importation: at the present time the UK manufacturers are supplying about 55 per cent of the country's requirements. A considerable proportion of the imported boards has always come from Scandinavia, but Belgium, Spain and Portugal are currently among the principal suppliers: most countries in Europe are on the list of suppliers.

When chipboard production first started, it was regarded as a means of utilising saw-mill waste, and while slabs and offcuts from the mills are still used and now supplemented by chipper canter material, the demand for raw material has far outstripped supply: other sources of waste such as peeler log cores, saw-mill edgings and planer chippings from furniture factories have done little to satisfy increased demand. Large quantities of roundwood are now used both in the UK and abroad, and despite the higher cost, roundwood (thinnings and small diameter trees) is frequently preferred, as it provides the mill with the opportunity to exercise more control over the size and shape of chip produced.

As the quality of chipboard continues to rise to meet particular requirements, so greater control on chip geometry is required, together with a higher degree of sophistication in the lay up and subsequent pressing of the board. Thus, it is normal to find that the rather coarse chips in the core of the board are produced from waste timber, while the thin flat particles in the surface layer originate from roundwood.

The chips after drying are sprayed with resin before being transported by air to the forming station. The average resin content in chipboard is about 8 per cent based on the dry mass of the board, but this represents a mean value of that in the surface layers (9–11 per cent) and that in the core (5–7 per cent). The increase in adhesive content of the surface layers is partly to compensate for the increased surface area of the thinner and smaller chips, and partly to increase the amount of bonding and consequently level of performance of these layers: this manifests itself in terms of higher bending strength and stiffness of the board.

The mat of chips is formed on a stainless steel caul or platen which is then fed batchwise into a press: generally the press can accommodate from six to eighteen platens and when completely filled the press closes, the rate of closure as well as the actual pressure and temperature having a profound influence on both the physical and mechanical properties of the boards. Some of the newer mills have a continuous press in which the board is pressed between two rotating cauls or platens.

Platen-pressed chipboard accounts for about 95-98 per cent of the market; the alternative method is to extrude the matt between parallel steel plates. This produces a board with the chips lying perpendicular to the board surface and, although low in strength and stiffness, it is cheaper and is quite adequate for the core of doors.

Although some platen-pressed chipboards possess a gradual change in chip size and density from the centre of the board to the outside, this profile (graded density) tends to have been superseded by the alternative three-layer profile which is now widely adopted. The skin layers are about 2 mm in thickness and there is a rapid change in density from skin layers to core layer.

The type of adhesive used, the density of the board, the level of strength and the degree of surface smoothness vary with the use to which the board is going to be applied. BS 5669: Part 2 specifies six grades of chipboard in terms of their mechanical properties and reaction to the moisture. These are:

C1   The lowest quality, used only for general purpose work, such as packaging and hoarding.

C1A   A chipboard with slightly improved strength, stiffness, surface soundness and resistance to axial withdrawal of wood screws. It is used mainly in certain types of furniture manufacture.

C2   This has mechanical properties that are an improvement on those of C1, plus a specified level of impact resistance. Its main use is for decking to suspended floors, or as a floating floor, in light industrial applications where there is no risk of wetting.

C3   This has better moisture resistance and mechanical properties than C1 board. Either PF resins or MUF resins, in which at least 40 per cent of the urea has been replaced with melamine, are used to satisfy the specification in terms of resistance to the effects of moisture. It recovers an acceptable level of strength on reconditioning after exposure to water or to high humidity for limited periods. However, while it has increased resistance to moisture, it is not 'moisture resistant', as is often claimed, because its durability is limited to that of the softwood from which the chips were prepared. It is used principally in the manufacture of kitchen work-tops.

C4   This type has the same degree of moisture resistance as grade C3, but also has a specified impact resistance. Its main use is in all types of flooring, especially where some degree of protection from accidental or irregular wetting is required.

C5   A chipboard with much better moisture resistance and mechanical performance than those of all other grades. It is a very high quality board for structural use where there is a risk of wetting during installation and of occasional wetting in service.

The suitability of chipboard for a particular use depends primarily on its long-term performance. Although initial levels of strength and stiffness must be appropriate to the use, it is important that these levels are maintained, particularly where adverse conditions may develop in service. One of the most significant factors affecting the performance of particleboard is moisture. High moisture conditions may be encountered in kitchens, bathrooms and roofs in the form of high relative humidity (which may fluctuate markedly) and actual soaking. The effect of these is accentuated with time and increasing temperature. The reaction of particleboards to such adverse moisture conditions depends on the type of resin used in the board's construction.

Urea–formaldehyde bonded boards, that is C1, C1A and C2, still comprise about 75 per cent of the chipboard used in the UK. When used under dry conditions they give most satisfactory performance, but their principal limitation is their marked reduction of performance in the presence of moisture which brings about deterioration of the resin. Thus, under conditions of high humidity, alternating high and low humidity, or wetting up by liquid water particularly at elevated temperature, urea–formaldehyde chipboard swells appreciably and, most important, these changes are largely irreversible. On drying, the thickness of the board is only slightly less than that in the swollen state (Figure 17.3) and there is considerable loss in all strength properties. These drawbacks can be reduced to a considerable extent with the use of either PF or MUF adhesives as used in grades C3, C4 and C5. The latter adhesive is widely used in the UK, France and Scandinavia and is either a urea–formaldehyde resin in which some of the urea has been substituted by melamine, or a blend of melamine formaldehyde and urea–formaldehyde resins. The principal difference in performance between UF chipboards on the one hand and the MUF and PF boards on the other is that the latter boards are capable of very considerable recovery in both strength and thickness on drying following a period of wetting. Their performance when wet under raised temperature is also vastly superior to that of UF boards (Figure 17.3).

In addition to physical deterioration, high moisture conditions can also encourage fungal degrade. Unless a fungicide is incorporated

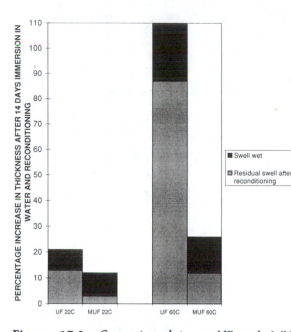

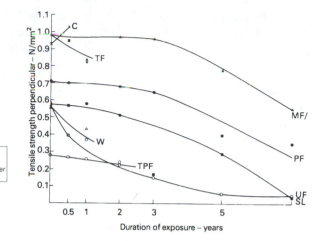

Figure 17.4 *Reduction in tensile strength perpendicular over seven years' natural weathering (Building Research Establishment, © Crown Copyright)*

Figure 17.3 *Comparison between UF and MUF chipboard of the percentage increase in thickness following 14 days' immersion in water at 22°C and 60°C, and the residual increase in thickness following reconditioning at 20°C/65 per cent relative humidity (Building Research Establishment, © Crown Copyright)*

during manufacture, resin-bonded chipboard is usually less durable than the wood species from which it is made.

It should be appreciated that not one of the six grades of boards is classed as an exterior board: in some countries MUF and PF bonded boards, when covered with certain finishes or treated with certain chemicals, are considered in their upgraded form as 'exterior chipboard'. In the UK, however, primarily because of difficulties in ensuring sufficient edge protection or treatment of the boards, no boards, whatever their treatment, are regarded as being suitable for external conditions. External weathering tests of a range of chipboards at the former Princes Risborough Laboratory have indicated not only a rapid fall off in performance of urea–formaldehyde bonded boards, as might be expected, but also a marked deterioration in performance of both the phenol and the melamine/urea–formaldehyde boards after seven year's external weathering (Figure 17.4).

About 40 per cent of the UK consumption of chipboard goes directly into the construction industry where it is widely used as a floor decking either suspended on joists, or floating on top of a concrete slab or a floor formed by beam and block construction.

A large proportion of the remainder is used in furniture manufacture; this includes fitted kitchen and bedroom units and therefore indirectly adds to the volume of chipboard used in construction. In addition to these major uses, there is a whole spectrum of other uses, most of which are included in the list of applications set out in Table 17.1; this also includes the principal relevant British Standards, the most frequently referred to being BS 5669: Part 5. The structure of the new draft European Standard prEN312 is similar to the BS with values of approximate equivalence or slightly lower; it is unlikely that this will replace BS 5669 before mid-1996.

The application of C5, structural grade chipboard by structural design in 'dry' conditions necessitates the use of the relevant grade stresses and time-modification factors set out in BS 5268: Part 2: 1993.

One of the problems in both the manufacture and use of UF and MUF bonded chipboard is the release of formaldehyde gas from the board. This

Table 17.1   *Selection of grades of chipboard for particular end uses (extract from BRE Digest 373)*

Application	Type of board as defined in BS 5669: Part 2	British Standard — application
Flooring — domestic	Type C4	BS 5669: Part 5 — section 4 BS 8201
Flooring — office	Types C, C4, C5	BS 5669: Part 5 — section 5 BS 8201 BS 5268: Part 2 — section 9 BS 6399: Part 1
— heavy duty	Type C5	BS 5669: Part 5 — section 5 BS 8201 BS 5268: Part 2 — section 9 BS 6399: Part 1
— raised	Types C2, C4, C5	BS 5669: Part 5 — section 5 BS 6399: Part 1
Roofing — pitched	Types C3, C4, C5	BS 5669: Part 5 — section 8
— flat	Types C3, C4, C5	BS 5669: Part 5 — section 7 BS 6399: Part 3
Sheathing (timber frame)	Types C3, C4, C5	BS 5669: Part 5 — section 6 BS 5268: Part 6
I Beams Box beams	Type C5	BS 5268: Part 2 — section 9
Industrial shelving	Type C5	BS 5268: Part 2 — section 9
Wall lining partitions and ceilings	Types C2, C3, C4	none
Furniture — normal exposure	Types C1, C1A, C2, C3, C4, C5	BS 5669: Part 5 — section 9
— hazard conditions	Types C3, C4, C5	BS 5669: Part 5 — section 9
Formwork	Types C4, C5	none

occurs in the initial pressing of the board, and can also take place after the board has been installed on site. Emission is primarily a function of the molar ratio of formaldehyde (F) to urea (U) or melamine (M), and great progress has been made over the last decade in the manufacture of adhesives with lower and lower F/U and F/M ratios. Emission is also a function of temperature and relative humidity, and is much more apparent in rooms where there is little or no air-change. The addition of certain chemicals in manufacture and the coating of boards following manufacture do much to reduce emission.

Sensitivity to formaldehyde varies considerably among individuals; in those sensitive to formaldehyde, irritation to the eyes, nose and throat occurs with attendant discomfort. Fortunately, the number of reported cases annually in the UK is exceedingly small, due in part to the formaldehyde levels set in BS 5669, and in part to the frequent number of air-changes that occur in UK buildings.

Guidance in the selection of chipboard and other materials is given in BRE Digests 323 1992a) and 373 (1992b).

Further information on chipboard is available from those organisations listed in Appendix III.

17.4.3 OSB (Oriented strand board)

It had been recognised for some time that the use in board manufacture of larger particles than those used in chipboard could lead to enhanced strength and stiffness. This concept came to fruition several decades ago in North America with the production of **waferboard**. Large, almost square wafers up to 75×75 mm in size and about 0.5 mm in thickness were bonded together using PF resins. Because all fines were excluded, it was possible to manufacture boards with resin amounts equivalent to between 2.0 and 2.5 per cent of the dry mass of the board.

Waferboard has been replaced by **oriented strand board** (OSB) in which the particles are narrower with a width about half their length, and more importantly, these strands are aligned either in each of three layers or only in the outer two layers of the board. In the latter case, the surface and core layers are oriented approximately at right angles to each other in order to produce a board which simulates the structure of plywood. The extent of orientation varies among manufacturers and the ratio of property levels in the machine to cross direction can vary from $1.25:1.00$ to $2.50:1.00$; these figures similar to that of plywood, but very much lower than the degree of anisotropy in timber. North American production in the north uses aspen, and in the south, pine; in France and the UK pine is used.

OSB manufacture is certainly a growth area with 43 mills worldwide producing in excess of $9 \times 10^6 \, \text{m}^3$ per annum. UK production by one mill is $240\,000 \, \text{m}^3$ per annum, a volume which currently exceeds UK demand and a considerable proportion of the production is exported to Europe.

In the UK, the values for two grades of boards are specified in BS 5669: Part 3. The F2 grade is generally stronger and has a much lower thickness swell than the F1 grade: equally as important, the F2 grade has higher moisture resistance, as measured by the residual bond strength following a 2-hour boil, as well as possessing higher durability as quantified by a decay susceptibility index. Values in the new draft European Standard pr EN 300 are similar to those in BS 5669: Part 3. In terms of mechanical properties, OSB is intermediate between C4 chipboard and structural softwood plywood; in terms of thickness swelling in the presence of moisture, OSB is equal to, or greater than that of C4 chipboard.

In North America, OSB has made substantial inroads into the structural plywood market, and that trend is beginning to appear in the UK as the cost of plywood continues to rise and quality to decline. Thus, one of the major attractions of OSB is its greater uniformity in performance compared with plywood.

In the UK, OSB has found wide acceptance as a sheathing material in timber frame construction according to BS 5268: Part 6: section 6.1 OSB is also used as a floor decking on joists where it is beginning to replace chipboard in certain parts of the country. Because of the higher strength and stiffness values of OSB compared with C4 chipboard, it has been possible to obtain the same performance with a 15 mm thick Type F2 (higher grade) OSB as an 18 mm C4 chipboard.

OSB also finds use as a roof decking material, in packaging and in temporary shuttering.

Further information on OSB is given in a BRE Digest (1994b).

17.4.4 Cement bonded particleboard (CBPB)

Cement bonded particleboard established itself in Switzerland and Central Europe in the mid-1970s and has been imported into the UK since the late-1970s. The number of plants worldwide is now over 40, with one in the UK.

Comprising by weight about 70–75 per cent Portland cement and 20–30 per cent wood chips, similar to those used in the manufacture of chipboard, the board is heavy with a density of about $1200 \, \text{kg/m}^3$. However, this amount of cement imparts four very significant bonuses to its performance; (1) with a pH of 11, this high degree of alkalinity ensures immunity to fungal attack; (2) the considerable amount of cement reduces very appreciably the dimensional movement of the wood chips in the presence of moisture; (3) the high cement content results in a panel which, though not totally incombustible, has very good resistance to spread of flame; and (4) the high density of the board ensures poor sound transmission.

In the UK the material is specified in BS 5669: Part 4. Although values for bending strength and impact resistance are very much lower and values for internal bond slightly lower than those for C4 chipboard, the stiffness of the board (modulus of elasticity), its dimensional movement with changing relative humidity, and its resistance to moisture are far superior to those of C4 chipboard. The values in the new draft European Standard pr EN 622-2 are similar to those in BS 5669: Part 4.

CBPB should really be viewed not as a competitor to chipboard, but as complementing its use by extending the application of so-called particleboards into hazard areas where there is risk of fire, moisture, or high noise levels. In the UK, CBPB has been used extensively as a lining for internal walls and partitions on account of its resistance to both fire and sound

transmission. Other uses have included ceilings, fire doors, ventilation and service ducts, soffits, overlayment of old floors in building refurbishment and as a core board for factory-applied finished cladding panels. Its external performance, even in an uncoated form, is far superior to that of chipboard and OSB, with strength values after seven years' exposure above the original unexposed strength (Figure 17.5).

Further information on CBPB is contained in a BRE Information Paper (Dinwoodie and Paxton, 1992); advice can be obtained from the relevant trade association (Appendix III).

17.4.5   Fibreboard

Fibreboard is the generic term for a wide spectrum of board types of differing density and hardness, made from wood in a fibrous state. Their properties, as well as their performance in service, depend mainly on the degree of pressing of the fibrous mat in manufacture, and on the incorporation of chemical additives. Fibreboard can be manufactured from most hardwoods and softwoods, either by the traditional 'wet' process, or by the newer 'dry' process which is becoming more widely used.

The raw material, usually small-diameter logs, but sometimes waste material from timber

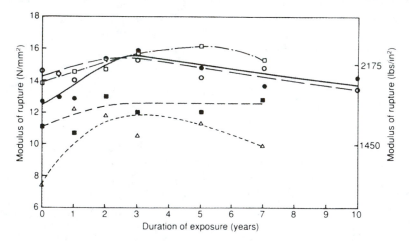

Figure 17.5   *Change in bending strength of unprotected Portland cement bonded particleboards exposed on the BRE weathering fence for up to 10 years. Each point is the mean value of samples taken from one to three panels (Building Research Establishment, © Crown Copyright)*

processing, is first reduced to small chips (25 × 25 × 4 mm); these are then defibrated to bundles of wood fibres, exceptionally to individual fibres, either by mechanical grinding (the 'Defibrator' method), or by the 'Masonite' process in which chips are explosively defibrated by expelling them from a heated pressurised chamber.

In the 'wet' process the defibrated wood is mixed with water to produce a pulp to which is added various chemicals depending on the board type. The wet pulp is then deposited on a moving wire mesh which encourages felting (interlocking) of the fibre bundles to form a sheet. In low density boards this mat is cut to length and allowed to dry: in boards of higher density the cut lengths are pressed at a temperature of 180°C. At these conditions the lignin component of the wood softens and flows, setting again when cool to bond together the fibre bundles; no adhesive is added. Boards produced by the wet process have a characteristic 'mesh' side.

One grade of hardboard, known as tempered hardboard, is produced by impregnating the hot sheets leaving the press with a drying oil to give a retention of about 8 per cent by weight.

In the 'dry' process the fibre bundles are first dried to a low moisture content prior to being sprayed with an adhesive and formed into a mat which is hot-pressed to produce a board with two smooth faces similar to the production of chipboard. Both multi-daylight and continuous presses are employed in the manufacture of MDF which is the principal (but not exclusive) board produced by the dry process.

BS 1142: 1989 defines 14 different types of boards in terms of density and property levels, but they can be classified in four broad groups. These groups and types are set out below:

1. **Hardboards** (density exceeding 800 kg/m³)
   Standard hardboard
   BS 1142 types SHA, SHB and SHC
   > Three types of standard hardboard are available
   > Type SHA is the highest grade, with greatest dimensional stability and strength.

   Tempered hardboard (density exceeding 960 kg/m³)
   BS 1142 types THE and THN
   > Two grades of oil-impregnated hardboard are available, THE having the higher specification.

2. **Mediumboards** (density 350–800 kg/m³)
   High-density mediumboards ('panelboard') – density 560–800 kg/m³
   BS 1142 type HMN – Standard high-density mediumboard
   BS 1142 type HME – High-density mediumboard with improved moisture resistance
   Low-density mediumboards – density 350–560 kg/m³
   BS 1142 type LMN – Standard low-density mediumboard
   BS 1142 type LME – Low-density mediumboard with improved moisture resistance

3. **Softboards** (density less than 350 kg/m³)
   Standard grade, BS 1142 type SBN
   – a standard softboard with no additives
   Impregnated boards, BS 1142 types SBI and SBS – these are bitumen-impregnated boards.
   Type SBS has the higher loading of bitumen and lower water uptake.
   Softboards are sometimes called insulating boards.

4. **Medium-density fibreboard** (MDF)
   BS 1142 recognises two grades of board: MDF and MDFMR, the latter grade having greater moisture resistance. Although not currently covered by the British Standard, the industry has also produced an MDF with further improved moisture resistance, suitable for 'protected' exterior use.

Many fibreboards can be treated with flame retardants. When this is done the boards carry special markings.

BS 1142 gives minimum strength requirement and maximum allowable dimensional movements for each of the fourteen types.

The consumption of fibreboard in the UK is about $0.5 \times 10^6 \, m^3$ of which 60 per cent is MDF, almost all of it home produced; total European production of MDF in 1993 was $2.68 \times 10^6 \, m^3$, the UK being the fifth largest producer. With such a wide range of board types, the range of uses of fibreboards is more extensive than with the other board types; the principal applications are shown in Table 17.2 together with the recommended grade(s) and relevant British Standards for each end use. All fibreboards have a very fine and homogeneous structure, and the medium and high-density boards are further characterised by having exceedingly smooth surfaces.

Table 17.2   *Guidance on the use of fibreboards in construction (extract from BRE IP 12/91 — Rodwell et al., 1991)*

Use	Recommended grades of fibreboard				Relevant British Standards for application
	Hardboard	Mediumboard	Medium-density fibreboard	Softboard	
Architectural mouldings	—	—	MDF MDFMR	—	BS 1186: Part 3
Box and I-beams	THE	—	—	—	BS 5268: Part 2
Ceilings	SHA	LME LMN		SBN	BS 476: Part 7 (Spread of flame)
External claddings	THE THN	HME HMN	—	—	BS 8200
Flat roof insulation	—	—	—	SBI	BS 6229
Formwork	THE THN	—	—	—	—
Furniture	—	—	MDF	—	—
Internal wall lining	SHA SHB	LME LMN	—	SBN	BS 476: Part 7 (Spread of flame)
Movement joint fillers	—	—	—	SBI	—
Overlay to structural floors	SHA SHN SHC THE THN	—	—	SBN* SBI	BS 8201
Roof sarking	THE THN	LME LMN	—	SBI SBS	BS 5534
Sheathing	THE	HME	—	SBI SBS	BS 5268: Part 6
Shop fitting	—	—	MDF MDFMR	—	—
Soffits and facias	THE THN	HME HMN	—	—	—
Staircases	—	—	MDF MDFMR	—	BS 585: Part 2
Windowboards	—	—	MDFMR	—	—

*Timber floors only.

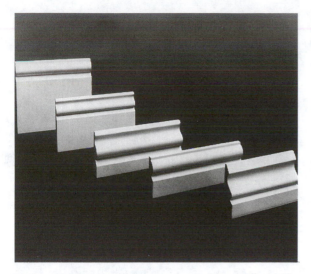

Figure 17.6 *Architraves and mouldings illustrating some of the newer applications for moisture-resistant medium-density fibreboard (Building Research Establishment, © Crown Copyright)*

Although the bulk of production of MDF is used in the manufacture of furniture, a not inconsiderable and increasing amount is being used in construction. Thus, 60 per cent of window boards and 10 per cent of skirtings and architraves in new build housing are MDF (Figure 17.6). It is estimated that there is between 150 and 250 m of skirting and architrave in each house.

The comments made in section 17.4.2 on formaldehyde emission from chipboard are also relevant to MDF, and the levels for free formaldehyde in the boards are set in BS 1142.

The new draft European standard pr EN 622 sets out the specification for the different types and grades of fibreboard: the levels are similar to those in BS 1142 after making allowance for their presentation as 5 percentiles.

Further information on the fibreboards is provided by Rodwell et al (1991) and can also be obtained from the relevant trade association, listed in Appendix III.

17.5 Comparative performance of board materials

With such a diverse range of board types each manufactured in several grades, it is exceedingly difficult to select examples in order to make some form of comparative assessment.

Table 17.3 *Strength properties of timber and board materials*

	Thickness (mm)	Density (kg/m³)	Bending strength (N/mm²)		E (N/mm²)	
			Parallel	Perpendicular	Parallel	Perpendicular
Solid timber − Douglas fir						
Small clear test pieces	20	500	80	2.2	12 700	800
Structural timber, SS grade	100	580	50	—	11 000	—
Plywood						
Douglas fir (three-ply)	4.8	520	73	16	12 090	890
Douglas fir (seven-ply)	19	600	60	33	10 750	3 310
OSB (oriented strand board) Type F1	18	670	28	12.5	4 000	1 500
Chipboard (BS 5669: 1989)						
Type C2	18	720	17		3 000	
Type C5 (structural)	18	740	24		3 750	
Fibreboard (BS 1142: 1989)						
Type SHA (standard hardboard)	3.2	900	45		—	
Type THE (tempered hardboard)	3.2	1 000	54		4 940	
Type MDF (medium density fibreboard)	18	790	30		2 500	

Table 17.4   *Dimensional stability of timber and board materials. Percentage change in dimensions from 65% to 85% relative humidity at 20°C*

	Direction to grain or board length		Thickness (%)
	Parallel	Perpendicular	
Solid timber			
Douglas fir	<0.1	0.8 (R)	1 (T)
Beech	<0.1	1.2 (R)	2.2 (T)
Plywood			
Douglas fir	0.15	0.15	2
OSB (oriented strand board)	0.20	0.20	15
Chipboard (BS 5669: Part 2: 1989)			
Type C2	0.25	0.25	7
Type C5 (structural)	0.20	0.20	4
Cement bonded particleboard			
Type T2 (BS 5669: Part 4: 1989)	0.18	0.18	0.5
Fibreboard (BS 1142: 1989)			
Type SHA (standard hardboard)	0.15	0.15	3.5
Type THE (tempered hardboard)	0.15	0.15	3.5
Type MDF (medium density fibreboard)	0.20	0.20	3

R = transverse radial direction; T = transverse tangential direction.

In general terms, the strength properties of structural softwood plywood are not only considerably higher than all the other board materials, but they are usually similar to or slightly higher than that of softwood timber. Next to plywood are the tempered and normal hardboards, followed by MDF and OSB. Chipboard is lower, but still ahead of the medium fibreboards. Table 17.3 provides some comparative data, but the reader must realise that considerable variability exists around each quoted value.

Comparison of the behaviour of these materials to changes in relative humidity (65 to 85 per cent) are set out in Table 17.4. In terms of thickness change, cement bonded particleboard is far superior to all other boards, while OSB, at the other end of the spectrum, shows very high thickness increase. There is less variation among board materials in terms of longitudinal movement, with plywood and the hardboards having slightly lower movement values.

References

(a)   *British Standards (BSI, London)*

BS 476   *Fire tests on building materials and structures.*
   Part 7: 1989   Method for classification of the surface spread of flame of products.
BS 585   *Wood stairs.*
   Part 1: 1989   Specification for stairs with closed risers for domestic use, including straight and winder flights and quarter and half landings.
   Part 2: 1985   Specification for performance requirements for domestic stairs constructed of wood-based materials.
BS 1088 and 4079: 1966   *Specifications for plywood for marine craft.*
BS 1142: 1989   *Specification for fibre building boards.*
BS 1186   *Timber for workmanship in joinery.*
   Part 1: 1991   Specification for timber.
   Part 3: 1990   Specification for wood trim and its fixing.
BS 1203: 1979   *Synthetic resin adhesives (phenolic and amino plastic) for plywood.*

BS 4978: 1988 *Timber grades for structural use.*

BS 5268 *Structural use of timber.*
Part 2: 1991 Code of Practice for permissible stress design materials and workmanship.
Part 6: 1988 Code of Practice for timber frame walls. Section 6.1: Dwellings not exceeding three storeys.

BS 5534 *Code of Practice for slating and tiling.*
Part 1: 1990 Design.
Part 2: 1986 Design charts for fixing roof slating and tiling against wind uplift.

BS 5669 *Particleboard.*
Part 2: 1989 Specification for wood chipboard.
Part 3: 1993 Specification for oriented strand board.
Part 4: 1989 Specification for cement bonded particleboard.
Part 5: 1989 Code of Practice for the selection and application of particleboards for specific purposes.

BS 6229: 1982 *Code of Practice for flat roofs with continuously supported coverings.*

BS 6399 *Loading for buildings.*
Part 1: 1984 Code of Practice for dead and imposed loads.
Part 3: 1988 Code of Practice for imposed roof loads.

BS 6566 *Plywood*
Part 6: 1985 Specification for limits of defects for the classification of plywood for appearance.
Part 7: 1985 Specification for classification of resistance to fungal attack and wood borer attack.
Part 8: 1985 Specification for bond performance of veneer plywood.

BS 8200: 1985 *Code of Practice for design on non-loadbearing external vertical enclosures of buildings.*

BS 8201: 1987 *Code of Practice for flooring of timber, timber products and wood based panels.*

(b) Draft European Standards (to be published 1996/97)

pr EN 300 *OSB: Definition and specification.*
pr EN 312 *Particleboard: specifications.*
Part 1 General requirements.
Part 2 Requirements for general purpose boards for use in dry conditions.

Part 3 Requirements for boards for interior fitments for use in dry conditions.
Part 4 Requirements for load bearing boards for use in dry conditions.
Part 5 Requirements for load bearing boards for use in humid conditions.
Part 6 Requirements for heavy duty load bearing boards for use in dry conditions.
Part 7 Requirements for heavy duty load bearing boards for use in humid conditions.

pr EN 314 *Plywood — Bonding quality.*
Part 1 Test methods.
Part 2 Requirements.

pr EN 622 *Fibreboards: specifications.*
Part 2 Requirements for hardboards.
Part 3 Requirements for medium boards.
Part 4 Requirements for softboards.
Part 5 Requirements for dry process boards.

pr EN 634 *Cement bonded particleboards: specifications.*
Part 1 General requirements.
Part 2 Requirements for ordinary Portland cement bonded particleboard used under dry, humid and exterior conditions.

pr EN 636 *Plywood requirements.*
Part 1 Plywood for use in dry conditions.
Part 2 Plywood for use in humid conditions.
Part 3 Plywood for use in exterior conditions.

(c) Other references

BRE (1992a) Selecting wood-based panel products. *Building Research Establishment Digest 323.*

BRE (1992b) Wood chipboard. *Building Research Establishment Digest 373.*

BRE (1994a) Plywood *Building Research Establishment Digest 394.*

BRE (1994b) OSB (Oriented Strand Board). *Building Research Establishment Digest 400.*

Dinwoodie J.M. and Paxton B.H. (1992) Cement bonded particleboard. *Building Research Establishment Information Paper IP 14/92.*

Rodwell D.F., Dinwoodie J.M. and Paxton B.H. (1991) Fibre building board types and uses. *Building Research Establishment Information Paper IP 12/91.*

Part 5

TIMBER IN SERVICE
– WHAT HAPPENS TO WOOD IN SERVICE?
– HOW CAN ITS LIFE BE EXTENDED?

18

Degradation of Timber in Service

18.1 Introduction

With the passage of time and the influence of the environment, all materials exhibit a loss in performance, a process which the materials scientist refers to under the all-embracing title of **degradation**. Undesirable changes occur to the material which, at best, are restricted to its surface layers, but which usually permeate the whole of the material and have a most pronounced effect on its mechanical performance, especially in terms of strength and toughness. The rate of the degradation is usually specific to a particular material and to specific environmental conditions.

Some timbers display good resistance to many degrading processes, while others have little resistance. There are six groups of degrading agencies – **biological**, **chemical**, **photochemical**, **thermal**, **fire** and **mechanical** – and these will be discussed in this and successive chapters.

18.2 Biological degradation

Wood is an organic material and therefore it is not too surprising to find that it is subject to attack by a whole host of biological agencies. Because of the significance of biological degradation to the performance of timber in service, and because of the variable and sometimes complex nature of the degradation, it has been decided to describe this form of attack in the following two chapters. Thus, readers are referred to

- Chapter 19, for a comprehensive account of the attack of wood by wood destroying fungi,
- Chapter 20, for a detailed description of the attack of wood by insects,

- Chapter 21, for means to eradicate both fungal and insect attack, and
- Chapter 22, for methods to prevent attack by fungi and insects.

18.3 Chemical degradation

Although significant advances in both plastics technology and metallurgy have resulted in new materials with higher chemical resistance, timber continues to be used in the manufacture of vats, pickling baths and other pieces of chemical equipment because it offers good resistance for the price. Additionally, in those processes where abrasion from suspended solids is inevitable, or where metal contamination such as staining needs to be avoided, timber has long been the preferred material.

The most important timber factors relating to good chemical resistance are impermeability and high density. The former property is largely due to the blocking of the intercellular pits and the cellular structure with natural deposits during heartwood formation (see Chapter 2, section 2.4), and the latter is usually associated with thicker wood cell walls and a lower proportion of voids. Sapwood must be eliminated as it is permeable.

There are many hardwoods of high density which are highly impermeable. Softwoods, because of their lower hemicellulose content, are intrinsically more chemically resistant than hardwoods. They also have the advantage of straighter grain. However, these advantages are generally offset by lower density and a greater degree of permeability.

The best service is obtained from vats and other chemical equipment when they are used for only one continuous operation, thus avoiding the shrinkage and contamination troubles caused by changing usage.

A vat constructed of impermeable timber has but a relatively thin layer of waterlogged wood on its inner surface during use. This swollen layer acts as a barrier to further liquid penetration. Exterior painting of vats should not be encouraged, as any barrier on the outside will raise the moisture content of the whole cross-section to a level which may be sufficient to promote fungal decay in all but extremely durable timbers.

Permeable timbers invariably give a shorter life as they are more prone to fungal decay, chemical attack throughout the wood and corrosion problems with metal fittings. Permeation with strong mineral acids leads to wood embrittlement and strong alkali can cause swelling/shrinkage problems owing to the dissolving of extractives and hemicelluloses.

The acid resistance of impermeable timbers is superior to that of most common metals; iron begins to corrode below pH 5, whereas attack on wood commences below pH 2, and even at lower values, proceeds at a very low rate. Elevated temperature and increased acid concentration will shorten the life of equipment. Wood has excellent resistance against acetic acid which is particularly destructive to most common metals. In alkaline conditions it has good resistance up to pH 11.

Dense, impermeable hardwoods are excellent for use under acidic conditions, but where alkaline solutions or solvents are involved the possibility of the leaching of coloured extractives should be checked. Softwoods are mostly free of coloured extractives and are therefore generally more suited for use with alkaline solutions and solvents, although excessively resinous material should be avoided.

Strong oxidising agents such as chlorine and nitric acid are destructive to wood. Both chlorine and sulphur dioxide attack the lignin component of wood, and since the bonding material between individual fibres is rich in lignin, they can cause defibration of the surface.

Another type of chemical degradation of timber can occur in sea-going wooden boats and is associated with corrosion of the metal fastenings, an effect which is often described as **nail sickness**. This is basically an electrochemical effect, the rate of the reaction being controlled by the availability of oxygen. Areas of different polarity are set up with the salt water, which has permeated the timber, acting as the electrolyte and the wet timber as the conductor. The reader will appreciate that the permeability of the timber is again a very significant factor.

Alkali is produced at the cathodic surface which causes the timber to become soft and spongy, impairing its ability to hold the fastenings, while in the anodic areas ions pass into solution and form with the negative ions of the electrolyte a soluble metallic salt. In the case of iron fastenings, the iron salts so formed cause further degradation of the timber, as noted above, resulting in considerable loss in strength and marked black staining of the timber.

Although as noted earlier, timber is remarkably resistant to acid, degradation can and does occur under conditions of very high acidity or elevated temperatures. An example of the latter is known to occur at times in the treatment of timber with the traditional type of flame-retardant solutions. These contain mono- and di-ammonium phosphates in aqueous solutions and, when the treated wood is rekilned to dry it, ammonia is liberated and the phosphates combine with the water to form phosphorous acid. This causes acid hydrolysis of the wood, destroying not only the middle lamellae, but also breaking the covalent bonds between the glucose units in the cellulose molecule. Strength and especially toughness are reduced and the timber fails with a brittle fracture. The propensity for this to occur can be reduced considerably by reducing the temperature at which the timber is dried following impregnation. It should be noted that the newer organic-based flame retardants appear to result in very little degradation of the timber.

18.4   Photochemical degradation

On exposure to sunlight, the colouration of the heartwood of most timbers, for example mahogany, oak, afrormosia, iroko and teak will lighten;

although a few, for example, Douglas fir and Rhodesian teak, will actually darken. Indoors, the action of sunlight will be low and the process will take several years before a noticeable change occurs. However, outdoors the change in colour is very rapid and noticeable change occurs in a matter of months.

Such colour changes are generally regarded as an initial and very transient stage in the whole process of the weathering of timber in which the action of not only light energy (photochemical degradation), but also rain and wind, results in a complex degrading mechanism. The weathering process, therefore, results in a loss of surface integrity of the timber which can be quantified in terms of the residual tensile strength of thin strips of the wood (Derbyshire and Miller, 1981). This loss in integrity embraces the degradation of both the lignin, primarily by the action of ultra-violet light, and the cellulose by shortening of the chain length, primarily by the action of energy from the visible part of the spectrum. The subsequent removal of the breakdown products by rain impacts a silvery-grey colour to the surface which consists of a thin layer of loosely matted fibres between a considerable number of small splits.

The rate of degradation from weathering agencies in the absence of biological attack is usually slow as the surface cells which are degraded serve to protect those at a greater depth; however, weathering frequently acts as a precursor for fungal attack.

The rate of photochemical degradation and weathering of exterior timber can be slowed down very appreciably by the careful selection and application of finishes which are described in detail in Chapter 23. Suffice it here to say that those surface layers showing loss in integrity must be physically removed prior to the application of any surface coating, otherwise this will not adhere to the wood.

18.5   Thermal degradation

It has already been mentioned in Chapter 10, section 10.1.5 how timber when subjected to

prolonged exposure to elevated temperatures results in a reduction in strength and a very marked reduction in toughness or impact resistance. There is some uncertainty as to the minimum temperature at which thermal degrade commences, but it would appear that temperatures as low as 60°C can induce degrade over very many years. The rate of degrade will rise markedly with increase in temperature and time of exposure: hardwoods appear to be more susceptible to thermal degrade than softwoods.

Considerable degrade occurs at temperatures below 100°C over a number of years, and at temperatures above 100°C in a number of months. Thus, tests on three softwood species subjected to daily cycles of 20°C to 90°C for a period of three years resulted in a reduction in toughness to only 44 per cent of its value on samples exposed for only one day (Moore, 1984).

Thermal degradation results in a characteristic browning of the wood with associated development of a caramel-like odour, indicative of burnt sugar. Initially this is due to degrade of the hemicelluloses, but with time the cellulose is also affected with a reduction in the degree of polymerisation of the cellulose molecule through scission of the $\beta1-4$ linkage. Thermally degraded wood breaks easily with a brittle type failure. Many examples can be found where, for example, timber has been in contact with hot pipes: one classic example, in the House of Lords, is described in detail by Dinwoodie (1989).

18.6   Degradation by fire

As an organic material, and in its dry state, wood is combustible. However, the thermal energy necessary to initiate combustion and to sustain it is quite considerable. Timber has the ability to form a **char** layer which reduces the rate of burning by acting as an insulation layer against both radiant and conducted heat. When the layer of char is sufficiently thick, burning may become so slow that insufficient heat is produced to maintain decomposition and the fire goes out: there are many examples where charred beams and props are still functioning as load-bearing

19

Decay and Sap-stain Fungi

19.1 General principles

Most forms of decay and sap-stain in timber are caused by fungi that feed either on the wall tissue or cell contents of woody plants. It is important to distinguish between wood-rotting fungi, responsible for **decay** in timber, and those that feed on the cell contents, causing **stains**. The former consume constituents of the cell wall, and lead to the disintegration of woody tissue, whereas the latter remove only stored plant food material in the cell cavities, leaving the cellular structure intact. Wood-rotting fungi seriously weaken timber, ultimately rendering it valueless, whereas sap-stain fungi spoil the appearance of wood, but do not affect most strength properties. Sap-stain is not a preliminary stage of decay, but such stained timber, exposed to suitable conditions, may later be attacked by wood-rotting fungi.

Wood-rotting and sap-stain fungi belong to a large group of organisms which includes edible mushrooms and toadstools. The visible mushroom or toadstool is the **fruit-body** of the fungus, the vegetative parts of the plant being out of sight, in the feeding medium. The fruit bodies of wood-rotting fungi are frequently flat, fleshy or woody plates, the undersides of which bear **spores**. The destructive part of a fungus is its vegetative system, or **mycelium**, made up of numerous exceedingly fine tubes called **hyphae**; these may become matted together to form a felt-like mass. Hyphae grow by elongating at their tips, passing from cell to cell of the host plant, feeding on the walls or cell contents in their path. The complete life cycle of a fungus is, therefore: (1) spore; (2) hyphae; (3) mycelium; (4) fruit body; and (5) spore.

All fungi feed on organic material of either plant or animal origin. Those of interest to the user of timber attack either living (often unhealthy) trees, or felled timber. One condition essential to the development of all fungi is the presence of sufficient moisture: initial infection will not occur in any timber below 20 per cent moisture content. Moreover, reduction of moisture below the critical minimum causes all fungi to cease growing. When growing vigorously, some fungi are, however, capable of extending their attack to adjacent timber near the critical moisture content; they will not ordinarily initiate attack unless the moisture content is about the fibre saturation point, and many fungi require appreciably higher moisture contents to initiate attack. Some fungi transport moisture from an outside source, or produce water from the breakdown of cell-wall substance in their path, enabling them to raise the moisture content of wood just below the 20 per cent level to this level. Wood-rotting fungi may survive infected wood for several years after drying.

The conditions essential to fungal growth are four in number: (1) food supplies; (2) adequate moisture; (3) suitable temperature; and (4) air (oxygen) supplies. In most circumstances, growth of fungi in wood under service conditions is dependent solely on the presence of adequate moisture. The wood itself, or the cell contents of the sapwood of certain species, constitutes the necessary food supply, and oxygen is always available under service conditions, except under completely waterlogged conditions. The heartwood of many timbers, however, contain extractives (see Chapter 2) that are poisonous to fungi. These substances render the wood less suitable as food, thus explaining why some timbers are 'naturally durable'; that is, they are resistant to attack, but not immune. Practically all fungal growth ceases at or below freezing point (0°C), and is very slow at temperatures below

5°C. Hence, in temperate regions with really cold winters, fungi may be quiescent in the winter months, and temperatures are a limiting factor in polar regions; optimum temperature conditions vary with different fungi, but generally are in the region 19–30°C.

The condition commonly accepted as decay is the final stage of attack. Beyond the decayed area and in the early stages of infection, a state of **incipient decay** exists: in the timber trade this is often referred to as **dote**. Even when only slight colour changes or softening may be detectable, strength properties, especially toughness, are usually reduced in the infected areas. Sterilisation at this stage will arrest decay, when provided attack was very slight, such timber is likely to be as good as sound stock, and it will be perfectly safe after sterilisation as far as the risk of spreading infection is concerned; for example, kiln-drying using initial temperatures adequate to ensure sterilisation is advisable for any green packaged timber from Western Canada that is intended for external joinery usage, with its attendant risks of re-wetting and continued spread of any infection that was present in the wood at the time of importation.

Kiln sterilisation is effected by raising the humidity and, to ensure a sterilising temperature of 65°C throughout the wood, a kiln temperature of 70°C is maintained for a period dependent on the thickness of the wood: 1 hour for 26 mm and 5 hours for 100 mm thick material.

19.2 Wood-rotting fungi

Some fungi attack primarily the heartwood of standing trees while others chiefly colonise and decay logs after felling, or sawn timber during the process of seasoning. Yet others, comprising the group of greatest economic importance, attack timber after it has gone into service. Though there are no hard and fast distinctions, there is a tendency for each group to consist of different species of fungi.

Actual decay in timber may be detected by the abnormal colour of the wood, by the transverse fractures of the fibres on longitudinal sawn faces,

and by lifting the fibres with the point of a penknife, when, if the timber is decayed, the fibres will break off short instead of pulling out in long splinters.

It has already been stated that the two main constituents of wood substances are cellulose and lignin. The **brown rots** feed mainly on the cellulose, while the **white rots** feed both on cellulose and lignin, but to a varying extent, depending on the particular fungus. The different wood-rotting fungi can be further subdivided, according to the form decay takes, into cubical, spongy, pocket, stringy rots, and so on. The terms **dry rot** and **wet rot** are misleading: a moisture content of about 20 per cent can be regarded as a critical minimum moisture content for the growth of all fungi. Most fungi require the moisture content of wood to be between 35 and 50 per cent for optimum growth, and some appear to prefer still wetter conditions. Precise figures for different fungi are virtually impossible to arrive at; it is known, however, that the minimum moisture content required for spores to germinate is higher than the figure for infection of wood adjacent to actively growing mycelium, or for fungi already present to continue growing.

Severely decayed timber, particularly when inspected after a perhaps temporary period of drying, may appear dry and friable. The term 'dry rot' is often mistakenly applied to such fungal attack. In the UK and Europe in general, the term **dry rot** is used to denote attack by one particular fungus, *Serpula lacrymans*, the 'dry-rot fungus'. This fungus causes decay of the brown rot type but occurs exclusively in buildings. It is of particular importance as it possesses an unusually high tolerance of alkaline conditions which allows it to spread away from decaying timber through lime mortars and plasters. This capability often results in rapid spread of the fungus in damp buildings and can result in difficulties in remedial treatment where masonry cannot be rapidly dried. (Dry rot is dealt with in more detail in section 19.2.3). The term **wet rot** is used to refer to the damage caused by all other wood-rotting fungi whether of the brown rot or white rot type. The term wet rot, however, is sometimes mistakenly applied to the slow disintegration of wood exposed to the weather, that is **weathering**, which is quite independent of fungal attack and is the result of superficial degradation caused by exposure to ultra-violet light together with repeated wetting and drying which leads to the formation of surface checks and their progressive development. Cross-checking may develop at a later stage as a result of secondary superficial fungal attack.

A number of fungi cause **pocket rots** either in the standing tree or in felled timber before completion of seasoning. The decay takes the form of small localised areas of affected wood with apparently sound wood surrounding them. The decay within the pocket may be of either the brown rot or white rot type.

A further entirely distinct group of fungi cause **soft rot** which is described more fully in section 19.2.3 (Fungi that attack wood in service).

19.2.1   *Standing tree fungi*

Fungi that attack standing trees are responsible for losses, principally to owners of the forest; they need not be a problem to the consumer of wood. Once seasoned, wood is usually safe from development of further decay. The principal fungi attacking standing trees are discussed below.

Phellinus pini (Trametes pini)
This fungus causes a white pocket rot in the heartwood of standing conifers. In Douglas fir the rot pockets are small, whitish, elliptical areas, 3–12 mm long with sound wood between them (Figure 19.1); in later stages the elliptical areas may develop into actual cavities. The fungus causes similar rots in Western hemlock and Baltic redwood. A number of other fungi also cause white pocket rots in both hardwoods and softwoods. Though in most cases the rot does not spread after conversion of the timber, affected wood is regarded as unsuitable for most purposes.

Laetiporus sulphureus (Polyporus sulphureus)
This brown rot causes heartwood decay in standing oak and other hardwoods, as well as in a few

Figure 19.1   *White pocket rot in Douglas fir (Building Research Establishment, © Crown Copyright)*

softwoods. In the later stages of decay, affected wood turns deep red-brown and breaks into large cubical portions: the checks in the wood often contain fungus in thin sheets which closely resemble a washleather. In the early stages the decay is difficult to detect and continues to develop if infected wood is used in situations, such as for boat construction, in which it remains wet in service. This fungus is the cause of the most extensive decay in the oak of *HMS Victory*.

19.2.2   *Log and timber fungi*

Delay between felling of the trees in the forest and extraction to the saw mill for conversion and drying may lead to the infection of logs by decay fungi. The problem is accentuated in tropical countries as both the development of fungi and the spread of infection by either bark-boring beetles or the ambrosia beetles is much more rapid under these conditions.

When converted timber is air-dried, periods of unfavourable drying weather can lead to infection. Such troubles are more likely to occur in those timbers, either softwoods or hardwoods, which do not have heartwood which is naturally resistant to fungal attack. Especially with the advent of 'packaging' of sawn timber for transportation prior to completion of seasoning, it is common practice to provide temporary protection of these timbers by the superficial application of a fungicidal solution as the boards leave the saw.

A number of fungi are responsible for attack of wood in the log or following conversion: the following are the more important ones.

Poria placenta (Poria monticola)
This brown rot fungus is a frequent cause of rot in softwoods imported from North America. In the early stages, decay takes the form of small, spindle-shaped pockets of soft, discoloured, dark-brown wood (Figure 19.2); when the wood has had time to dry out, attack of this nature is often more apparent since shrinkage cracks develop in the affected zones. Timber containing such visible decay will have lost mechanical strength and should obviously not be used.

Antrodia serialis (Trametes serialis)
This fungus causes a brown pocket rot in imported European spruce very similar to that caused by *Poria placenta*, and indeed rot caused by that fungus was originally wrongly ascribed to A. *serialis*.

From time to time various fungi, often unidentified, have been responsible for incipient decay in imported non-durable hardwoods. The decay often appears as a white flecking on the surface in small oval areas that can be removed by planing; the sound wood beneath is then acceptable for use. It is more troublesome when the decay occurs below the superficial layer of sound wood protected by anti-stain treatment. In a more advanced stage, attacked timber is worthless. Attack of this nature is sometimes accompanied by the formation of fine black lines which run irregularly through the wood (Figure 19.3). Even when these 'zone lines' or 'pencil lines' are the only visible form of attack, the wood has probably suffered loss of toughness and should

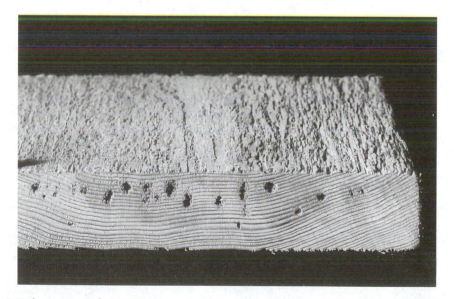

Figure 19.2 *Initial stages of brown pocket rot in imported softwood* (Poria placenta) *(Building Research Establishment,* © *Crown Copyright)*

Figure 19.3   *Zone lines in Beech caused by incipient rot (Building Research Establishment, © Crown Copyright)*

not be used for any purpose in which shock resistance is important.

There is nothing that the importer or ultimate consumer of wood can do to combat these fungi, other than to arrest development of such infection by either kiln-drying or proper stacking of green timber on receipt; it is presumed that timber identified as showing unacceptable levels of deterioration will be destroyed. Provided these two points are observed, decay will not be a problem in timber yards.

19.2.3   Fungi that attack wood in service

By far the most important fungus destructive to wood in buildings is the 'dry-rot fungus', *Serpula lacrymans*. Because of the necessity for adopting

appropriate remedial measures promptly, if *Serpula lacrymans* is the causal agent of decay, it is important for all who are responsible for the care of buildings to be able to identify the fungus correctly. It must not be confused with the 'cellar' fungus, *Coniophora puteana*, which, while capable of severe localised decay, cannot spread through masonry and is therefore not usually so serious a problem. Nor should *Serpula lacrymans* be confused with mould growths that frequently appear as tiny green or black tufts on damp wood. Such moulds do not cause decay, but their presence is indicative of damp conditions, favourable to the attack of wood-rotting fungi.

The most effective method of preventing wood-rotting fungi from attacking interior woodwork is to use sound, seasoned timber, free from fungal infection in the first place, and to ensure that timber will not be exposed to damp conditions subsequently. Timbers must not be built into damp or potentially damp brickwork, masonry or concrete; where timbers need to bear on, for example, damp brickwork, they should be isolated by damp-proof courses or membranes. Constructions which place timbers in unventilated situations, where condensation or minor leakage or penetration of rain or groundwater could accumulate, should be avoided. In particular, suspended timber floors should be provided with efficient cross-ventilation. In older buildings where ventilation of grouped floors may be less generous, installation of floor coverings which restrict moisture vapour loss from the floor surface (such as vinyl or rubber-backed carpet) have resulted in problems of decay in floor timbers. If timber is to be laid directly onto concrete, it should be laid in a bituminous mastic. This is standard practice with woodblock floors, but strip floors are frequently nailed to fillets let into an upper concrete screed above a damp-proof membrane, or alternatively the damp-proof membrane may be a bituminous layer laid above the fillets; in either case the fillets must be preservative treated under pressure.

Once rot makes its appearance, it is imperative to take active measures immediately. It is of utmost importance that, even before eradicating the decay, the building defects which permitted entry of the water which caused the decay should be sought for and remedied, and simultaneously measures should be taken to dry out the moisture that has already entered. Ideally, all decayed timber, together with any fungal growths, should be cut out and removed from the building immediately. Where drying of affected timbers can be achieved rapidly, it may be permissible to retain timbers showing evidence of slight decay but only after specialist advice has been sought and appropriate *in situ* treatment with wood preservative specified. Those timbers which do not show evidence of decay but seem likely to remain damp for more than a few weeks may also require localised *in situ* application of wood preservatives.

Replacement timbers must be treated with wood preservative; the use of timber pre-treated by pressure or double vacuum impregnation is to be preferred.

Wood in contact with the ground, exposed to the weather, or otherwise exposed to persistent dampness, as in coal mines, must be recognised as potentially having a relatively short life. It is then a simple matter of economics to decide whether the use of more resistant timbers is the correct answer, or whether the use of wood preservatives is the better solution. If the latter course is adopted, then adequate treatments are essential (see Chapter 22). In general, when the decay hazard is unavoidable, adequately pressure-treated non-durable timbers are usually more economical, and give a longer service life, than the most durable timbers untreated.

The descriptions of the different fungi, given in the notes on different species, have been culled from publications by the Building Research Establishment, to which reference should be made for more detailed information.

Serpula lacrymans (Merulius lacrymans)
The decay caused by the dry rot fungus, *Serpula lacrymans*, is a brown cubical rot. The appearance, both of the fungus and of infected wood, depends on the stage attack has reached, and on the growth conditions of the fungus. In conditions of high humidity, the fungus develops as white, fluffy, cotton-wool-like masses spreading over the surface of attacked wood (Figure 19.4). In drier conditions, the mycelium forms a grey-white felt over the wood, usually with small

Figure 19.4   *Dry-rot mycelium which has developed between the flooring timbers and the lathe and plaster ceiling, now removed (Building Research Establishment, © Crown Copyright)*

Figure 19.5   Serpula lacrymans: *portion of a decayed joist showing two fruit-bodies, mycelium, and deep cracks along and across the grain (Building Research Establishment, © Crown Copyright)*

Figure 19.6 *(Top) Typical cubical breakdown following* Serpula *attack; (bottom) similar breakdown in a piece of timber caused by* Coniophora *(Building Research Establishment, © Crown Copyright)*

patches of bright yellow or lilac. The mycelium filaments (or hyphae), though individually microscopic, can penetrate the mortar of a brick wall and cross concrete and steelwork to reach as yet uninfected wood. Behind the advancing fungus, hyphal strands develop varying in thickness from coarse threads to strands as thick as a pencil. The strands contain conducting hyphae which enable the fungus to grow over inert materials despite increasing distance from their food base. The fruit-bodies are soft, fleshy plates, with white margins (Figure 19.5). Numerous folds or shallow pores occur on the surface of a fruit-body and contain the rust-red spores. These spores are microscopic, and so light that they are easily blown about; they are sometimes produced in such quantities that a whole room may be covered with a rust-red layer of spores. The fruit-bodies sometimes grow vertically, in the form of a thick bracket, when the pore-bearing surfaces become elongated like small stalactites. Water

may be exuded in drops by the fruit-bodies, hence the name *lacrymans* or weeping.

The fruit-bodies, which grow out into the air and light, are frequently the first indication of dry rot in a building. Slight waviness on the surface of panelling, or the sinking of a floor, may be the first warning of extensive damage. Infected wood is soft when tested with the blade of a pen-knife, and it will not 'ring' when struck. Wood beneath a coating of mycelium may be wet and slimy to the touch, but in the final stages of attack it is dry and friable, brown in colour, and breaks up into cube-shaped pieces (Figure 19.6).

Fibroporia vaillantii (Poria vaillantii)
This is a brown cubical rot, responsible for decay in buildings, but also common on timbers in coal mines. The final stages of attack are

similar in their effect on wood to the action of *Serpula*. Unlike *Serpula*, the hyphae and mycelium remain white and soft; the fruit-body is plate-shaped, covered with fine pores, and also white.

Coniophora puteana (Coniophora cerebella)

The 'cellar' fungus is also a cubical brown rot, but sometimes the decayed wood develops longitudinal splits or cracks (see Figure 19.7). Cubical breakdown, virtually indistinguishable from *Serpula* attack, may occur (see Figure 19.6). Ultimately the wood is extremely brittle, and can be powdered in the fingers. Any strands which form are always fine; they rapidly turn brown or almost black. The fruit-bodies, which are rarely seen, are thin plates, olive-green in colour. *Coniophora* favours decidedly wet conditions; it is very liable to occur where there is persistent water leakage or condensation.

Paxillus panuoides

This is a brown rot generally associated with softwoods in very moist conditions. The strands are paler than those of *Coniophora*, the mycelium is rather fibrous, and yellow or violet. The fruit-bodies, which are often bell-shaped, are dingy-yellow with deep gills on the under surface.

Donkioporia expansa (Phellinus megaloporus)

This is a white rot, usually found attacking oak timbers and occasionally on associated softwood; it is often found where there has been persistent water leakage, for example in the ends of beams embedded in damp walls. The hyphae are white and fibrous; the fruit-bodies are thick, leathery plates or brackets, buff-coloured, with darker brown pores. In the final stages of attack, the wood is reduced to a soft, white mass.

Lentinus lepideus

This is a brown cubical rot, requiring moist conditions. Mainly it attacks timber out-of-doors, for example, telegraph poles, railway sleepers and paving blocks. The fruit-body is a brown, woody mushroom. The fungus and decayed wood have a characteristic aromatic odour. Cartwright and

Figure 19.7 Coniophora puteana: *portion of a decayed joist showing dark strands of mycelium, and longitudinal cracks in the wood (Building Research Establishment, © Crown Copyright)*

Findlay (1958) record that this fungus 'occurs quite frequently on worked timber which has been imperfectly creosoted'.

Amyloporia xantha (Poria xantha)

This is a brown cubical rot, requiring moist, warm conditions; it is commonly found in greenhouses. The fruit-body is a thin plate with small pores, yellow in colour and, when fresh, smelling of lemons.

Phellinus contiguus (Poria contigua)

This is a white rot which has caused much decay in external joinery, especially in softwood window frames, of houses built since the war. Decayed wood finally breaks into fibrous strings; the soft, woolly, light-brown fungus mycelium may develop in voids in joints and hard, dark-brown fruit-bodies sometimes appear, often along the lower sides of the affected members. The National House Builders' Council and many local authorities now demand preservative treatment of exterior doors and windows to combat this and other joinery decay fungi. Oddly, preservative treatment of non-durable softwoods used as external cladding, a situation usually presenting less decay risk than framed external joinery, has for long been mandatory under the Building Regulations.

The types of decay distinguished above are all caused by the higher fungi, the wood-rotting *basidiomycetes*. There has been an increasing awareness of the significance of decay caused by lower fungi, the wood-rotting microfungi (for example, *Chaetomium globosum*). It was first observed in the wooden slats used to fill water cooling towers. There, and in other situations under water, the wood surface darkens and this superficial layer, up to 3–4 mm deep, becomes very soft. For this reason the name **soft rot** was applied to this type of decay.

Wood from such underwater situations is usually sound immediately below the decayed surfaces; when allowed to dry out the softened surfaces harden and shrink to produce numerous superficial cross-cracks. While not of much importance in buildings, soft rot occurs notably in timber in ground contact such as fence posts or transmission poles; it may then penetrate much deeper from the wood surface and, as it then does not shrink and cross-crack upon drying, may be difficult to detect before the wood fails under stress.

Preservative treatment notably with copper/chromium/arsenic (CCA) formulations has given good protection for all softwood species for many years; records show that CCA has given successful protection of softwood in water cooling towers over the past thirty-five years. However, CCA has been found to be less effective for some hardwood species. Creosote applied in a high-vacuum pressure treatment has been shown to give effective protection to softwood poles and fence posts.

Mention should also be made of two fungi, unimportant in themselves, which are nevertheless an indication that dangerously damp conditions exist in buildings. Elf cups (*Peziza* sp), which are pale-brown cups about 25 mm in diameter, not infrequently develop on plaster ceilings following flooding from defective plumbing or frost damage. Unless steps are taken to secure rapid drying out of the affected areas, there is a risk of subsequent 'wet rot' or 'dry rot' infection in timbers in the wet areas. Another fungus that should similarly be regarded as a warning that dangerously wet conditions have become established is a species of ink cap (*Coprinus* sp). This fungus produces small soft toadstools that dissolve into an inky fluid. Fruit-bodies of this species often appear on damp walls, but they may also appear on the underside of ceilings saturated by persistent plumbing leaks or defective internal gutters.

19.3   Sap-stain fungi

Sap-stain or **bluestain** in timber is caused by several species of fungi. These fungi are distinct from those that cause decay; hence, 'bluestain' is not an incipient stage of decay, but its presence may be an indication of conditions favourable for the attack of wood-rotting fungi. Moreover, badly blued timber should be suspected of possibly also containing incipient decay (or dote).

All staining of wood is not necessarily 'bluestain'. Green timber rich in tannin that comes in contact with iron, as in sawing, may become stained blue-black. This is the result of chemical actions, and the stained areas are usually superficial and easily planed off. Similarly coloured stains, caused by sap-stain fungi, penetrate wood deeply and rapidly (Figure 19.8).

Some wood-rotting fungi cause discoloration, but staining from this cause is accompanied by softening of the wood, whereas blue-stain fungi have little or no effect on strength properties, other than reducing resistance to impact bending, sometimes by as much as 40 per cent, which is of material importance in timber for tool handles and sports goods. On the other

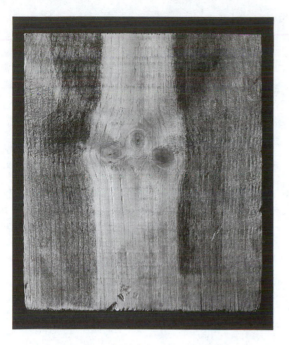

Figure 19.8 *A board of Scots pine showing discoloration caused by sap-stain fungi (Building Research Establishment, © Crown Copyright)*

The discoloration caused by blue-stain fungi is not a stain in the true sense of the word: it is the presence of numerous dark-coloured hyphae in the translucent cells of the wood that produces a tinting visible on the surface. The fruit-bodies of *Ceratocystis* species are small, black flask-shaped **perithecia**, as large as a pin head, often with long necks, and containing numerous spores.

As with wood-rotting fungi, four conditions are necessary for these fungi to grow: (1) sufficient moisture; (2) food supplies, in the form of starch and sugars stored in the cell cavities, but not the wood-substance of which the walls are composed; (3) suitable temperatures; and (4) oxygen (obtained from the air). The right type of food material in sufficient quantities is a limiting factor: in most softwoods these requirements are only found in the sapwood. The presence of bark saturated with moisture also inhibits growths by want of air. Relatively high temperatures are necessary for active growth: the optimum is between 21°C and 26°C; below 10°C, growth is very slow. In temperate regions the most favourable temperature conditions occur in the summer months. Reduction in moisture content in the surface layers of wood can rapidly become a limiting factor: the fungi require moisture contents above the fibre saturation point of wood to initiate attack.

Susceptible timbers will often become infected by sap-stain in the forest after felling, either from the ends of logs or through places where the bark has been removed or damaged in felling. Especially with valuable tropical hardwood logs of stain-susceptible species, spraying with mixtures of insecticides and fungicides coupled with the use of end coatings are acceptable, temporary measures, which should be regarded only as auxiliary to rapid extraction from the forest, followed by immediate conversion at the mill. Suitable proprietary end-coating materials are available although creosote or tar may be used. Any areas where the bark has been removed should also be dressed with the material used for end coating.

Unless infection has occurred in the forest, rapid reduction of surface moisture in converted timber is the simplest method of inhibiting

hand, 'bluestain' is responsible for degrading large quantities of susceptible timbers, because their value is reduced for decorative purposes, and, if heavily stained, they may be unsatisfactory for paint finishes.

Beside the fungi responsible for 'bluestain', there are several other fungi that stain wood green, pink, purple and, more rarely, brown; the majority of these produce a powdery or downy growth of mould that is easily brushed or planed off.

'Bluestain' in softwoods is caused by several species including members of the genus *Ceratocystis*; attack is confined to the sapwood. 'Bluestain' in the light-weight, light-colour tropical hardwoods, such as obeche and ramin, is commonly the result of *Diplodia* infection; attack is not confined to the sapwood, but may extend right through a log. Several fungi attack the light-coloured temperate hardwoods, not necessarily confining their attack to the sapwood. Ash and poplar are liable to be discoloured a dark brown, and oak a pale yellow.

growths. Kiln-drying immediately after conversion is the surest safeguard. Piling in properly built stacks, with stickers of maximum thickness, does not always ensure sufficiently rapid drying to prevent staining of particularly susceptible timbers. With these timbers the use of chemical dips is standard commercial practice in parts of Europe and North America (see Chapter 21).

With the increasing tendency to apply clear finishes and exterior wood stains to external joinery instead of paint, new problems of superficial 'bluestain' of timber service have arisen. They are being countered by the development of new persistent fungicides for application before the clear finishes or for incorporation in the exterior wood stain.

References

Bravery A.F., Berry R.W., Carey J.K. and Cooper D.E. (1987) Recognising wood rot and insect damage in buildings. *Building Research Establishment Report*. HMSO: available only from the Building Research Establishment Bookshop.

Cartwright K.St.G. and Findlay W.P.K. (1958) *The Decay of Timber and its Prevention*. HMSO, London.

20

Worm in Timber

The damage referred to as **worm** in timber is the result of insect activity but, in salt water, teredo or ship worm and a wood-louse-like animal belonging to the crustacean family are responsible for damage of this type. Insects tunnel in timber, spoiling the appearance of exposed faces and, if the tunnels are numerous, they may so reduce strength properties as to make the wood valueless. Some insects only attack living trees or newly felled logs, some only seasoned wood, and others only the sapwood of certain species. In consequence, the presence of insect damage is not in itself necessarily a cause for alarm: the damage may be of the first type and therefore of no consequence in seasoned timber, beyond the disfigurement caused. Moreover, some insects and crustaceans commonly associated with timber are of no importance because they do not attack it. For example, the land form of wood-louse is to be found under any piece of wood that has been left in contact with the ground, in sheds, or in the open for any length of time: these are the small oval-shaped crustaceans that roll up into balls when touched. Although probably the most familiar creature associated with timber, wood-lice are of no practical importance, as they do not attack sound wood. At most, they are an indication that storage conditions are not good, and may lead to infection by wood-rotting fungi. On the other hand, dangerous pests are often overlooked because their insignificant appearance results in their escaping notice.

By far the most important of the insect types causing deterioration of timber are the beetles. A very large number of species are found attacking timber in its various conditions; as standing trees, felled logs, seasoned timber or when softened by fungal decay. Each beetle type is usually associated with only one timber condition; for example, those that attack freshly felled logs cannot normally attack seasoned timber.

The typical **life cycle** of a beetle is very similar to that of butterflies and moths. The adult beetles lay eggs from which hatch **larvae** that feed and grow and eventually enter a resting stage, the **pupae**. From the pupa emerges the new generation of adult beetles.

Although with most beetle types it is the larva which normally damages the wood by its tunnelling, in some species the adults may also feed to a limited extent. In some types which attack standing trees or green logs, the adult only may carry out the tunnelling and the larvae feed only on mould fungi growing on the tunnel linings. The length of the life cycle may also vary between species; some, particularly those feeding on green timber, may complete their life cycle in a matter of months, whereas others may take several years.

This range of timber condition requirements and life cycle length makes it important that insect damage is correctly identified. Incorrect diagnosis can lead to inappropriate or unnecessary remedial treatments.

The different types of insects, the means of their identification and the methods of control are discussed fully in various publications issued by the Building Research Establishment, the Forestry Commission and other Government organisations. Below, the groups of insects of importance to timber users are briefly described.

20.1 Forest and mill-yard pests

20.1.1 Longhorn beetles

These beetles belong to the family *Cerambycidae* and lay their eggs in crevices, or just under the bark, of living but usually unhealthy trees, or newly felled logs. The adults do no tunnelling, the damage being done by the larvae, which feed on the wood substance. The tunnels or galleries are 3–25 mm in diameter, and oval in cross-section; they are packed with coarse **bore dust** or **frass**. The damage done is considerable but, except for the house longhorn borer, infestation occurs only in green timber; the larvae may continue feeding in relatively dry wood, but they will not migrate to adjacent seasoned stock. Rapid extraction of felled logs, immediate removal of bark of susceptible timbers and heat-sterilisation of infested wood will secure adequate protection against most longhorn borers.

Rhagium bifasciatum F. is one of the commonest longhorns in the UK, but it is of little economic importance; it attacks decayed softwoods. *Tetropium gabriele* Weise is the larch longhorn, which confines its attack almost entirely to *Larix* spp (larch). It is of common occurrence in England and Wales, and has been responsible for appreciable damage when the simple precautions mentioned above have been omitted. The beetles may lay eggs on unhealthy trees, but more usually on sound, felled logs that have been on the ground throughout a summer. The larvae feed beneath the bark, and enter the sapwood for pupation. Infestation can be entirely prevented by barking logs when felled.

20.1.2   Pin-hole borers (ambrosia beetles)

These pests belong to the families *Scolytidae* and *Platypodidae*; they feed on a mould fungus, introduced by the adult beetles, and not on wood; the fungus grows on the walls of the galleries, which are wholly constructed by the adults and not by the feeding larvae. The fungus is usually referred to as the **ambrosia fungus** (hence ambrosia beetles, the name first-coined by Schmidberger in 1836). The adult beetles tunnel at right angles to the grain, into living trees and newly felled logs, and lay their eggs in specially constructed 'egg-chambers'. At the time of egg-laying they also introduce a fungus into their galleries on which the larvae feed when hatched. The galleries of different species vary from 0.5 to 3 mm in diameter; they are usually oriented at right angles to the grain of the wood; that is, at right angles to the vertical axis of the tree. The galleries are usually empty, but they may become plugged with resin or other compounds; the walls of the galleries are stained black, and the tunnels themselves may be surrounded by an elongate-oval area of tissue discoloured by the fungus. Pin-hole borers ruin the appearance of considerable quantities of timber, and the galleries may be so numerous as to reduce strength properties appreciably, but attack does not continue in, nor can it spread

Figure 20.1   *'Pin-hole' borer damage in white chuglan (note direction of the galleries at right angles to the grain of the wood) (Building Research Establishment, © Crown Copyright)*

to, seasoned timber, because the fungus on which the larvae feed requires moisture. Prompt conversion and drying of felled logs is therefore important in minimising this type of insect attack. Since the damage is usually done before the timber gets to the mill, and sometimes even before the tree is felled, the timber merchant can purchase infested timber without fear of the attack becoming any worse than it is at the time he makes the purchase. Wormy grades of mahogany, lauan, seraya and meranti contain damage of this type. The 'shot-holes' referred to in the Malayan grading rules are caused by the same type of beetle, referred to as 'pin-hole' borers here. Typical 'pin-hole' borer damage in a tropical hardwood is illustrated in Figure 20.1; the diameter of the galleries can vary considerably among the different species of 'pin-hole' borers.

20.2 Pests of seasoning yards

The most serious pests of timber yards in temperate regions are the so-called **powder-post beetles**, belonging to the families *Bostrychidae* and *Lyctidae*. The larvae of these two families do not live on cell-wall substance, but on the starch content of the sapwood of certain timbers. In this respect, the food requirements of powder-post beetles resemble those of the sapstain fungi but, unlike fungi, the larvae have to devour the cells to obtain the starch they seek.

20.2.1 Lyctus *or powder-post beetles*

These beetles lay eggs in the vessels of wood, and the larvae tunnel about, feeding on the starch contained in the storage cells. The attack is confined to the sapwood of certain hardwoods;* it does not occur in the heartwood of any species, although, in emerging, adults may tunnel through heartwood immediately adjacent to sapwood (see Figure 20.2). The size of the vessels is

---

* *Lyct* attack has been reported in the sapwood of a softwood, *Pinus canariensis* C. Sm., grown in South Africa.

a limiting factor: they must be large enough to admit the **ovipositor** (egg-laying tube) of the adult female beetle, since it is in the vessels that the eggs are almost invariably laid. The fine-textured timbers, such as beech, with vessels below 0.1 mm in diameter, are ordinarily immune. Additionally, the starch content of the sapwood must be sufficiently high for the powder-post beetle to select the timber for egg-laying purposes. Several timbers, with large enough vessels, may fail in this respect and, consequently, are immune to attack. Infestation occurs in partially or fully seasoned timber.

The life cycle of *Lyctus* from egg to adult beetle is normally about one year, but the period may be as short as ten months and, where food supplies are deficient, the life cycle may be considerably extended: to two, or even three to four years. Adults normally emerge from April to September, appearing in largest numbers in June, July and August. Immediately on emerging, the adults mate, and the female begins egg-laying, being most fastidious in regard to the suitability of the particular piece of wood selected for egg-laying: it must be rich in starch.

There are several species of *Lyctus*, and the related genus, *Minthea*, the former being cosmopolitan, and the latter tropical. There are four species of *Lyctus* in the UK, L. *brunneus* Steph. (Figure 20.3) being the most abundant. *Minthea rugicollis* Walk, is the commonest species worldwide of *Minthea*; adult beetles have emerged in the UK, but the species is not known to have bred here.

The galleries of powder-post beetles are similar in diameter to those of the smaller 'pin-hole' borers. They run along the grain; that is, parallel with the vertical axis of the tree but, as attack progresses, the separate galleries become merged, and all the attacked wood, with the exception of a thin outer skin, is eventually reduced to a flour-like powder (Figure 20.4). Powder-post beetles can be responsible for enormous damage to the sapwood of susceptible timbers, such as ash, oak and agba, probably causing heavier financial losses in UK yards than any other insect pest. Particularly heavy losses can occur with some of the more lightweight tropical

Figure 20.2   *Stacked oak planks showing* Lyctus *attack confined to the sapwood (Building Research Establishment,* © *Crown Copyright)*

Figure 20.3   Lyctus brunneus *Speth (×8) (Building Research Establishment,* © *Crown Copyright)*

hardwoods which are used increasingly as soft-wood substitutes in the framing of furniture.

Control of powder-post beetle attack presents many difficulties. The total exclusion of sapwood of susceptible species will secure 100 per cent immunity from attack; this course of action is probably only economically justified in first-class joinery, flooring, panelling and furniture. Where susceptible timbers are used for structural purposes, sapwood should either be excluded or the timber should be pressure treated with a preservative.

Prolonged storage of logs in water results in starch depletion and is effective in securing immunity from *Lyctus* attack, but the storage period for timber in log form is too long for the method to be practicable. Alternatively, prophylactic measures may be resorted to in an endeavour to

Figure 20.4 Lyctus *frass (Building Research Establishment,* © *Crown Copyright)*

secure immunity from infestation during the seasoning and storage period, and prior to manufacture, since it is in these stages that infestation usually occurs. To reduce the risk of infestation in timber yards and factories it is recommended that stocks be inspected twice yearly in March and October. Yards should be kept clean and free from accumulating sapwood waste, and only softwood or heartwood piling sticks should be used. Chemical control of *Lyctus* in the UK is normally achieved by spraying stored timber with an insecticide emulsion, the treatment being annual and timed to take place immediately before the emergence of the adult beetles in June.

Since the treatments are of a superficial nature they are not effective once the treated timber is dressed or re-sawn. The transitory nature of the protection is, however, of high commercial importance, where all the evidence points to infestation occurring prior to manufacture: if timber could reach the joinery and woodworking shops immune from infestation, losses from *Lyctus* attack would be of negligible importance in the UK.

20.2.2 Bostrychid *powder-post beetles*

Except for differences in size of galleries, the damage done by these insects is similar to that of the previous group. That is, the larvae tunnel in partially, or recently, seasoned sapwood of certain hardwoods, reducing the wood to a flour-like dust, which is slightly coarser in texture than that of the *Lyctus* group; the galleries and exit holes are commonly up to 3 mm in diameter, but may be considerably larger. As with *Lyctus* attack, damage is confined to the sapwood.

Bostrychid beetles are typically larger than those of *Lyctus* species, and they are pests of tropical rather than of temperature regions, although there are a few species in temperate climates, including *Apate capucina* L., found in European oak. The adults are characterised by a hooded, roughened thorax covering the head, and with a three-jointed club at the end of the antennae. The life cycles of the different species have not been the subject of such critical study as those of *Lyctus*, but it is apparent that the adults will infest timber in an appreciably wetter condition than that favoured by *Lyctus*: stacks of mersawa (*Anisoptera* spp) in Malaysia have been seen to become heavily infested by *Bostrychid* beetle within a few days of the timber being sawn, whereas *Minthea* (the commonest of the *Lyctus* group in Malaysia) would be unlikely to infest timber until it had been in stick for several weeks.

Bostrychid beetles also differ from the *Lyctidae* in their egg-laying habits: the adults bore into the wood, constructing a Y-shaped egg-tunnel, which is kept free from dust, and in which the female lays her eggs. When the eggs hatch, the larvae continue burrowing but longitudinally (as do *Lyctus* larvae), packing the gallery system with fine, flour-like dust. The methods of control of *Bostrychid* beetles are identical with those for *Lyctus* beetles.

20.3   Pests of well-seasoned wood

Furniture beetles, and the death-watch beetle, belong to the family *Anobiidae* — important pests of timber in buildings although also occurring naturally in decaying stumps out-of-doors (Bravery *et al.*, 1987). There are several species of furniture beetles, belonging to more than one genus, but the most frequently encountered 'indoor' species is *Anobium punctatum* De G; the common furniture beetle. The death-watch beetle is *Xestobium rufovillosum* De G.

20.3.1   Common furniture beetle

The natural home of this pest is out-of-doors, in decayed trees and posts, but it is better known as a pest of well-seasoned softwoods and hardwoods; it is commonly referred to as 'woodworm'. The damage is done by the larvae, which hatch from eggs laid in cracks in the wood, in joints of made-up woodwork, and in old flight holes. The larvae travel along the grain, but as they feed and grow they tunnel in all directions, filling their galleries with loosely-packed, granular frass, which feels gritty when rubbed between the fingers (Figures 20.5 and 20.6); the pellets are appreciably thinner than those in longhorn borer frass.

The life cycle and biology of the common furniture beetle has been extensively investigated. The adults (Figure 20.7) emerge in May, June, July and August, and mate, when the female lay their eggs in suitable places; they will not lay on smooth surfaces. The eggs hatch shortly after they are laid, and the larvae commence tunnelling into the wood, on which they feed. The length of the cycle in the UK is usually three to five years, but may well be considerably extended when food supplies are not entirely suited to the pest's requirements.

The common furniture beetle is widely known as a pest of old furniture, and of hardwood constructional timbers in period houses, but since the 1940s it has been recognised as a common pest in the sapwood of softwoods in buildings of all ages. It was formerly thought that initial attack did not occur until the timber had been in service for several years, and that it was necessary for the timber to have 'matured' in some way for it to become attractive to the beetle. Entomologists now conclude that initial infestation may occur as soon as the timber has become seasoned, but the presence of attack may not be discovered until several life cycles have been completed, and flight holes are quite numerous.

Damage is confined to the sapwood of timbers except those that display poor differentiation between sapwood and hardwood (such as

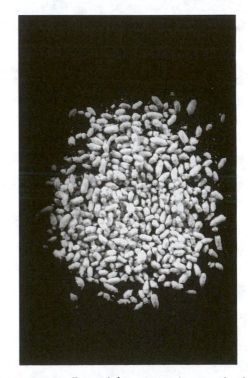

Figure 20.5   *Frass of the common furniture beetle (note typical elongate pellets) (magnification ×8) (Building Research Establishment, © Crown Copyright)*

Figure 20.6 *Damage by the common furniture beetle in a structural timber of Scots pine, commonly referred to as 'woodworm' attack (Building Research Establishment, © Crown Copyright)*

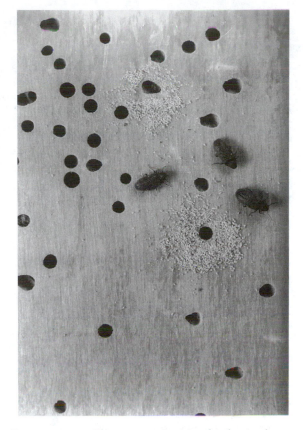

Figure 20.7 *The common furniture beetle, Anoboium punctatum De G (×3 approximately) (Building Research Establishment, © Crown Copyright)*

beech and spruce); in these timbers, damage may occur throughout the wood. In furniture and small wooden articles, the damage done may be quite serious, as, for example, attack in the leg of a chair, but in structural timbers damage is only serious when the amount of sapwood is abnormally high. Attack in softwood timbers may develop occasionally to the extent of causing some timbers to collapse, if sufficient sapwood is present, but the presence of a beetle population is probably of more importance because of the risk of subsequent infestation of furniture.

Control of existing infestations can be effected by application of suitable preservatives, particularly those of the solvent type, and by sterilisation or fumigation; neither of the last two mentioned methods confers immunity from fresh infestation. To reduce the risk of infestation, it is recommended that care be exercised in the purchase of second-hand wooden articles as these may well be infested, thereby constituting a source of infestation for the spread of furniture beetles to sound timber. Fuel logs and garden woodwork in close proximity to the house are likely breeding grounds of furniture beetles from which infestation can spread indoors.

20.3.2 Death-watch beetle

Death-watch beetles lay eggs in crevices, cracks, or old exit holes, and the larvae do the damage by tunnelling in, and feeding on the wood. Attack is usually confined to old timbers of several species of hardwoods, but it has been known

Figure 20.8 *Portion of an oak wall plate decayed by fungal activity and attacked by the death-watch beetle (Building Research Establishment, © Crown Copyright)*

to spread to adjacent softwood timbers.* Attack is not confined to the sapwood, but it is more likely to begin in sapwood than in heartwood. Adequate moisture, and the presence of fungal decay, are necessary conditions for infestation (Figure 20.8). The galleries made by the larvae are about 3 mm in diameter; they are filled with coarse frass, containing bun-shaped pellets (Figures 20.9 and 20.10). Removal of decayed wood, and the causes of decay, are the first essential steps in eradicating death-watch-beetle infestation. All too frequently, however, major structural damage has occurred before the infestation is discovered, when replacement of the attacked timbers, rather than *in situ* chemical treatments, is the only practicable course.

Wherever possible, it is preferable to use pressure-treated softwood timber in repairs but, if for aesthetic reasons hardwoods, and usually oak, must be used, such timber must be well seasoned and free from sapwood. Not infrequently, the elimination of dampness responsible for the initial fungal decay, the replacement of decayed and

Figure 20.9 *The death-watch beetle (×2 approximately) (note characteristic bun-shaped pellets in bottom right-hand corner) (Building Research Establishment, © Crown Copyright)*

heavily attacked timbers, and the introduction of steel straps and the like to restore structural stability, will suffice, because any continuing attack will die out when dry conditions, adverse to the rapid development of the beetle, are established.

The life cycle and biology of the death-watch beetle have been exhaustively studied between 1937 and 1941 by the late Dr R.C. Fisher, who concluded that the length of the life cycle 'is dependent upon the moisture content of the timber, the presence and extent of fungal decay, and also upon temperature'. Under optimum conditions the life cycle may only be one year, but in less favourable circumstances it is prolonged over two to several years. Heavy infestation is often accompanied by *Korynetes*

---

* One case is known of widespread and quite heavy death-watch beetle infestation in softwoods in an old house at Surbiton that contained no structural hardwoods, joinery or flooring.

Figure 20.10 *Bun-shaped pellets of death-watch beetle frass (magnification ×8) (Building Research Establishment, © Crown Copyright)*

caeruleus Deg, a steely blue, hairy beetle, which is predatory on the death-watch beetle, but which does not attack the timber. The work is fully described in Fisher (1937–41).

20.3.3 House longhorn beetle

The house longhorn (*Hylotrupes bajulus* L) (see Figures 20.11 and 20.12), a member of the family *Cerambycidae*, is a serious pest in many parts of Europe, particularly France and Germany. The pest has long been known here, and has become of sufficient importance in some parts of Surrey to necessitate by-laws and subsequently a Building Regulation requiring softwood timber used in new work in roofs to be treated with approved wood preservatives. The pest attacks the sapwood of softwoods, and usually only timbers in the roof space. As the

Figure 20.11 *The house longhorn beetle (×2) (Building Research Establishment, © Crown Copyright)*

life cycle is up to 10 years, serious damage may result before the first flight holes bring the attack to light. The frass contains pellets that resemble truncated cylinders when examined with a pocket lens. The finding of such frass or oval flight holes is not proof of house longhorn infestation; still less that there is continuing activity. Other longhorn borers produce similar frass and oval flight holes. Active house longhorn borer infection is concentrated in the UK in the Walton and Weybridge area and Camberley, although isolated cases are found throughout

Figure 20.12 *Softwood roofing timber attacked by the house longhorn (Building Research Establishment, © Crown Copyright)*

the southern counties (Lea, 1994). Even within these areas of high concentration, active infestations are not usually found in buildings more than 50 years old, except where new timber has been recently used for repair.

20.4   Other timber beetles

Several other beetles may sometimes be responsible for causing damage to timber in service, but the only two likely to be encountered at all frequently in the UK are the wood-boring weevils, family *Cossoninae*, and *Ernhobius mollis* L. The wood-boring weevils are essentially secondary infestation, following on fungal decay, and, in tackling this, the secondary pest is also eliminated. *Pentarthrum huttoni* Wollaston is probably the commonest wood-boring weevil, and the damage done superficially resembles that caused by the common furniture beetle. The exit holes are rather ragged, the frass is rather finer, usually round, and the conspicuous 'snout' of the weevil is a final clue to the pest at work (Figure 20.13). *Ernobius mollis* L. is of no economic importance; it has not as yet acquired a popular name. It is a reddish or chestnut-brown beetle, up to 6 mm long, which leaves flight holes resembling those of the furniture beetle. The frass consists of small bun-shaped pellets, resembling those of the death-watch beetle but appreciably smaller; the frass is characteristically a mixture of red-brown and white particles because the larvae feed on the bark of softwoods, often just penetrating the outer sapwood. Attack is quite common when the inner bark has been left on the waney edges of carcassing timbers. Removal of the bark brings an attack to an end, and there is no need for any additional chemical treatment.

It is important to stress that when dealing with insect attack in buildings, mere discovery of a flight hole, or even the finding of bore dust, is not necessarily a cause for alarm. It is important to identify the type of insect responsible and the extent of activity before having to resort to expensive *in situ* chemical treatments, which, so often, are quite unnecessary because attack has ceased. Careful inspection is also important as substantial structural repairs may be called for rather than *in situ* chemical treatments.

The wharf borer, *Nacerdes malanura* L., should also be mentioned; it is a large beetle, superficially similar to some longhorn beetles, which leaves a flight hole about 3 mm in diameter; attack is confined to very damp decayed timber.

20.5   Termites

The insect pests discussed in the preceding pages are those that commonly occur in temperate regions; the same species, or close relatives, and other insects unknown in cooler regions, are pests of timber in the subtropics and tropics. In these regions **termites** or **white ants** are the most serious insect pests; they probably cause more damage to timber annually than do all other insect pests together. Simple control measures exist, which would reduce the termite problem in the tropics to negligible proportions, but all too often effective precautions are not taken.

There are many species of termites, but those that attack timber may be classified into one or other of two broad groups; **subterranean termites** and **dry-wood termites**. The former live in large colonies in the ground, and must retain

Figure 20.13   *The common wood-boring weevil,* Pentarthrum huttoni, *Wollaston (×10)*

Figure 20.14 *Subterranean termite damage: wood hollowed out and filled with mud (Building Research Establishment, © Crown Copyright)*

unbroken covered runways which they construct from soil particles from the colony to their feeding grounds. Dry-wood termites live in small colonies in seasoned wood on which they feed; they require no access to the soil.

Termites do not make definite tunnels in timber, but tend to excavate large cavities that, in the case of subterranean termites, become packed with mud as attack progresses (Figure 20.14). Naturally resistant timbers tend to be gnawed by termites, but the soft, non-resistant species may be completely hollowed out, except for an outer skin of wood. Contrary to popular statements, no timber is immune to subterranean termite attack, but the range in resistance of different timbers is appreciable: exposed to conditions of equal intensity of attack, one timber may last less than six months and another more than ten years, with many other timbers with a variable serviceable life between these extremes. Moreover, resistance to fungal decay is not necessarily an indication of resistance to subterranean termite attack. For example, such naturally decay-resistant timber as oak heartwood does not show up well if exposed to termite attack. Conversely, some tropical timbers that are not regarded as particularly durable in their countries of origin, where termites are a more serious problem than fungi, may prove exceptionally

resistant in temperate regions, where fungal decay is the serious hazard; examples are kempas and kapur. Hardness of a timber is no criterion of its power of resistance to termite attack.

Dry-wood termites invariably feed just below the surface of wood, and in most timbers attack is more or less confined to the sapwood; they produce granular dust, appreciably coarser than that of furniture beetles, which showers out when the sound skin of wood left on the surface of an attacked piece of timber is broken. The feeding termites also push out granular dust through their exit holes, the mounds of this dust often being the only evidence of attack in progress.

Termites need not be nearly so serious a timber problem as they are generally supposed to be, and, of the two types, dry-wood termites are more troublesome than subterranean termites, because of difficulties of control: differences in the habits of the two groups result in precautions for rendering buildings proof against subterranean termites being ineffective against dry-wood termites. Effective control against the former is secured by proper design and construction of buildings: barriers must be provided at ground level. Where walls are contiguous with the foundations, a strip of metal, extending 50–75 mm from the wall face, and projecting downwards at an angle of 45°, let into the damp-proof course below the ground-floor floor joists, is effective. The use of chemical barriers consisting of insecticide emulsions sprayed into foundations prior to the placement of foundation concrete and floor slabs is also to be recommended. (Details of such procedures are described in the Building Research Establishment's *Overseas Building Note No. 170.*) Where these barriers cannot be provided, as, for example, with much timber used in contact with the ground (railway sleepers, fence posts and poles), no timber will last indefinitely, and the choice is between the naturally resistant timbers or the less resistant ones adequately treated with wood preservatives. Chemical control of subterranean termites, usually involving the introduction of insecticidal dusts into the runways, is effective in keeping down the termite population, but it is a palliative and not a curative measure, in spite of claims to the contrary. Brush applications of wood preservatives are not effective in rendering wood immune from subterranean termite attack. Pressure processes are in a different category: with adequate absorptions of suitable preservatives, wood can be made to outlast its mechanical life, that is, a properly treated railway sleeper will fail from rail-cutting or spike-killing rather than from termite attack, or, for that matter, fungal decay.

Dry-wood termites cannot be excluded from buildings in the same way as subterranean termites. In areas in which dry wood termites are known to be a hazard, there is therefore a strong case for the use of timber pressure impregnated with a suitable wood preservative. The pests can be positively eliminated from building timbers and other indoor woodwork by screening buildings with fine metal gauze, a course that is extremely expensive and usually impracticable. Alternatively, reasonable precautions can be adopted that will minimise the risk of dry-wood termite attack, with dependence on curative measures when attack occurs. A US authority recommends the painting of wooden surfaces as an effective method of denying dry-wood termites entry into timber, and where this is practicable for example on joinery, the course should be adopted. It seems probable that the total exclusion of sapwood may appreciably

delay dry-wood termite attack, even if it does not confer complete immunity.

The nature of the damage caused by dry-wood termites is apt to be misleading; at first sight the damage appears devastating, but closer inspection usually reveals the destruction to be less severe than was thought. In furniture, panelling and high-class joinery, even a small amount of damage may be serious, because the appearance is spoiled, but in carcassing timbers it is necessary for the damage to be sufficiently serious to weaken the structure before alarm need arise. In practice, dry-wood termite attack is localised; some wooden members may be attacked while adjacent ones are quite free. Moreover, it is usually only parts of such members that are infested, and then often only to a depth of about 12 mm. If the infested zone is removed, and the remainder of the timber is liberally dressed with an oil-solvent wood preservative, attack will often cease, and the reduced member is often still strong enough to carry the load required of it.

20.6   Marine borers

Although not insects, several marine organisms, of which the teredo or ship-worm is probably the best known, are responsible for heavy losses of timber used in salt water. Intensity of attack by ship-worm varies in different regions, but is generally much more severe in tropical than in temperate climates: even the naturally resistant species such as greenheart and belian may have a very short service life in some tropical waters. The damage done takes the form of tunnelling, either vertically or horizontally, in the wood, which may be so extensive as to destroy the strength properties of a timber member completely.

A second form of damage is caused by the marine organism known as the 'gribble' which closely resembles the terrestrial wood-louse. This animal creates shallow tunnels approximately 1 mm in diameter in the surface layers of exposed timbers. In combination with fungal decay and weathering, deterioration may be extremely rapid.

Any wood used in salt or brackish water* is liable to attack by these marine organisms, and only in situations where infestation is known to be slight is it economical to depend on naturally resistant timbers. Pressure treatments with capacity absorptions of creosote or other good preservatives have been found effective in temperate waters, but metal sheathing is likely to prove more economical wherever marine borers are particularly active. Preservation aspects are discussed in Chapter 22. Many, but not all, of the timbers that are resistant to a greater or lesser degree to marine borer attack have been found to contain silica deposits in their storage tissue. However, some timbers containing appreciable quantities of silica have not revealed any particular resistance when exposed to attack.

* There is some evidence to suggest that water can be too salty for optimum development of the teredo, whereas very low concentrations have been found to be associated with exceptionally heavy infestation.

References

Bravery A.F., Berry R.W., Carey J.K. and Cooper D.E. (1987) Recognising wood rot and insect damage in buildings. *Building Research Establishment Report.* HMSO: available only from the Building Research Establishment Bookshop.

BRE (1976) Termites and tropical building. *Building Research Establishment Overseas Building Note, OBN 170.* HMSO: available only from the Building Research Establishment Bookshop.

BRE (1992) Identifying damage by wood-boring insects. *Building Research Establishment Digest 307.* HMSO: available only from the Building Research Establishment Bookshop.

Fisher R.C. (1937–41) Studies of the biology of the death-watch beetle *Xestobium rufovillosum* De G, Parts I–IV. *The Annals of Applied Biology* **XXIV(3)** 600–613; **XXV(1)**: 155–180; **XXVII(4)**: 545–557; **XXVIII(3)**: 244–260.

Lea R.G. (1994) House longhorn beetle: geographical distribution and pest status in the UK. *Building Research Establishment Information Paper IP8/94.*

21

Eradication of Fungal and Insect Attack

Timber used as railway sleepers, fence posts, power-line poles and the like, is inevitably exposed to conditions favouring attack throughout its service life. By contrast, certain conditions of service ensure that timber never becomes attacked: piling timbers in deep water or timbers in the abnormally dry atmosphere of the Egyptian tombs will remain free from attack indefinitely. Much more timber is used in circumstances between these extremes of certain attack and complete immunity, giving rise to the need for prophylactic measures or problems of eradication.

Timbers predisposed to attack by sap-stain fungi, 'pin-hole' borers, or 'powder-post' beetles are very liable to suffer deterioration unless appropriate precautions or prophylactic measures are taken. The sapwood of those timbers susceptible to 'mould' fungi are liable to become stained unless logs are covered immediately after felling, and the converted material is then dried so rapidly that the fungal 'spores' are not given the opportunity to germinate. It is often not possible to arrange for sufficiently rapid drying, in which case the use of chemical dips (anti-stain treatments) provides a practicable solution.

'Pin-hole' borers present a somewhat similar problem to that of sap-stain fungi, in that attack is dependent on the existence of sufficient moisture in the wood to support the growth of the ambrosia fungus on which the pin-hole borer larvae feed. Some species attack standing trees, so that the damage is done before the trees are felled, and no methods have as yet been devised for dealing with infestation at this stage. Most infestation, however, occurs after felling, while logs are lying in the forest or at the mill awaiting conversion. Extraction immediately after felling, followed by immediate conversion at the mill,

and rapid drying of the converted timber either in a kiln or by proper stacking, will unquestionably minimise the damage done by pin-hole borers. The rapidity with which these borers attack the logs of some species makes dependence on rapid extraction and conversion to avoid infestation uncertain, particularly in tropical countries where infestation occurs most rapidly and extraction and conversion can be problematic. Under such circumstances, spraying of freshly felled logs with insecticides is sometimes done to reduce the risk of infestation. Emulsion-based formulations of persistent insecticides such as permethrin are normally used for this purpose.

The countering of 'powder-post' beetle attack presents a problem intermediate in complexity between elimination of sap-stain fungal infection and 'pin-hole' borer attack on the one hand, and eradication of fungal or insect attack in timber in service on the other. Attack does not normally occur in the log, but *Bostrychid* beetles will attack timber very soon after conversion, and *Lyctus* as the timber becomes drier. Good yard hygiene plays the most important part in minimising the depredations of the powder post beetles. In the UK, the available evidence points to the need for purely transitory protection against *Lyctus* (apart from eradication of outbreaks when these occur): where joinery, furniture, etc., is delivered free from infestation, the likelihood of attack occurring in service would be remote. In effect, precautions are especially necessary while timber, containing a significant proportion of susceptible sapwood, is in stick for drying, or awaiting manufacture. For these circumstances, it has been found that the use of emulsion-based formulations of permitted insecticides such as permethrin, applied by spraying to piled timber, is effective; complete coverage of the sapwood must be secured. The spray is prepared from miscible oil concentrates containing insecticide which is then diluted with water to the concentration required. No penetration of the timber is aimed at, or secured, with this spray, so when timber is worked or re-sawn it will require spraying anew if it is likely to be exposed to infestation before being made up.

Longer-term protection against damage by *Lyctus* beetles may be achieved by the use of a variety of wood preservatives, for example:

(a) pressure applied aqueous solutions of copper-chrome-arsenic salts (Tanalith or Celcure);

(b) diffusion of the green timber with aqueous solutions of disodium octaborate tetrahydrate (Timbor);

(c) organic solvent solutions of persistent insecticide such as permethrin applied either as a three-minute dip or by the double vacuum process (see Chapter 22).

Availability of moisture is, in practice, the factor that governs liability to fungal infection, although different fungi vary in regard to their 'total' moisture requirements. With insect pests, favourable factors are more varied: the death watch beetle, the wood-boring weevil and the wharf borer thrive only on timber that has first been attacked by fungi, although the death-watch beetle will spread to sound wood. The common furniture beetle is less exacting, but attack is usually confined to sapwood. The *Lyctus*, or powder-post beetle, is the most exacting of all, confining itself to the sapwood of certain hardwoods, and then only if adequate supplies of starch are present, if the timber is sufficiently seasoned, and if the pores or vessels are large enough to permit egg laying.

21.1   Eradication of fungal decay

Recommendations for dealing with 'dry-rot' are to be found in the Bible (Leviticus xiv. 34–38). The fungal origin of the 'plague' or 'leprosy' was not known until centuries later, but the passage is of interest in that it brings out two important points: the need for establishing that the attack is still active, and the need for drastic remedial measures — rather too drastic in the light of modern knowledge. The fungus would not have been *Serpula lacrymans* as the temperatures in Palestine are too high for this species. The Old Testament writer overlooked an all-important point: namely, the importance of tracing the source of moisture that gave rise to fungal infection in the first place. By comparison, it is less important to identify the species of fungus, although it is, of course, essential to determine whether *Serpula lacrymans* is involved, since this fungus can require more drastic remedial measures than with fungi of the 'wet rot' type.

The three most common faults when dealing with outbreaks of fungal attack are to do too little 'site' investigation, to do too extensive replacements and to ignore the fundamental cause, namely a supply of moisture. It is essential to determine the full extent of infection, but knowing what to look for, and where to look for it, will minimise the amount of opening up to be done. Time spent on the careful examination of the exterior of a building is usually well repaid. The more obvious points in regard to damp-proof courses, and levels of flower-beds and paths, are generally understood. The importance of an adequate number of air bricks to provide through flow of air in sub-floor voids is also usually appreciated; although it is sometimes overlooked that their number may be adequate and yet the air bricks are ineffective: they may be obscured by a plate or joists on the inside. Many air bricks have, at best, only 50 per cent of open area and a floor joist or plate can reduce this 50 per cent to almost nothing. Water from above — from defective gutters, rainwater heads, or down pipes — can cause as widespread devastation as indifferent ventilation below the ground floor (Figure 19.5). Hence, the condition of the rainwater disposal arrangements, and their efficacy, warrant close examination: staining of external walls, the growth of algae and the condition of pointing, or evidence that such matters have recently received attention, will often indicate where the search for fungal infection should be directed inside the building. Parapet gutters, 'internal' valley gutters, and lead or asphalt 'flats' are other fruitful sources of trouble, and any defects in these features, or signs of past patching, call for particular attention to be given to the condition of timber beneath such weak spots. Past history can be most relevant: flooding from burst pipes, when the property is unoccupied, or water used in putting out fires, can cause the most extensive dry rot. Areas of new slates or tiles should arouse suspicion. Armed with clues on the lines discussed above, the tracing of decay is simplified. Corrugations in skirtings and panelling are obvious defects to look for, but even bowing of such timbers, if on the opposite side of a wall where 'defects' exist outside, should not be given the benefit of any doubt: removal of such timber often reveals, surprisingly extensive, 'unsuspected' decay.

Having located decay and its extent, it is essential to determine where the wood obtained a supply of moisture sufficient to render it fit to support fungal growth. Very often an outbreak has more than one 'focal' point, which means more than one source of moisture, and it is all-important that these should be detected. Moisture will not travel upwards nearly so far as it will travel downwards: infection in basement or ground-floor rooms will be sustained by a different source of moisture from infection in the same premises but on the first or higher floors.

Until the source of moisture is traced, the appropriate remedial measures cannot be laid down. For example, moisture arising from constructional defects can only be eliminated if these defects are corrected or, when this is impracticable or too costly, by recognising that no timber should be used in repairs that will not be isolated from the source of moisture by a water-impervious barrier.

Badly ventilated basement floors, with no over-site concrete and no damp-proof course, present just such a problem. However thoroughly the wood is stripped out and replaced by new wood, fungal infection and subsequent decay will reappear unless moisture can be excluded from the underfloor space, and good ventilation provided; brush, spray or dip treatment of replacement timbers with wood preservatives will, at best, defer the onset of decay. Even timber pressure impregnated with preservatives of the copper–chromium–arsenic type will not remain immune from decay indefinitely under persistently damp conditions. Ideally, where basements are to be used as habitable space, a complete waterproof lining of walls and floor is required before timber can be reinstalled.

When the floor area is large, as in a gymnasium or concert hall, decay has been known to occur, in spite of the existence of damp-proof courses in all walls, and provision of the normal number of air bricks; it is sometimes impossible to ventilate the underfloor space of large areas adequately, and moisture must be excluded by providing a waterproof barrier in the over-site concrete.

Cure of the cause of dampness, once identified, presents no great difficulty, and such action alone will, in many cases, arrest further decay. It is, of course, necessary to cut back affected wood to sound material, and to use well-seasoned timber in repairs; if there is any risk of dampness recurring from neglect of maintenance in the future, pressure-treated timber should be used for all replacement work. Timber that has been exposed to attack and is only very slightly decayed should only be retained if rapid drying can be guaranteed. Application of preservatives of timbers likely to remain damp for perhaps a few months can provide temporary control of decay while drying is achieved. Only penetrating treatments, such as emulsion paste formulations or drilling and insertion of diffusible fungicides, are likely to provide any significant degree of control. When panelling or valuable flooring has become sufficiently damp to support fungal growth, it is usually advisable to dismantle such timber and to dry it, otherwise splits may develop as the timber dries *in situ* or, with floors, compression set may have been induced (Chapter 9).

Where *Serpula lacrymans*, the true 'dry rot' fungus, is the causal agent, additional measures to those outlined above are likely to be essential; this fungus penetrates masonry and brickwork. In practice, walls that have become saturated are likely to retain sufficient moisture to sustain *Serpula* for some years after the original source of water has been cut off, particularly if there are even quite small pieces of timber hidden in the wall to provide the essential food material: fixing blocks for down pipes in the outside of a wall may suffice, and bond timbers will, of course, ensure vigorous continuing attack. The time it takes for thick walls to dry out explains the recurring attacks sometimes experienced after an outbreak of *Serpula lacrymans*: as soon as the new timber has had time to absorb sufficient moisture from the wall to raise its moisture content to about 20 per cent, conditions are ripe for active fungal hyphae to attack the new timber provided. The heat sterilisation of walls with a blow-lamp commonly practised in the past is totally inadequate: raising the temperature of one face of a 100 mm brick wall to 900°C, and holding that temperature for 4 hours, will only raise the temperature of the opposite face to about 50°C. The surface application of fungicides is unlikely to be any more effective; elaborate irrigating of walls with fungicide is costly in labour and in the large quantities of chemicals absorbed, besides being somewhat uncertain in efficacy. Moreover, extensive boring of holes to permit of irrigation treatments may well weaken old brickwork.

An additional problem associated with large-scale irrigation of walls with water-based fungicides is the efflorescence of soluble salts during drying of the wall, resulting in disruption of plasterwork. These difficulties have resulted in irrigation treatments becoming less popular and usually restricted to localised areas of wall to provide a *cordon sanitaire* of sterilised masonry around embedded timbers.

In general, remedial control of dry-rot outbreaks necessitates careful identification of the area affected, followed by removal of all decay-susceptible timbers whether decayed or not, in contact with the damp, infected masonry. For example, fixing blocks for joinery can be replaced with plastic plugs, joist ends resupported on joist hangers or metal brackets and timber lintels replaced with concrete or steel. Timbers pretreated with preservative by pressure or double-vacuum impregnation should be used to replace or repair existing timbers in contact with infected masonry. Considerable caution is necessary if *in*

situ preservative treatment with emulsion pastes or other penetrating treatments is to be attempted as an alternative to removal of timbers from infected masonry. Such treatments should only be contemplated where rapid drying of the masonry can be assured and where the efficacy of the treatment can be monitored at intervals following completion of remedial works.

Fungicidal plasters or renders may be used to create a barrier between infected brickwork and re-instated joinery. Proprietary products based on zinc oxychloride (ZnOCl) fungicide are available from specialist suppliers.

While in many buildings the normal processes of opening up to determine the extent of the outbreak facilitate ventilation and the consequent drying out of the moisture which entered the fabric before defects were repaired, the possibility that deliberate measures may be needed to expedite drying should not be overlooked. This need is particularly evident in flood-damaged buildings or in larger buildings in which massive walls have become saturated through long neglect of maintenance. 'Crash' drying can be effected in such cases by the use of hot air dryers, but control can be difficult and it is not always easy to obtain the co-operation of all concerned to provide the ventilation which must be provided simultaneously if this form of drying is to be successful. A more convenient method of drying is provided by the 'dehumidifiers' which can be hired from some larger builders merchants. These circulate air dried by refrigeration and require that the area to be dried shall be sealed off to prevent ventilation from outside or from other dry parts of the building.

To sum up: control and eradication of fungal infection within buildings is first and foremost dependent on tracing and eliminating the source of moisture. Subsequent steps depend on whether the fungus is *Serpula lacrymans* or one or other of the less virulent fungi. Fungicides have a part to play, but in a secondary role. The work of eradication is likely to necessitate the assistance of carpenters, bricklayers and plasterers, often the plumber and tiler as well, and is, therefore, essentially a job for the building contractor, fully aware of the importance of each step. Thoroughness is the keystone of success.

The particular case of decay in external softwood joinery deserves separate consideration. The Building Regulations require the preservative treatment of non-durable timber used for external cladding, but have made no such provision for doors and windows in which decay of the wet-rot type became increasingly prevalent after the late 1950s. However, in 1970 the National House Builders' Council, to which most private sector builders subscribe, specified that the softwood window joinery of houses built by its members should receive preservative treatment; the requirement has since been extended to external doors. Similarly, the Department of the Environment now requires that all new external softwood joinery in dwellings for which it is responsible should be preservative treated and Local Authorities are tending to follow suit. Nonetheless, there are still many houses with untreated joinery at risk of decay. To deal with this problem, a number of the firms which specialise in remedial treatment of decay and insect attack now provide a service of treatment for existing window joinery by either drilling and insertion of plugs of diffusible fungicide, or injection of liquid preservative, into the wood near the joints which are most at risk. These *in situ* treatments have been demonstrated to provide a considerable degree of protection of sapwood in Baltic redwood, the primary window joinery timber.

21.2 Eradication of insect infestation

In temperate climates, insect infestation of timber on land means beetle attack; in salt or brackish water, marine borers are the destructive agent. Control and eradication of beetle infestation are dependent on an understanding of the life cycle of different species and their food requirements. The life cycle of any species is liable to be prolonged if its preferred food supply is deficient. In dealing with insect attack, it is all-important to determine which pest is at work before attempting eradication.

If the damage is the work of the 'pin-hole' borer (ambrosia beetle), no remedial measures are necessary: such 'wormy' timber is perfectly safe to use, since the damage will not get worse, and cannot affect other wood. Damage caused by these pests is most common in tropical timbers, infestation and the full extent of the damage having occurred before such timbers are exported. Control of pin-hole borer infestation rests with those responsible for felling and extraction of timber from the forest and the appropriate measures are discussed in Chapter 20, section 20.1.2. 'Pin-worm' infestation can usually be differentiated from other forms of insect attack by the galleries running at right angles to the grain; the galleries of different species vary from 0.05 to 3 mm in diameter.

Longhorn borers are also mainly forest pests, although one species, the house longhorn (*Hylotryupes bajulus*), attacks converted softwoods. With the exception of the last-mentioned species, attack must be dealt with in the forest: logs must not be left lying on the ground, and if extraction is likely to be delayed, logs should be barked immediately after felling. The house longhorn is a serious pest in parts of Sweden, Denmark and Germany, and in the UK has caused damage to buildings, in particular, in parts of Surrey. It attacks the sapwood of seasoned softwoods (the heartwood is not completely immune). Extensive damage is likely to have occurred before the first flight holes are detected, because the life cycle is relatively long (3 to 11 years), and indications of attack, other than flight holes, are often wanting, or are easily overlooked (for example, blister-like swellings on the surface of infested wood). Attack generally originates in the roof timbers and attics, from which it may spread to other timbers throughout the building. Remedial measures are likely to involve replacing appreciable quantities of timber, coupled with application *in situ* of wood preservatives. *In situ* heat sterilisation and fumigation are used on the European continent in combating outbreaks, but facilities for heat sterilisation are not available in the UK, and fumigation is impracticable in most houses, particularly in built-up areas. The use of wood preservatives calls for very thorough applications to ensure that all timber surfaces are adequately treated. Inspection for signs of renewed infestation in ensuing years is advisable. Oil-solvent wood preservatives are appropriate: most commercially available remedial fluids consist typically of a persistent insecticide, such as permethrin, dissolved in an organic solvent such as white spirit or odourless kerosene. The most recently introduced emulsion pastes contain the same type of insecticide, but can provide far deeper penetration than the conventional fluids; they are, however, more expensive and time consuming in application. Where the pest is prevalent it is wise to use only pressure-treated timber in repairs and for all new work.

The use of home-made wood preservative formulations cannot be too strongly discouraged owing to the hazardous and flammable nature of some of the constituents used. In the UK, only wood preservative products approved under the Control of Pesticides Regulations (1986) can be advertised, sold, stored or applied (BRE, 1992). Use of non-approved products is an offence under the Regulations.

The death-watch beetle is almost always associated with decay, although, once established, the beetle may extend its attack to relatively sound wood. Hence, in dealing with death-watch beetle infestation, it is imperative to deal with the decay too, since the cause of this will have been responsible for the subsequent and 'secondary' beetle infestation. It is also important to decide whether the beetle infestation is still active. It is a common feature of damage by the death-watch beetle for attack to cease before all the available timber has been destroyed, and this is no doubt due to absence of the conditions of moisture and fungal decay now known to be necessary for attack and which must once have been present in the building. It is not always easy to determine whether attack is continuing, although clear-cut rims to the flight holes, from which bright-coloured frass is spilling out, indicate the recent emergence of adult beetles. A search of the ground beneath attacked timber during the emergence period (April to June) is helpful: live beetles will usually be found if attack is still active. The presence of large numbers of live, steely blue beetles (*Korynetes caeruleus*), predatory on the death-watch beetle, also provides evidence of continuing attack. If investigation shows that attack has ceased, the use of wood preservatives is obviously unnecessary. In many cases of continuing attack, cutting off supplies of moisture, which bring decay to an end, may be as effective, in the long term, as attempting to directly eliminate the beetle infestation. In dealing with serious devastation, the first step is to conduct a thorough check for any signs of decay, the cause of which must be eliminated; next, replacement of all structurally weakened timber; and, finally, treatment *in situ*, with an oil-solvent preservative, of infested timber to destroy any remaining infestation. With larger timbers surface applications should be supplemented by pressure injection. It is particularly important to detect and treat by injection any internal pockets of decay which may form a breeding ground for the insects. Eradication may necessitate 'repeat' treatments in succeeding years, but these can be confined to areas of continuing active infestation.

Insecticidal smoke generators (BRE, 1986)* have been successfully applied in many buildings infested by death-watch beetle. An annual treatment, just prior to the emergence season, provides an insecticidal deposit which should ensure that none of the emerging beetles is able to complete its mating and egg-laying activities. Repeated over a number of years, the procedure should therefore gradually reduce the population to insignificant proportions and possibly eradicate it altogether.

This form of treatment is particularly attractive where large, high roof structures are involved and where the cost of scaffolding is prohibitive.

Two other beetles infest decayed building timbers: the common wood-boring weevil, often found in basement floors, and the much larger wharf borer. Both these pests are secondary to fungal decay, and their control by chemical means (wood preservatives) should not be attempted. Exclusion of the source of moisture, coupled with the cutting out of decayed and

*Insectidal smoke generators are manufactured by Octavius Hunt Ltd, 5 Dove Lane, Bristol, and ICI Ltd.

infested wood, may well suffice. Damage done by the wood-boring weevil is sometimes mistaken for furniture-beetle attack; the galleries run longitudinally, and are of about the same diameter as those of the furniture beetle, and the frass excreted by the feeding larvae is gritty but finer than that of the furniture beetle. If the galleries are searched, or heavily attacked timber is broken up, it is usually possible to find, if not whole beetles, at least a snout, which puts identification as weevil damage beyond doubt.

The common furniture beetle or 'woodworm' is a pest of sound, seasoned timber but, even with softwoods, infestation is largely confined to the sapwood unless decay is also present. It was formerly thought that softwoods were unlikely to be attacked until they had been in service for about 15 years, and hardwoods not within the first 20 to 30 years. As a result of more detailed studies of the life cycle of the common furniture beetle, entomologists have revised their earlier views, concluding that initial infestation may occur as soon as the timber is seasoned, although it may not be discovered until attack has been underway for some years (see also Chapter 20). The larvae excrete a gritty frass, which contains elongated pellets. In dealing with outbreaks, it is important to determine whether the infestation is still active, because only then are remedial measures necessary; inspections should be made in the early months of the 'flight' season, that is May and June. With experience, fresh infestation can be detected by the bright colour of the frass, which appears to be spilling out of the flight holes, and the margins of the holes are clear-cut. With age the frass becomes discoloured, and the rims of the flight holes are burred over.

Where the treatment of small portable articles such as furniture is being considered, there is no doubt that the most effective control measures are heat sterilisation and fumigation.

Where kiln facilities exist, and the timber is such that it will not suffer from a kiln-sterilisation treatment, the method is to be recommended because of its 100 per cent efficacy when properly carried out. A temperature of 70°C and 100 per cent relative humidity, held for 2 hours, would be lethal to fungal infection and

insects in timber up to 75 mm in thickness, but these conditions would be too severe for most manufactured woodwork. Temperatures as low as 50°C and a relative humidity of 60 per cent, held for 36 hours, have been found effective for sterilising *Lyctus*-infected timber; allowing for the lag period in heating up the wood to the temperature of the kiln, and a margin for safety, 46 hours for 25 mm material, and up to 50 hours for 75 mm timber, are suggested as maximum periods. It seems probable that a temperature of 55°C and 80 per cent relative humidity, with a total exposure period of 2.5 hours for 35 mm and up to 7 hours for 75 mm material, will prove adequate against any form of insect infestation, but this has not yet been conclusively tested. Neither French polish nor turpentine varnish finishes are appreciably affected by a temperature of 55°C and a relative humidity of 80 per cent, nor is plywood bonded with resin, casein, or blood-albumin glues for the periods necessary for sterilisation. A table of alternative temperatures and relative humidities for the successful sterilisation of beetle infected timber is given in BRE Digest 327 (1993).

Next to kiln sterilisation, fumigation is the most effective means of eliminating active insect attack, whether in the egg, larval, or beetle stage. Portable articles are fumigated in special chambers; skilled operators are required to carry out the work because of the poisonous nature of the fumigants. The fumigant in general commercial use in the UK is methyl bromide, although in other countries gases such as phosphine and sulphuryl fluoride are also used. Fumigation is essentially a task for the expert since a considerable degree of knowledge and skill is required regarding the concentration of gas and exposure times required. A knowledge of the possible deleterious effects of the fumigant on associated non-woody materials is also essential. (The use of methyl bromide is a specialist operator procedure and may only be undertaken by professional operators in compliance with the HSE Guidance Note CS12).

Both of these methods will provide complete eradication but will not prevent subsequent reinfestation if the articles are replaced in an infested building. Thorough application of an organic solvent-based preservative containing a persistent insecticide such as permethrin following sterilisation is advisable in these circumstances.

Use of the sterilisation technique is not considered practicable in the treatment of carcassing timbers in buildings in the United Kingdom. *In situ* treatments of common furniture beetle in such timbers have traditionally consisted of the spray or brush application of solutions of permitted insecticides, such as permethrin, in organic solvents, such as white spirit or odourless kerosene.

These will provide both an initial mortality of a large proportion of the larvae within the timber together with a layer of persistent insecticide which will kill any surviving insects attempting to emerge as adults in the years following treatment. This insecticide layer will also prevent the successful establishment of larvae hatching from eggs laid by beetles from other sources.

Formulations of these insecticides as emulsion pastes can be used for eradication of common furniture beetle although the extra cost and time involved normally restricts their use to treatments of buildings in which cost is a low priority.

Since the 1970s there has been a trend away from organic solvent insecticidal liquid and towards those based on liquid emulsion systems. There is no record of any level of failure of these products but they are known to achieve poor initial mortality, and thus rely more heavily for their effectiveness on the toxicity of the persistent insecticide layer to emerging adults. Since these products usually contain a very high proportion of water, they do have the advantage over the organic solvent based types of very low fire hazard.

Timber should be brushed down prior to treatment, or a powerful vacuum cleaner can be used; any heavily infested edges or surfaces should be cut back to sound wood. The preservative should not be allowed to run down on to plaster surfaces. In theory, all timber within, for example, an infested roof or floor should be treated, but this is both economically impracticable and undesirable from an environmental viewpoint. Treatment usually can be confined to actually attacked

timber, and immediately adjacent members; it is advisable to take up flooring to permit the treatment of joists and plates beneath. In most buildings there is likely to be much hidden timber requiring treatment, for example joists and plates, and roof timbers behind sloping plaster ceilings. Some opening up is essential to determine the severity of attack in such timbers. Treatment of painted timber such as skirtings and joinery is a difficult process since the treating fluids will not pass through the paint film. Wherever possible, such infected items should be replaced completely although repeated injections of fluid into flight holes may provide some measure of control.

Unless the amount of sapwood in carcassing timbers is excessive, the likelihood of the timbers being structurally weakened by furniture-beetle attack is remote. There is, however, no justification for ignoring active infestation, but careful consideration should be given to the necessity for carrying out general insecticidal treatments. If the treatment is really thorough, the cost of 'making good' after the complete treatment may be appreciably greater than any real damage the pest could possibly do in the next half century or more. Further, it is quite common to find ample evidence of old attack, but no continuing active infestation, although by no means all sapwood in adjacent timbers has become affected, indicating that even widespread infestation may die out completely while there are still ample food supplies available to support active attack of the pest. Structural damage is more likely to be found in old period cottages, where the ceiling joists and rafters were little more than half-round poles. With the passage of time all the sapwood may have been destroyed, when there is insufficient sound wood left for structural purposes. In such circumstances an insecticidal treatment is pointless, and the attacked timbers have to be renewed, although this may necessitate stripping and recovering roofs, and taking down and renewing ceilings.

There are certain ancillary precautions that are worth observing, even by residents of modern blocks of flats containing little or no structural timber. Second-hand furniture should be carefully inspected for any traces of flight holes before it is brought into the home. Wicker-work articles, such as linen baskets and waste-paper baskets, old plywood packing cases and discarded articles stored in lofts, should always be suspect because these are ideal breeding grounds for the common furniture beetle. A surprising number of outbreaks can be traced to such sources, and hence the frequency with which attacks are found to be concentrated around access hatches to roof voids, and in cupboards under stairs. The articles mentioned should be carefully inspected at the end of May or early in June each year, and any infected article is best burned or otherwise disposed of.

The *Lyctus* beetle is the common powder-post beetle, the larvae of which feed on the starch stored in the sapwood of some timbers. Only hardwoods are attacked, and then only species of timber with pores large enough for egg-laying (about 0.1 mm in diameter): seasoned, or nearly fully seasoned timber is selected by the adult for egg-laying. In effect, only some timbers are attacked, and then only the sapwood of such timbers, provided it contains sufficient stored starch. The frass is very fine powder, which feels like flour when rubbed between the fingers.

Eradication of powder-post beetle infestation is rather more difficult than dealing with furniture beetle attack, because the damage is usually more deep-seated and extensive by the time it is detected. Where practicable, heat sterilisation is the most effective method of eradication, and, provided no timber is excluded from sterilisation, further precautions are likely to be unnecessary. Heat sterilisation is only practicable for 'portable' articles or for stocks of sawn timber or components at mills or factories.

Lyctus infestation in timbers more than about 10 years after felling is rare, as the starch content becomes reduced with time. In buildings, infestations are usually either in old hardwood timbers, such as oak, and are extinct, or are in recently installed joinery. In the latter case, treatment is rarely practical and since infestation is usually initiated before installation, compensation from the supplier is usually sought.

To sum up: identification of the particular pest at work is important, but it is of no less importance to establish whether or not there is continuing *active* infestation. In the past, many thousands of pounds have been expended in applying wood preservatives indiscriminately to the timbers of our churches and other buildings, often when infestation was no longer active, and even may have been 'dead' for upwards of a century or more. With some insects, it is often sufficient to concentrate on eradicating the decay, which will dispose of the beetle pests too. Heat sterilisation is the most certain method of killing all stages of infestation, and is recommended for portable items, sawn timber stocks or stocks of components. The *in situ* use of wood preservatives has an important place in dealing with areas of continuing, active infestation, and where it is essential to prevent re-infestation. It must not be overlooked that preservatives do not restore the strength properties of attacked timbers. The proper cleaning of wood prior to treatment, and the thoroughness with which the preservative is applied, are all-important.

Practical advice on safety measures that will reduce the risk to individuals and the environment before, during and after the application of wood preservatives in remedial work is set out in BRE Digest 371 (1992).

References

BRE (1986) Controlling death watch beetle. *Building Research Establishment, IP 19/86.* HMSO: available only from the Building Research Establishment Bookshop.

BRE (1992) Remedial wood preservatives: use them safely. *Building Research Establishment Digest 371.* HMSO: available only from the Building Research Establishment Bookshop.

BRE (1993) Insecticidal treatments against wood-boring insects. *Building Research Establishment, Digest 327.* HMSO: available only from the Building Research Establishment Bookshop.

HSE (1991) Fumigation using methyl bromide. *Health and Safety Executive Guidance note CS12.*

Preservation of Timber

22.1 General principles

The principal causes of deterioration of wood in service, as distinct from deterioration during seasoning, are attack by wood-destroying fungi, insects and marine borers, fire, and mechanical abrasion and weathering. Mechanical abrasion and weathering will not be considered here, but the resistance of timber to the other agents of destruction may frequently be enhanced by the application of suitable chemical formulations. Various substances were used for this purpose by the ancient Egyptians and the Romans, but the extensive use of wood preservatives is essentially a development of the last hundred years or so. This chapter will be devoted mainly to the preservation of timber against attack by biological agencies; methods used to decrease the risk of fire damage are dealt with briefly at the end of the chapter.

Wood preservatives are usually applied to non-durable timbers so as to render them more resistant to biological attack. This may be because they are the only materials available or because they can then be used to replace a naturally durable, but more expensive timber. Modern wood preservation uses relatively few different biocides and methods of treatment. The correct selection of the most suitable wood preservative and method of treatment for a particular timber component is of the utmost importance and must be based on a thorough understanding of the scope and limitations of preservative treatments: none confers complete immunity to the treated timber, and a treatment suitable for one particular end-use situation may be useless for others.

Wood preservatives are also used remedially to control or eliminate an existing infestation. This requires a different approach from that used for the pretreatment of timber since the infested timber will be in an existing structure, such as a house, and the timber or the organism within the timber may be difficult to reach. In such instances, surface application or injection of biocidal preparations are used to control both insect attack and rot.

22.2 Properties of preservatives

It is important to keep in mind the special circumstances of each particular job when selecting a wood preservative. For example, a substance that is readily soluble in water may be excellent for indoor use but worthless for outside work and, conversely, a substance with a pronounced odour may be quite satisfactory for outdoor work but totally unsuitable for indoor use. The better preservatives can protect properly treated timber for decades, but extravagant claims made for some proprietary products should always be treated with caution; even the best preservative does not confer immunity for ever.

The ideal wood preservative has yet to be found, but the properties desirable in such an agent may be enumerated, and are useful as a basis for comparison. It should be:

(1) highly poisonous (toxic) *only* to the target organism(s), such as fungi, insects or marine borers;
(2) easily able to penetrate wood;
(3) chemically stable (that is, not readily decomposed or altered);
(4) permanent (that is, not readily volatilised nor leached out by liquids);
(5) non-deleterious to the timber when treated;
(6) cheap and readily available; and
(7) not liable to increase the flammability of wood.

The following additional qualities are sometimes important, depending on the end use of the treated timber:

(8)   odourless;
(9)   colourless and free from effect on subsequent painting or finishing processes;
(10) non-corrosive to metals or other materials.

The first four properties are essential qualities of any wood preservative, and the remainder are of diminishing importance, according to circumstances. Guidance is available in BS 1282.

22.3   Testing and approval of preservatives

As successful preservation of timber is a long-term process dependent upon the thoroughness of treatment as well as the preservative, the most reliable evidence of the likely performance of a preservative in service can only be obtained through the establishment of field trials. In such trials, fully recorded and controlled experiments are carried out in which different timber species, treated with different preservatives using different methods of application, are placed in the field and can be inspected at regular intervals to assess the benefits of the treatment. Originally, the trials consisted primarily of noting the performance of treated stakes planted in the ground with secondary trials in particular locations, such as of railway sleepers in the track and test pieces in the sea or in water cooling towers. Since the 1960s and coincident with the growth or preservative treatment for external joinery, an increasing number of trials have included treated joints (for example, simulating the corner of window frames) exposed out of doors and out of contact with the ground. Experimental buildings have also been constructed so that model window frames may be exposed to natural outdoor climates and controlled indoor climates of enhanced temperature and humidity.

Because field trials take many years to complete, short-term laboratory tests have been developed to test the important characteristics of wood preservatives, such as toxicity to the intended target organisms and permanence within the treated timber. European Standards organizations through their membership of the Comité Européen de Normalisation (CEN) have developed European Standards covering many of these requirements. Data generated from these Standards are often demanded by the various approvals bodies that some European countries have established to control the use of preservatives. More recently, the European Economic Community (EEC), following the passing of the Single Market Act, has asked for European Standards to be harmonised so that a common approach can be taken by all EEC members in assessing and marketing wood preservatives throughout the Community.

There is increasing awareness of the dangers to human health and to the environment posed by widespread use of toxic chemicals. Within the UK, the sale of wood preservatives is controlled by the Pesticides Regulations which are administered by the Health and Safety Executive (HSE) of the Department of Employment. Under these regulations, no wood preservative can be used within the UK without first receiving clearance from the HSE, and containers used to distribute a preservative must bear labels carrying information adequate to ensure its safe handling and usage. Developments within Europe, associated with the Single Market Act, will again result in an agreed common approach throughout Europe on the control of wood preservative use. A list of the new European Standards is given in section 22.9.

22.4   Classes of wood preservatives

Conventional wood preservatives which give protection by virtue of their toxicity to wood-destroying organisms are classified as:

(1)   tar oil preservatives, such as creosote,
(2)   water-borne preservatives, which are salts or mixtures of salts in water, and
(3)   organic solvent preservatives, in which the active ingredients are dissolved in petroleum distillates.

Other types of wood preservatives are available; these are discussed briefly at the end of this section.

22.4.1   *Tar oil group of preservatives*

The principal representative of this group is creosote which is derived from the distillation of coal tar, usually at between 200°C and 400°C. It is a complex mixture of well over 200 substances, making it difficult to assign the role that these various substances play in preserving the wood.

The heavy creosote oils, such as those defined in British Standard BS 144: Part 1 (1990) as type 1 and type 2, are suitable for use in vacuum/pressure treatment processes. Heating during the treatment process is needed to improve their liquidity and penetration into the wood; BS 144: Part 2 (1990) specifies 65°C to 100°C. The lighter creosote oils, such as those defined in BS 144: Part 1 (1990) as type 3, are used for application by immersion, spraying and brushing.

Creosote is mainly used in the treatment of large-dimensioned timbers for exterior use, such as railway sleepers, transmission poles and harbour works. It is precluded from many indoor and other building uses because of its smell, the risk of tainting food stuffs, its diffusion into adjacent plasterwork and the difficulties in overpainting. Exudation or 'bleeding' of creosote from timber treatment may be a problem, but this has been reduced as a problem by modified treatment methods and more precise creosote specifications.

22.4.2   *Water-borne preservatives*

The water-borne formulations used today are mainly based on copper/chromium mixtures of salts or oxides, with copper/chromium/arsenic or CCA (marketed under the 'Tanalith' or 'Celcure' trademarks) probably the most common (see BS 4072). Other combinations include copper/chromium itself, copper/chromium/boron and copper/chromium/silicofluoride. In all of these,

chromium acts as an oxidising agent so that, once deposited in the wood, the water-soluble mixture becomes water-insoluble ('fixed') and thus not liable to leaching. As with any preservative, however, an adequate treatment is essential: the firms manufacturing these chemical mixes ensure this by treating the timber under pressure in their own plants or supplying the chemicals only to treatment plants that conform to the manufacturer's process requirements.

The application of CCA, and similar treatments, is an alternative to creosote, especially when freedom from odour and a paint finish are required. While, as with most industrial pre-treatments using creosote, the timber must be dry before application of the preservative, it must not be overlooked that timber pressure treated with water-borne salts has a very high moisture content immediately after treatment and should be dried again before being put into service; for many purposes this may entail kiln-drying after treatment to obviate delays in a building or repairs programme. Apart from their use in buildings where creosote cannot be easily employed, water-borne preservatives have found increasing use for treatment of posts and poles and for protecting timber in water cooling towers. The CCA mixtures have been outstandingly successful. Despite this, it is generally accepted that they do not give such long-term protection against soft rot to some hardwoods in tropical countries. Difficulties in achieving adequately uniform penetration of preservative into the wood structure have been cited as the possible cause.

Inorganic boron compounds have long been recognised for their fungicidal and insecticidal properties. However, they do not become fixed within the wood following treatment and are, therefore, liable to be leached from the wood during service. For this reason they are confined to those environments where leaching will not take place. The most important application of a boron compound as a preservative is by the diffusion process (see later) using disodium octaborate tetrahydrate which is much more soluble in water than boric acid, allowing higher loadings to be achieved.

Two further applications of water-borne fungicidal preparations, which are not preservation in the strictest sense, lie in the control of sap stain and in the sterilisation of walls infected by the dry rot fungus, *Serpula lacrymans* (see Chapter 21).

The most effective of the anti-sapstain formulations have been those containing the sodium salts of the chlorinated phenols, particularly sodium pentachlorophenoxide (NaPCP). In the tropics, for example to prevent *Lyctus* attack on susceptible hardwoods, oil-soluble insecticides are incorporated into the treatment by using dilute oil-in-water emulsions. The use of NaPCP is diminishing because of possible risks to the environment, and new anti-sapstain fungicides are being sought. Research so far has led to the appearance of such compounds as 2-(thiocyanomethylthio) benzothiazole (TCMTB) and methylene bis-thiocyanate (MTB).

Stain control can be difficult on thicker dimension (50 and 75 mm) boards if rapid air drying leads to surface splitting. When this occurs, 'bluestain' spores can gain access to moist untreated wood below the superficial layer of fungicide, leading to internal stain which may not be discovered until the dried wood is re-sawn. This trouble has occurred sporadically in Finland with Baltic redwood, where it is known as 'secret' stain, and also in Malaya with ramin.

With the growth of the practice of packaging and transporting unseasoned timber, usage of the anti-stain treatments, originally developed to prevent sap-stain during seasoning, has been extended to protect the packages from both stain and decay (see Chapter 19).

Water-based formulations for the treatment of walls infected with dry rot are based on NaPCP, sodium 2-phenylphenoxide or, more recently, mixtures of dodecylamine salicylate and lactate. Again, it is likely that the use of NaPCP will diminish as new formulations appear on the market.

22.4.3 *Organic solvent wood preservatives*

These consist essentially of fungicides or insecticides dissolved in petroleum distillates, such as white spirit. They are more expensive than water-borne preservatives because of the higher cost of the solvent. However, the penetration of this group of preservatives is usually good and most are reasonably permanent for both interior and exterior use. They do not cause dimensional changes or distortion to the wood on treatment, and the treated surfaces can generally be painted without difficulty when dry. Organic solvents are not so benign as water and can affect other materials, so care should be taken when using them indoors, for instance polished surfaces should be protected. If treating by spray, goggles and face masks should be used to protect against inhalation or eye contact. They are also flammable, calling for especial care in storage and handling.

A variety of biocides are used in formulating these preservatives. Probably those with the longest use are the fungicides pentachlorophenol (PCP) and tri-n-butyltin oxide (TnBTO), PCP being used at 5 per cent m/m (mass-to-mass basis) and TnBTO at 1 per cent m/m when they are the only biocide present. Copper and zinc naphthenates also have a long history of use in this area, although more recently other carboxylic acid derivatives have appeared, because the naphthenates are becoming more expensive and more difficult to specify. Thus copper and zinc versatate, zinc octoate, and the preparations known as acypetacs zinc and acypetacs copper, are now approved and used in the UK. Since the chemical mix within the napthenic acids used originally to produce the copper and zinc salts was complex, it has become normal to express the concentrations of these biocides in terms of the equivalent metal content. For both copper and zinc this is about 2–5 per cent m/m for current formulations.

The insecticides traditionally used in these formulations were gamma-HCH (lindane) and HEOD (dieldrin). Dieldrin is now banned in many countries and, although lindane is still used, the insecticides of choice are now the synthetic pyrethroids permethrin and cypermethrin. Lindane is used at 0.5–1.0 per cent m/m permethrin at 0.1–0.2 per cent m/m and cypermethrin at 0.05–0.1 per cent m/m.

An organic solvent preservative may contain a fungicide, or an insecticide, or may be a 'dual purpose' formulation containing both types of biocide. It is therefore imperative that the correct type of preservative is obtained for the particular use envisaged. In addition, these formulations can contain a range of additives, such as waxes, oils and resins. One of the main reasons for including such materials is to improve the water repellency of the treated wood, but improvements in penetrability and drying time, and the prevention of 'blooming' (that is, the appearance of the biocide as crystals on the surface of the treated wood during drying) are also associated with the inclusion of these materials.

Part 1 of BS 5707 specifies certain organic solvent formulations used in the UK, while Part 2 describes suitable methods of treatment; Part 3 specifies PCP formulations that may be used without subsequent painting.

22.4.4   *Other preservative types*

Most of these are used principally for the remedial or *in situ* fields of preservation. Oil-in-water emulsions, where the oil phase carries the biocide, are available as thick pastes or as liquids. The pastes are spread locally on the surface of the wood, where they slowly diffuse into the infested zones. They are useful where large-dimensioned timbers in buildings are difficult to treat by other means. Dilute oil-in-water emulsions are now popular for remedial insecticidal spray treatments, replacing many of the organic solvent formulations used for this purpose, because water is more environmentally acceptable and cheaper. More recently micro-emulsion formulations have been developed for this purpose.

Solid inserts are also used in remedial treatment; commonly these are fused or compressed preparations of water-soluble boron compounds. These are placed in holes drilled into the affected wood and subsequently sealed in with a pretreated wooden plug. Emulsion pastes are sometimes injected into drilled holes in a similar fashion. Both of these treatments are best used as remedial applications to those timbers which are likely to

begin decaying in the near future, rather than those that are already being degraded.

Gases may be used in diffusion treatments, particularly for the remedial treatment of standing poles, where capsules are inserted and sealed into drilled holes and then broken so that the released gas can permeate the wood. Pretreatment with gaseous boron compounds is currently being developed.

22.5   Application of wood preservatives

Unless the method of treatment is specifically designed to treat freshly felled, green timber, all timber must be sufficiently dry before treatment. This is because the objective of a preservative treatment is to introduce the preservative solution into the wood via the voids within the wood's cell structure. At high moisture contents, these voids will be filled with water and it is only by reducing the moisture content of the wood to 28 per cent or below (when only the wood substance will contain water) that the voids will be free of water and available to the incoming preservative solution. The importance of adequate drying before treatment cannot be overstressed. If the wood is too wet, it will be impossible to introduce the appropriate amount of preservative, leading to inadequate penetration and retention, and premature failure in service. The moisture content of the timber should be checked before commencing treatment.

Inner and outer bark should be removed from the wood (this is not necessary for the sap displacement method – see section 22.5.7) as they are both impervious to preservatives. The timber should be free from dirt and surface water, and whenever possible, all machining, cutting, boring, etc. should be completed before preservative treatment.

The principal methods of applying wood preservatives to building timbers are briefly described below. Most of these are available worldwide, although in some developing countries the choice of commercial processes using industrial plant may be restricted or non-existent for the building industry. Where choice is restricted, it is usually confined to the less effective treatment methods. If these have to be used, it is essential that the timber is in a proper condition for treatment and that the most appropriate preservative formulation is employed. Different timbers vary widely in their response to preservative treatment; BRE Digest 296 (1985) lists the treatability of over 200 species. BS 5268 Part 5 and BS 5589 provide general guidance on the treatment of timber components.

22.5.1   *High pressure/vacuum processes*

The main advantage of a high pressure/vacuum process is that it provides a relatively deep, uniform penetration together with a high absorption of preservative, thus ensuring long-term effective protection. The process is normally used for the water-borne preservatives, such as copper/chromium/arsenic (CCA) formulations, and creosote. This type of treatment is particularly suited to the protection of timber where it is exposed to a high risk of biological attack; it is recommended where timber will become and remain wet during its service life, as occurs in ground contact or in water.

High pressure/vacuum processes all require the same basic plant and equipment. This includes a treating cylinder capable of withstanding vacuum and high pressure; storage, measuring and mixing tanks for the preservative; vacuum, circulating and filling pumps; an air compressor or pressure pump; tramways and bogies for carrying the timber into and out of the treating cylinder; heating equipment in certain cases; and associate pipework, valves and instruments.

Cylinders vary widely in dimensions. A common size is about 1.5 m diameter and 12 m in length, but some of the larger cylinders can be 2 m in diameter and 36 m in length. Normally the cylinder has a door at one end, although for quicker throughput, a door at each end allows treated stock to be removed at one end while fresh material is introduced at the other. Recent environmental legislation within Europe has required a greater degree of retainment of the plant and the treated timber. Bunding and concreted storage areas have been introduced to prevent loss of the preservative to the surrounding environment, especially if a major accident results in the release of a large quantity of preservative fluid. Much of the operation is now carried out under cover so that rain does not wash preservative from freshly treated timber or the immediate environs of the plant.

The essential step in the high pressure/vacuum type of treatment is that the timber is placed in a closed cylinder and immersed in the preservative fluid. A relatively high over-pressure is applied, usually 10–14 bar held for between 1 and 6 hours, which forces the preservative into the wood. This results in the sapwood of many timbers being completely penetrated, but penetration into the heartwood is not normally as deep and may vary considerably between different timber species. The CCA preservatives are applied at ambient temperatures and never above 40°C, while creosote is used at temperatures between 65°C and 100°C.

There are two main types of process in this category, **full cell** and **empty cell**. In a full cell (or **Bethell**) process, the intention is to produce the maximum possible retention and depth of penetration, and this results in the available voids in the wood becoming more or less completely filled. This is achieved by first applying a vacuum to the closed cylinder after it has been charged with timber. While maintaining the vacuum, the cylinder is filled with preservative fluid and then the vacuum is released. Since the pressure inside the timber is still below atmospheric pressure, this causes the preservative to move into the timber. The ingress of preservative is enhanced by the application of the high over-pressure previously mentioned. On completion of the pressure period, the cylinder is drained of preservative and a final vacuum applied which withdraws a small amount of preservative (generally referred to as the '**kickback**') and leaves the surface of the treated timber dry and in a reasonable state for handling.

In the empty cell treatment, the intention is to recover more of the preservative in the 'kickback' stage, thus decreasing the retention. However, the penetration is generally similar to that obtained with the full cell process. This results in the internal walls of the wood cells being coated with the preservative, but the voids in the wood being only partially filled. This is achieved by either applying a pressure to the loaded cylinder before it is filled with the preservative (the **Rueping** process) or filling the cylinder at atmospheric pressure (the **Lowry** process). In all other details, these processes are the same as the full cell process.

22.5.2 Double vacuum process

This method introduces light organic solvents into timber and is particularly useful for the treatment of made-up components, such as windows, and other exterior joinery, since there is no distortion nor change of dimension associated with the treatment. It may also be used to treat roofing timbers and studding against insect and termite attack. For permeable timbers, the process allows control over the quantity of preservative introduced into the timber and the depth to which it penetrates.

The plant associated with this process is similar to that required for the high pressure/vacuum process, except that the pressures involved are much less. The timber is placed in a treating cylinder and a partial vacuum created and held for several minutes before the cylinder is filled with preservative while still under vacuum. The vacuum is released and the timber allowed to remain in the preservative for up to an hour under either atmospheric pressure or an applied pressure of up to 2 bar. After the pressure is released and the cylinder drained, a second vacuum is created to recover a proportion of the preservative from the wood and to provide a dry surface.

22.5.3 Diffusion with aqueous solutions

The diffusion process can result in complete penetration even in timbers resistant to penetration by pressure methods. Since the preservative remains water-soluble even after drying, timber treated in this way cannot be used where leaching is likely to occur, for example in ground contact or where it is directly exposed to the weather. However, it may be used out of doors when protected by paint or design. The preservative usually introduced in this way is a highly soluble boron compound, such as disodium octaborate, which acts as both a fungicide and an insecticide/termiticide. These boron compounds are relatively harmless to man and other mammals, and may therefore be used in treating internal building timbers where contact with the occupants is possible.

The diffusion method is unusual in that it is applied to 'green' timber, that is undried timber freshly felled or fresh from the saw; in fact, the higher the moisture content of the timber, the better the treatment. The timber is immersed for a short period in a concentrated (sometimes hot) solution of a water-soluble wood preservative and then close-stacked under cover to prevent drying. Over a period of several weeks the preservative diffuses into the wood to give complete penetration. The timber is then dried to the required moisture content. Because the treatment must be carried out before any drying has taken place, the diffusion process is usually carried out close to the felling operation, for instance, at the saw-mill.

A variation on the method is the double diffusion process; this provides the opportunity to incorporate a non-leachable preservative system in timber using simple diffusion techniques. This is achieved by first introducing one water-soluble chemical into the wood, following the methodology described above, and then introducing a second in the same way, which reacts with the first chemical as it diffuses through the wood to produce a water-insoluble preservative. For example, the first solution could contain copper sulphate and the second a mixture of sodium dichromate and sodium arsenate, in which case an insoluble preservative could be formed similar to that ultimately obtained by treating with a CCA formulation. Although simple to carry out, the process is time-consuming and there is little control over the depth of penetration and the retention achieved.

Recent developments in preservation include pretreatment by gaseous diffusion of wood and panel products using trimethylborate. This is introduced into a sealed, evacuated chamber containing timber or panel products at a relatively low moisture content. The gaseous trimethylborane permeates the wood or board material and reacts with the moisture to produce the biocide boric acid which is deposited where the reaction occurs.

22.5.4   Immersion treatment

Usually this type of treatment method is suitable only for the more permeable timbers where the risk of biological degrade is relatively low, such as in internal joinery. Immersion treatments include dipping, soaking and steeping. The term 'dipping' usually applies to short immersion periods (up to 10 minutes), while 'soaking' and 'steeping' are associated with longer periods (hours or days). Only light organic solvent preservatives and certain grades of creosote are suitable for immersion treatments; water-borne preservatives achieve relatively little penetration, even after prolonged steeping. The process involves merely submerging and retaining the timber below the surface of the preservative fluid for the required period.

Care should be taken in processing the timber in this way, for the open tanks of preservative and the handling of treated timber must be considered a significant health risk to operatives.

22.5.5   Brushing, spraying and deluging

Although brushing and spraying are probably the easiest and most widespread methods of applying wood preservatives, their effectiveness is slight compared with other methods, particularly for protection in the tropics. If the method has to be used, the woodwork should always be flooded with as much preservative as possible, particularly at the end-grain and where joints and cracks

appear. If brushing, the application should be liberal and not brushed out as with painting. A second treatment after a few days and a regular retreatment every few years are recommended, always when the timber is very dry.

Deluging is essentially a sophisticated form of spraying. The timber is passed through a tunnel on a conveyor system during which preservative is sprayed or flooded over all the surfaces of the timber. The process is generally regarded as equivalent to a 10-second immersion.

As with immersion, all these treatments are only useful with light organic solvents and certain grades of creosote, although emulsion systems are also applied in this way, especially for remedial applications (see section 22.4.4).

22.5.6   Hot and cold open-tank process

This simple process is probably the most effective method of introducing preservatives into timber without the use of expensive equipment and, as such, is especially attractive to the developing countries. The penetration and retention of preservative achieved with this process results in treated timber that can be used for most situations where the risk of biological degrade is high. However, the use of open-tanks of preservative fluid and the application of heat require that all possible safety precautions are taken when using this type of treatment.

The hot and cold open-tank process can be used with any preservative that does not decompose with heating, although it is not recommended for light organic solvent formulations and similar volatile preparations because of the fire risk. Creosote or creosote-type preservatives are probably the most commonly used. Where the preservative is liable to heat degradation, the timber can be heated in hot water, hot air or steam, and then quickly transferred to a tank of the cold preservative where it is left to cool.

The process relies on the changes in volume that occur when a fixed quantity of air is heated or cooled. Seasoned wood is immersed in a tank of cold preservative which is then heated to

85–95°C for up to 3 hours. During this period, the air within the voids in the wood expands, causing much of it to be expelled. The heat source is removed and the tank contents are allowed to cool, thus causing the residual air within the wood to contract and draw preservative into the vacated spaces. The treatment times and temperatures are varied to suit the treatability, and the size of the timber. If a relatively large retention is required, the process can be terminated at this step. However, similar penetration but less retention can be achieved by re-heating the preservative for 1 to 3 hours, thus inducing a 'kickback', and removing the wood while the preservative is still hot. The hot timber is then allowed to cool in the air.

The plant required for this process can be simple and inexpensive. As with the immersion process, all that is required is a tank capable of supporting the weight of the timber and preservative together with a system to keep the timber submerged, for example holding-down bars or a weighted cradle or cage. The heat source can be an open fire built beneath the tank: particular care should be taken when using this approach to minimise the risk of fire and spillage.

22.5.7 Sap displacement method

This method is used mainly to treat unseasoned 'green' poles, essentially removing the sap by forcing a preservative solution in to replace it. The level of treatment is good, for the whole of the sapwood band is saturated with the preservative. There are several variants to the process, although all involve water-tight caps being attached to the ends of the poles. The **Boucherie** process uses hydrostatic pressure; the preservative is led into one end of the pole from an elevated tank, thus forcing the sap out at the other end. In the **Gewecke** process, the poles are placed in a cylinder which is filled with preservative and pressurised. This ejects the sap from the poles, which is led away through the caps and the tubes to the outside of the cylinder.

22.5.8 Treatment with limited resources

Where effective and long-lasting protection of timbers is required and few resources are available, the methods most likely to achieve this are those based on the principles underlying the diffusion process, the hot and cold open-tank process or prolonged steeping.

22.5.9 Drying of treated timber

Timber treated with CCA preservative requires a period of storage without drying to ensure fixation (that is, conversion to insoluble forms) of the preservative within the wood. In temperate climates this will take 7–14 days but in the tropics a shorter fixation period may be expected. Recent developments in plant technology include accelerated fixation systems where the preservative is fixed in the timber before it leaves the cylinder. Following the fixation period, the treated timber may be brought to its service moisture content either by open-stacking to allow air-drying, or by kiln-drying.

Timber treated with light organic solvent preservatives (LOSP) or creosote are normally open-stacked immediately after treatment and left for 2–7 days to allow the solvent to evaporate. Creosote is used undiluted and, by definition, does not have a drying solvent. Nevertheless, a period of storage does allow the more volatile components to evaporate and the timber to acquire a relatively dry surface.

Concern in certain countries over the uncontrolled emission of volatile organic compounds and impending legislation has increased interest in forced drying systems which incorporate solvent recovery.

22.5.10 Retreatment of cut surfaces

Most preservative treatments are best carried out after all machining, cutting and boring operations are complete. If such action is unavoidable, all freshly cut surfaces should be liberally coated with the preservative fluid originally used or a formulation especially manufactured for this purpose. However, in general such action cannot reinstate the original quality of the treatment.

22.6   Health and safety

All preservatives should be handled with care, paying great attention to the manufacturer's recommendations and instructions. Protective gloves should be worn at all times when handling treated timber which is wet or contains solvents, although this is not normally necessary once the timber is dry. After any operation involving the handling of treated timbers, hands should be washed thoroughly, especially before handling food. If treated timber is machined or sanded, an efficient dust extraction system should be used. If this is not available, dust masks should be provided and used.

22.7   Properties of treated timber

22.7.1   Strength

In general, the treatment of timber with wood preservatives does not significantly affect its strength properties and can be ignored in design calculations.

22.7.2   Corrosion of metal fasteners and fittings

Certain types of metal fixings will be liable to corrode if the wood in which they are embedded becomes wet. This effect is exacerbated if salt-type wood preservatives are present (for instance, CCA formulations) and if the ambient conditions are hot and humid. There are several ways of minimising this effect:

(i)   If CCA preservatives are to be used, select an oxide, rather than a salt formulation, and do not attach the fittings until the chemical

fixation process is complete and the moisture content of the wood is below 20 per cent.

(ii)   If the timber is likely to become wet during service and a long life is required, fittings of austenitic stainless steel (excluding free machining grades), or copper, or silicon bronze should be chosen.

(iii)   If only occasional dampness is expected, coated low carbon steel (for example, sherardized, galvanized or cadmium plated) fittings may be used. Better performance will be achieved with thicker coatings. And

(iv)   For CCA treated wood, aluminium alloys containing copper should not be used. Sheet aluminium roof coverings or claddings should not be used in direct contact with CCA-treated timber.

22.7.3   Surface finishes

A normal paint system or exterior wood finish can be applied to timber treated with waterborne and most light organic solvent wood preservatives provided the timber has been allowed to dry properly. Exceptionally porous wood may need an extended drying period before painting.

Timber painted with creosote, or pentachlorophenol in heavy oil cannot be painted.

22.8   Validation of treatments

It may be possible to ascertain by general visual inspection whether a preservative has been applied to timber, but if the preservative is colourless and has no distinct odour, it will be necessary to apply chemical tests. The British Standard BS 5666: Part 2 lists methods which can be used to determine the presence (but not the quantity) of certain preservatives in wood. These methods include the use of spray reagents to reveal the penetration patterns of preservatives within the wood, and the separation and identification of preservatives in solvent extracts obtained from treated wood.

Current British Standards specify or recommend wood preservative treatments by defining the process to be used. However, the new European Standard, EN 351-1, instructs specifiers to describe the treatment they require in terms of the results of the process, that is by defining the penetration and retention achieved, rather than by the process itself (BRE, 1994). This will require a greater emphasis on chemical analysis of the treated timber to determine whether the appropriate level of treatment has been achieved. This standard will eventually replace the existing approach taken in current British Standards. While BS 5666: Parts 3 to 7 inclusive specify methods of analysis for treated wood containing certain types of wood preservative, not all formulations are included in this set of standards. Where no standard method of analysis exists the manufacturer's advice must be sought.

22.9 European standards

The following is a list of the new European Standards for wood preservation already published (or to be published in the near future) by the British Standards Institution:

BS EN 350-1 *Durability of wood and wood-based products. Natural durability of solid wood.*
Part 1 Principles of testing and classification of the natural durability of wood.
Part 2 Guide to natural durability and treatability of selected wood species of importance in Europe.
BS EN 351-1 *Durability of wood and wood-based products. Preservative-treated solid wood.*
Part 1 Classification of preservative penetration and retention.
Part 2 Guidance on sampling for the analysis of preservative-treated wood.
BS EN 355-1 *Hazard classes of wood and wood-based products against biological attack.*
Part 1 Classification of hazard classes.
BS EN 355-3 *Durability of wood and wood-based products. Definition of hazard classes of biological attack.*
Part 3 Application to wood-based panels.

BS EN 460 *Durability of wood and wood-based products. Natural durability of solid wood.*
Part 3 Guide to the durability requirements for wood to be used in hazard classes.
BS EN 599-1 *Durability of wood and wood-based products. Performance of wood preservatives as determined by biological tests.*
Part 1 Specification according to hazard class.

22.10 Flame-retardant solutions

The reaction of timber and board materials to fire is described fully in Chapter 10, section 10.1.6. In this section, attention is focused on the treatment of these products to reduce their ignitability and subsequent rate of flame spread; by effective treatment it is possible to achieve a class 1 spread of flame rating. However, it must be appreciated that these treatments have little effect on changing either their combustibility or the fire resistance rating of a component or timber structure fabricated from them. These treatments are no more than **flame-retardant** processes and their applicability to only one single aspect of fire behaviour must be clearly understood.

Flame-retardant treatments take the form of either surface applications or the impregnation of the timber with certain chemical solutions. Surface applications are used for *in situ* treatments and comprise either intumescent materials, or flame-retardant paints; these are described in Chapter 23, section 23.5.

Flame-retardant chemicals may also be applied to timber by impregnation, but this must be done prior to site construction. The chemical salts most commonly employed in the UK at the present time are mono- and di-ammonium phosphates, ammonium sulphate, boric acid and borax. Most of the proprietary flame-retardants currently on the market are mixtures of these constituents, formulated to give a balance between performance and cost. Treatment of the timber involves high pressure/vacuum processes (see section 22.5.1) with an aqueous solution, thereby necessitating redrying of the timber following impregnation.

While the application of this type of flame-retardant effectively decreases the ignitability and spread of flame ratings of timber, thereby permitting the use of timber in otherwise restricted applications, it should be appreciated that there are a number of implications in specifying such treatments especially in terms of structural efficiency. These implications are:

(1) *Reduction in strength and toughness.* It is now well established that the re-drying of the impregnated timber at normal kilning temperatures results in a loss of strength, and a marked loss in toughness or impact resistance of the timber, when solutions containing ammonium salts have been employed. This results from the loss of ammonia on heating and the subsequent attack of the cellulose by the acid residue (sulphurous and phosphorous acids). The extent of degradation is dependent primarily on the kilning temperature: using a temperature of 65°C, the resulting loss in most strength properties lies between 15 and 20 per cent, while toughness is reduced by 30–50 per cent. It is therefore imperative that timber impregnated with this type of fire retardant be re-dried at rather low temperatures.

(2) *Equilibrium moisture content.* There is a general tendency for timber impregnated with fire retardants to have a higher equilibrium-moisture content, which is of paramount significance when treated timber is exposed to high humidities for a long period of time. Such wet conditions affect the structural performance of the treated wood and favour the corrosion of metals.

(3) *Effect on metal fasteners.* Metals in contact with dry retardant-impregnated timber will remain free from corrosion; however, as the wood becomes wetter, corrosion commences and intensifies with increasing wetness of the wood: it is therefore imperative that treated wood containing metal fasteners be kept dry continuously. Alternatively, metal fixings of the types listed in section 22.7.2 may be used.

(4) *Bonding and painting.* Problems can and do arise in both bonding and painting flame-retardant-treated timber. Adhesives based on phenol and resorcinol formaldehyde or casein are unsuitable, while urea-formaldehyde should only be employed when the moisture content is below 12 per cent. Painting must also be restricted to timber with a moisture content below this level.

Conscious of the limitations of those flame retardants based on ammonium salts, a number of companies have developed retardants of markedly different chemical composition which appear to display much reduced or even zero degradation of the timber at normal drying temperatures; furthermore, they are non-corrosive against metal and result in normal equilibrium moisture contents. Such products, however, tend to be more expensive than their ammonium based counterparts. They are listed by TRADA (1991).

References

(a)   *British Standards (BSI, London)*

BS 144   *Wood preservation using coal tar creosotes.*
   Part 1: 1990   Specification for the preservative.
   Part 2: 1990   Methods for timber treatment.
BS 1282: 1975   *Guide to the choice, use and application of wood preservatives*
BS 4072   *Wood preservation by means of copper/chromium/arsenic compositions.*
   Part 1: 1987   Specifications for preservatives.
   Part 2: 1987   Method for timber treatment.
BS 5268   *Structural use of timber.*
   Part 5: 1989   Code of Practice for the preservative treatment of structural timber.
BS 5589: 1989   *Code of Practice for preservation of timber.*
BS 5666   *Methods of analysis of wood preservatives and treated timber.*
   Part 2: 1980   Qualitative analysis.
   Part 3: 1991   Quantitative analysis of preservatives and treated timber containing copper/chromium/arsenic formulations.
   Part 4: 1979   Quantitative analysis of preservatives and treated timber containing copper naphthenate.

Part 5: 1986   Determination of zinc naphthenate in preservative solutions and treated timber.

Part 6: 1983 (1990)   Quantitative analysis of preservative solutions and treated timber containing pentachlorophenol, pentachlorophenyl laurate, γ-hexachlorocyclohexane and dieldrin.

Part 7: Quantitative analysis of preservatives containing bis(tri-n-butyltin) oxide: determination of total tin.

BS 5707   *Solutions of wood preservatives in organic solvents.*

Part 1: 1979   Specification for solutions for general purpose applications, including timber that is to be painted.

Part 2: 1979 (1986)   Specification for pentachlorophenol wood preservative solution for use on timber that is not required to be painted.

Part 3: 1980 (1990)   Methods of treatment.

(b)   European Standard already published by the British Standards Institution

BS EN 351-1   *Durability of wood and wood-based products. Preservative-treated solid wood.*

Part 1   Classification of preservative penetration and retention.

Part 2   Guidance on sampling for the analysis of preservative-treated wood.

(c)   Other references

BRE (1985) Timbers: their natural durability and resistance to preservative treatment. *Building Research Establishment Digest 296.*

BRE (1993) Wood preservatives: application methods. *Building Research Establishment Digest 378.*

BRE (1994) Specifying preservative treatments: the new European approach. *Building Research Establishment Digest 393.*

TRADA (1991) Flame retardant treatments for timber. Appendix – Tables of treatments. *Wood Information Section 2/3, Sheet 3. Appendix March 1991.*

Wood Finishes

23.1 Introduction

Unless very durable timbers are employed, the service life of wooden components in contact with the external climate will be determined primarily by the efficacy of the applied finish. This in turn is determined not only by its chemical formulation but also by the nature of the substrate.

Difference in chemical formulations will be reflected in differences in the finish in terms of:

- the elasticity of the cured finish, which determines its ability to contend with dimensional change in the substrate without cracking or losing adhesion
- light transmission, which determines its ability to protect the substrate from the degrading effects of ultra-violet light
- viscosity, which in turn determines the degree of penetration of the substrate, one of the factors affecting the bond of the finish to the wood.

The nature of the substrate has a most important effect in determining the durability or service life of the finish resulting from:

- poor surface quality due either to bad machining, or to prior weathering
- instability, due to changes in moisture content
- poor penetration by the finish in high density timbers, or in those with a very marked contrast in density between early and latewood, and
- uneven penetration between bundles of fibres and large diameter vessels.

Wherever possible, the type of finish should reflect the nature of the substrate. The desire to use clear finishes externally for aesthetic reasons

must be resisted, unless very frequent maintenance is acceptable.

Finishes are of two basic types, each of which can be subdivided: the first group comprises the pigmented paint systems, either solvent, or more recently, water-borne. The second group contains the natural finishes which permit the colour and figure of the wood to be seen at least to some degree. This group contains the wood stains, varnishes, oils and preservatives.

In addition to these two major categories of finishes, there can be added two specific types of wood coatings; the first are the floor seals which, though for inside use, have to withstand high levels of wear, and the second are the flame-retardant paints which combat the inherently poor spread of flame resistance of the substrate.

23.2 Paint systems

These pigmented systems can be conveniently subdivided into two basic types, **solvent-borne** and **water-borne**, the former of which can be further subdivided into **conventional** and **exterior** quality.

Simple classification and description of exterior wood coatings are difficult. Conventional ('general purpose') paints can still be considered in terms of the different primer types, undercoats and gloss topcoats, but many of the new exterior paints cannot be classified in this way. Indeed, with the wide range of technologies now used in paint formulation, many are unique to a single manufacturer. Consequently, conventional and exterior paints will be treated separately, the latter embracing both solvent and water-borne types. A comparison of their properties is given in Table 23.1. Both systems result in a coating thickness of from 80–100 μm.

Table 23.1 *Comparison of properties of the three types of paint system (reproduced from BRE Digest 354)*

Property	Conventional paints (solvent-borne) for exterior use	New exterior-quality paints	
		Solvent-borne	Water-borne
Adhesion	← Good over well-prepared surfaces →		
Long-term extensibility	Poor	Moderate	Good
Flow	Good	Good	Moderate
Gloss levels*	High	High	Moderate
Colour stability	Moderate	Moderate	Good
Moisture permeability	Low	Low to moderate	Moderate to high
Maintenance interval	3–4 years	4–6 years	5–8 years
Redecoration procedure	Difficult	Moderate	Easy
Compatibility with putty	Good	Good	Moderate
Tolerance to adverse weather (during application)	Good	Good	Moderate
Blocking resistance†	Good	Good	Moderate

*Low-gloss level finishes available in all paint types.
†Self-adhesion between contacting surfaces.

23.2.1 Conventional paints ('general purpose')

The traditional approach to painting exterior wood is the application of a three-part system comprising primer, undercoat and gloss topcoat. If these conventional systems fail prematurely, it is usually because the balance of properties has been inadequate to prevent or control the kind of substrate conditions which can develop in service. Lack of coating flexibility has been a fundamental problem, especially in undercoats.

When specifying a conventional paint system, it is essential that as much of the system as possible is formulated to meet the requirements for durability on exterior wood. For primers, which provide the all-important foundation for the rest of the system, this may be achieved by using products to BS 5082 or BS 5358, which are performance based. Most of the traditional undercoats based on varnish or alkyd resins are too hard and brittle to produce maximum durability on wood outdoors. However, emulsion-based primer/undercoats applied in two consecutive coats are more flexible and potentially more durable. Table 23.1 gives the typical properties for conventional paints.

23.2.2 Special exterior paints

The new range of exterior quality paints is available as either solvent-borne or water-borne formulations (see Table 23.1). Some of these have a rather higher level of moisture permeability than the conventional paint systems and have been described as 'microporous', 'breathing' or 'ventilating'. They are generally claimed to resist the passage of liquid water, but to allow the passage of water vapour, thereby permitting the wood to dry out. These paint systems do perform well in service, but all the evidence suggests that this is because these coatings have been formulated specially for external use rather than the fact that they are microporous (BRE, 1990).

Solvent-borne exterior paints
These can be subdivided into gloss and low-sheen finish: the former is usually based on flexible types of alkyd resins with separable primer, undercoat and topcoat. The low-sheen finish paints are generally applied as coat-on-coat systems (that is, two coats of the same paint), thereby having the simplicity and convenience of one-can products.

Water-borne exterior paints

These are based on acrylic, or alkyd–acrylic emulsions. Acrylics normally have a gloss finish while alkyd–acrylics are of low sheen: these paints are normally applied in two or three coat systems depending on formulation. All these paints have a higher level of moisture permeability than the equivalent solvent-borne types. However, they also have high levels of film extensibility (elasticity) which is retained in ageing, a most important attribute (Miller and Boxall, 1994) – see Table 23.1. Site experience over many years has shown these to be generally more durable than solvent-borne alkyd paints. However, they do have the disadvantage that they are slow to dry in cold, wet weather.

23.3   Natural finishes for exterior wood

Most natural finishes are less durable than well-formulated opaque paint coatings, so maintenance intervals may be shorter and overall costs higher. The principal reason for using a natural finish for exterior wood lies in maintaining the decorative features of the wood substrate in terms of its colour and figure. There are four main types; their principal uses are set out in Table 23.2 (BRE, 1993).

23.3.1   Exterior wood stains

These form the predominant type of natural finish; they are pigmented resin solutions which produce a range of coatings that can vary in build, opacity and sheen. They are also semi-transparent and available in a wide range of colours. Most contain organic-based fungicides to provide a preservative function and other additives of benefit to their durability.

There are three main types of wood stains:

- **Zero-build** which penetrate the substrate almostly entirely and are used where dimensional movement of the wood is permitted.
- **Low build** which penetrate the substrate to some degree leaving a thin film on the surface (10–15 µm): these are widely used for cladding and on other sawn surfaces.
- **Medium build** which are more viscous and result in a noticeable film, a two-coat system giving a film thickness up to 40 µm. Because these are more resistant to moisture transmission, the wood is less liable to move and this type of wood stain is recommended for joinery components.

Wood stains generally erode by weathering and colour change may be pronounced. Maintenance is normally simply by the application of further coats, but if the original colour and semi-transparency are to be retained, old treatments

Table 23.2   *The principal applications for the different types of natural finishes on exterior wood (extract from BRE, 1993)*

Application	Wood stain			Varnish	Oil	Creosote	CCA
	Zero	Low	Medium				
Cladding/boarding	—	√	√	—	—	—	√
Plywood	—	√	—	—	—	—	—
Joinery: windows	—	—	√	—	—	—	—
: doors	—	—	√	√	—	—	—
Gates	—	√	—	—	—	√	√
Fences	√	—	—	—	—	√	√
Sheds	√	√	√	—	—	√	√
Benches	—	√	—	—	√	—	√
Handrails	—	√	√	√	—	—	√

will have to be removed and the wood surface restored by sanding (Boxall, 1991).

23.3.2 *Varnishes*

Varnishes are unpigmented resin solutions which give a clear, high build coating about 80 μm in thickness. The clarity and depth of finish of newly varnished wood is difficult to equal, but this appearance is often short-lived, especially on full exposure to the weather. Failure occurs because most varnishes embrittle during weathering, so that in time they crack, lose adhesion and flake. Light passing through the film causes the substrate to bleach and degrade, thereby becoming more sensitive to the ingress of moisture.

23.3.3 *Oils*

Finishing wood with simple drying oils such as linseed or modified types like teak oil, though initially attractive, has several disadvantages; the finish is sticky, holds dirt and supports the growth of staining fungi. Treatments based on oil should only be used if they have been formulated for exterior use by the inclusion of pigments and fungicides.

23.3.4 *Preservatives*

Some preservative treatments, such as copper/chrome/arsenic (CCA) and creosote give protection against weathering degrade and have some limited decorative effect.

23.4 Finishes for wood floors

Wood is an excellent flooring material: not only is it aesthetically pleasing, but it is also extremely hard-wearing. However, to maintain these attributes the floor must be sealed.

Traditionally, wooden floors have been coated with wax or a solution of shellac (button polish).

While providing an attractive finish, neither of these surface treatments is hard-wearing, nor resistant to the spillage of water or other liquids.

Such disadvantages are not present with the new range of synthetic seals for wooden floors. These both penetrate and coat the wood, providing a hard-wearing surface that resists the penetration of spilled water. A number of types are available on the market with a range in performance. At the lower end of the range are the **oleoresins** or clear varnishes, which are widely used but easily scratched, and give only fair wearing resistance. However, if these oleoresins contain a small amount of urethane resin, their wearing performance is increased though they still scratch readily.

One-can epoxy esters produce a harder wearing and glossier surface as do the **one-can urea–formaldehydes** which, however, can give off very strong odours on curing. Even higher wear and stain resistance can be obtained with **two-can urea–formaldehyde**, **two-can epoxy resin** and **two-can polyurethane** which self-cure, and **moisture curing polyurethane** which cures by reaction with atmospheric moisture. More information on these seals, on the choice of seal and on its maintenance is given in a TRADA Wood Information sheet (TRADA, 1993).

23.5 Flame-retardant coatings

These coatings are applied to the surface of wood or wood-based panel products in order to achieve either Class 1 rating in the surface spread of flame test set out in BS 476: Part 7, or a Class 0 rating complying with the requirements for both the fire propagation and surface spread of flame tests set out in BS 476, Parts 6 and 7 respectively, and in accordance with the Building Regulations, 1985. The most commonly used flame-retardant coating are the intumescent ones; in the presence of heat, these coatings rapidly expand (intumesce) to produce a thick barrier which protects the substrate to the spread of flame.

There are on the market a few intumescent coatings, either emulsion or solvent-based, which satisfy the Class 0 requirements, and a larger number of coatings, both clear and pigmented, either water- or solvent-based, which satisfy the Class 1 requirements (TRADA 1980, 1991). Care has to be exercised in the selection of proprietary intumescent coatings to ensure that they are compatible with the substrate: some are recommended only for softwood or fibreboard, while others have the potential for wider application.

References

Boxall J. (1991) Exterior wood stains. *Building Research Establishment Information Paper IP 5/91.*

BRE (1990) Painting exterior wood. *Building Research Establishment Digest 354.*

BRE (1993) Natural finishes for exterior wood. *Building Research Establishment Digest 387.*

BS 476   *Fire tests on building materials and structures.*
    Part 6: 1989   Method of test for fire propagation for products.
    Part 7: 1987   Method for classification of the surface spread of flame of products. BSI, London.

BS 5082: 1986   *Specification for water-borne priming paints for woodwork.* BSI, London.

BS 5358: 1986   *Specification for solvent-borne priming paints for woodwork.* BSI, London.

Miller E. R. and Boxall J. (1994) Water-borne coatings for exterior wood. *Building Research Establishment Information Paper IP4/94.*

TRADA (1980) Flame retardant treatments for timber. *TRADA Wood Information, Section 2/3, Sheet 3.*

TRADA (1991) Flame retardant treatments for timber. Appendix — Tables of treatments. *TRADA Wood Information, Section 2/3, Sheet 3.*

TRADA (1993) Seals for timber floors. *TRADA Wood Information, Section 2/3, Sheet 14.*

Selected Bibliography
– additional to that listed at the end of most chapters

Part 1   STRUCTURE OF WOOD

Boyd J.D. (1985) *Biophysical control of microfibril orientation in plant cell walls: aquatic and terrestrial plants including trees*. Martinus Nijhoff Dr W. Junk Publishers, Dordrecht, The Netherlands.

Browning B.L. (1963) *The Chemistry of Wood*. Interscience Publishers, New York.

Carlquist S. (1988) *Comparative Wood Anatomy: Systematic Ecological and Evolutionary Aspects of Dicotyledon Wood*. Springer-Verlag, Berlin.

Côté W.A. (Ed.) (1965) *Cellular Ultrastructure of Woody Plants*. Syracuse University Press.

Cutler D.F., Rudall P.J., Gasson P.E. and Gale R.M.O. (1987) *Root Identification Manual of Trees and Shrubs*. Chapman & Hall, London.

Farmer R.H. (1972) *A Handbook of Hardwoods*, 2nd edn. HMSO: available only from the Building Research Establishment Bookshop.

Harding T. (1988) British softwoods: properties and uses. *Forestry Commission Bull 77*. HMSO, London.

Harris J.M. (1988) *Spiral Grain and Wave Phenomenon in Wood Formation*. Springer-Verlag, Berlin.

Preston R.D. (1974) *The Physical Biology of Plant Cell Walls*. Chapman & Hall, London.

Young R.A. and Rowell R.M. (Eds) (1986) *Cellulose – Structure, Modification and Hydrolysis*. Wiley–Interscience, New York.

Zinnermann M.H. (1983) *Xylen Structure and the Ascent of Sap*. Springer-Verlag, Berlin.

Zobel B.J. and van Buijtemen J.P. (1989) *Wood Variation: Its Causes and Control*. Springer-Verlag, Berlin.

Part 2   PROPERTIES OF WOOD

Anon (1979) Wood moisture content – temperature and humidity relationships. *Proc Symposium Virginia Polytechnic Institute and State University, 29 October 1979: USDA Forest Service, North Central Expt. Stn.*

Barrett J.D. and Foschi R.O. (Eds) (1979) *Proc First Int Conf Wood Fracture, Banff, Alberta (1978)*. Forintek Canada Corp., Vancouver.

Barrett J.D., Foschi R.O., Vokey H.P. and Varoglu E. (Eds) (1986) *Proc International Workshop on Duration of Load in Lumber and Wood Products (1985)*. Special Pub. SP-27. Forintec Canada Corp., Vancouver.

Bodig J. and Jayne B.A. (1982) *Mechanics of Wood and Wood-composites*. Van Nostrand Reinhold, New York.

Cheremisinoff N.P. (1980) *Wood for Energy Production*. Ann Arbor Science Publishers, Michigan, USA.

Goldstein I.S. (Ed.) (1977) *Wood Technology – Chemical Aspects*. ACS symposium series 43, American Chemical Soc., Washington, DC.

Kollmann F.F.P. and Côté W.A. (1984) *Principles of Wood Science and Technology. I: Solid Wood*. Springer-Verlag, Berlin.

Meyer R.W. and Kellogg R.M. (Eds) (1982) *Structural Uses of Wood in Adverse Environments*. Van Nostrand Reinhold, New York.

Rowell R.M. and Youngs R.L. (1981) Dimensional stabilization of wood in use. *Forest Products Laboratory, Madison, Research Note FPL-0243*.

Siau J.F. (1984) *Transport Processes in Wood*. Springer-Verlag, Berlin.

Part 3   PROCESSING OF TIMBER

Koch P. (1985) *Utilization of Hardwoods Growing on Southern Pine Sites. Vol II: Processing*. US Dept of Agriculture, Forest Service, Agriculture Handbook 605: 1419–2542.

Walker J.C.F., Butterfield B.G., Harris J.M., Langrish T.A.G. and Uprichard J.M. (1993) *Primary Wood Processing, Principles and Practice*. Chapman & Hall, London.

Part 4   UTILISATION OF TIMBER

Dinwoodie J.M. (1979) The properties and performance of particleboard adhesives. *J Inst Wood Sci* **8(2)**: 59–68.

Gibson E.J., Laidlaw R.A. and Smith G.A. (1966) Dimensional stabilisation of wood. I: Impregnation with methyl methacrylate and subsequent polymerisation by means of gamma radiation. *J App Chem* **16**: 58–64.

Farmer R.H. (1967) *Chemistry in the Utilisation of Wood*. Pergamon, Oxford.

Farmer R.H. (1972) *A Handbook of Hardwoods*, 2nd edn. HMSO: available only from the Building Research Establishment Bookshop.

FPR (1987) *Wood Handbook – Wood as an Engineering Material*. Agriculture Handbook 72. US Dept of Agriculture, Forest Service.

Harding T. (1988) British softwoods: properties and uses. *Forestry Commission Bull 77*. HMSO, London.

Kollman F.F.P., Kuenzi E.W. and Stamm A.J. (1975) *Principles of Wood Science and Technology. Pt. II: Wood-based Materials*. Springer-Verlag, Berlin.

Maloney T.M. (1977) *Modern Particleboard and Dry-process Fibreboard Manufacturing*. Miller Freeman Publications, San Francisco.

Pizzi A. (1983) *Wood Adhesives. Vol. 1*. Marcel Dekker, New York.

Pizzi A. (1989) *Wood Adhesives. Vol. 2*. Marcel Dekker, New York.

TRADA (1982) Wood based sheet materials for formwork linings., *Trada Wood Information, Section 2/3, Sheet 17*.

TRADA (1983) Edge sealants for wood-based boards. *Trada Wood Information, Section 2/3, Sheet 20*.

TRADA (1987) Low flame spread wood-based panel products. *Trada Wood Information, Section 2/3, Sheet 7*.

Part 5   TIMBER IN SERVICE

Berry R.W. (1994) Termites and tropical building. *Building Research Establishment Overseas Building Note OBN 201*.

BWPDA (1986) *Manual*. The British Wood Preserving and Damp-proofing Association (available from BWPDA, Building 6, The Office Village, 4 Romford Road, Stratford, London E15 4EA).

Cartwright K.St.G. and Findlay W.P.K. (1958) *Decay of Timber and its Prevention*, 2nd edn. HMSO, London.

Eaton R. and Hale M. (1993) *Decay, Pests and Protection*. Chapman & Hall, London.

Findlay, W.P.K. (Ed.) (1985) *Preservation of Timbers in the Tropics*. Martinus Nijhoff Dr W Junk Publishers, Dordrecht, The Netherlands.

Laidlaw R.A. and Pinion L.C. (1977) Metal plate fasteners in trussed rafters treated with preservatives or flame retardants – corrosion risks. *Building Research Establishment Information Sheet IS 11/77*.

Meyer R.W. and Kellogg R.M. (Eds) *Structural Uses of Wood in Adverse Environments*. Van Nostrand Reinhold, New York.

Norimoto M. (1982) Structure and properties of wood used for musical instruments. 1. On the selection of wood used for piano soundboards. *J Jap Wood Res Soc* **28(7)**: 407–413.

Richardson B.A. (1993) *Wood Preservation*, 2nd edn. Chapman & Hall, London.

Le Van L. and Winandy J.E. (1990) Effects of fire retardants on wood strength: a review. *Wood and Fibre Science* **22(1)**: 113–131.

Appendix I: List of Botanical Equivalents of Common or Trade Names used in the Text

In many cases it is not possible to give a single botanical name, because the latter is often applied to the timbers of more than one species or even genus; such cases are given below as 'spp.', no attempt being made to list all the species that may provide commercial supplies.

Reference should also be made to BS 7359: 1991 *British Standard nomenclature of commercial timbers, including sources of supply.*

abura = *Mitragyna ciliata* Aubrev. and Pellegr.

afara = *Terminalia superba* Engl, et Diels

afrormosia = *Pericopsis elata* van Meeuwen (syn. *Afrormosia elata* Harms)

afzelia = *Afzelia* spp.

agba = *Gossweilerodendron balsamiferum* Harms

albizia = *Albizia ferruginea* and spp.

alder = *Alnus glutinosa* Gaertn.

alerce = *Fitzroya cupressoides* F.M. Johnston

American whitewood, *see* whitewood, American

apitong = *Dipterocarpus* spp. (Philippines)

araracanga = *Aspidosperma* spp.

ash = *Fraxinus* spp.

ash, American = *Fraxinus americana* L., F. *pennsylvanica* var. *lanceolata* Sarg., F. *nigra* March.

ash, European = *Fraxinus excelsoir* L.

ash, mountain, *see* oak, Tasmanian

balau = *Shorea* spp.

balsa = *Ochroma pyramidale* Urb. (syn. O. *lagopus* Sw.)

beech = *Fagus sylvatica* L.

birch = *Betula* spp.

blackwood, African = *Dalbergia melanoxylon* Guill. et Perr.

blackwood, Australian = *Acacia melanoxylon* R. Br.

box, Cape = *Buxus macowani* Oliv.

box, European = *Buxus sempervirens* L.

boxwood, Ceylon = *Canthium dicoccum* Merr.

camphorwood, Borneo, *see* kapur

camphorwood, East Africa = *Ocotea usambarensis* Engl.

camphorwood, Formosan = *Cinnamomum camphora* Nees et Eberm.

canary whitewood, *see* whitewood, American

cedar = *Cedrus* spp.

cedar, Central American = *Cedrela odorata* L. (syn. C. *Mexicana* Roem.)

cedar, cigar-box, *see* cedar, Central American

cedar, pencil, African = *Juniperus procera* Hochst. ex Endl.

cedar, pencil, Virginian = *Juniperus virginiana* L.

cedar, Port Orford = *Chamaecyparis lawsoniana* (A. Murr.) Parl.

cedar, South American = *Cedrela* spp. including C. *fissils* Vell.

cedar, western red = *Thuja plicata* D. Don

chengel = *Balanocarpus heimi* King

cherry = *Prunus avium* L.

chestnut, American = *Castanea dentata* Borkh.

chestnut, sweet = *Castanea sativa* Mill.

coachwood = *Ceratopetalum apetalum* D. Don

cocus wood = *Brya ebenus* D.C.

cornel, Turkish = *Cornus* spp.

cryptomeria = *Cryptomeria japonica* (L.f.) D. Don

dahoma = *Piptadeniastrum africanum* (Hook. f.) Brenan (syn. *Piptadenia africana* Hook. f.)

deal, Baltic, *see* redwood, European
deal, red, *see* redwood, European
Douglas fir = *Pseudotsuga menziesii* (Mirb.) Franco (syn. *P. taxifolia* Brit., *P. douglasii Carr.*)

ebony = *Diospyros* spp., *Maba* spp.
ekki = *Lophira alata* Banks ex Gaertn.
elm = *Ulmus* spp.
elm, common = *Ulmus procera* Salisb.
elm, Dutch = *Ulmus hollandica* Mill. var. *holandica* Rehd.
elm, European = *Ulmus procera* Salisb.
elm, wych = *Ulmus glabra* Hudson (non Miller)

fir = *Abies* spp.
fir, Douglas, *see*, Douglas fir
fir, silver = *Abies alba* Mill.

gaboon = *Aucoumea klaineana* Pierre
gedu nohor = *Entandrophragma angolense* C. DC.
greenheart = *Ocotea radiaei* Mez
guarea = *Guarea cedrata* Pellegr., *G. thompsonni* Sprague et Hutch.
guarea, scented = *Guarea cedrata* Pellegr.
gum, American red = *Liquidambar styraciflua* L.
gum, spotted = *Eucaluptus maculata* Hook. and *E. citriodora* Hook.
gurjun = *Dipterocarpus* spp. (Andamans, Burma)

hazel = *Corylus avellana* L.
hemlock = *Tsuga* spp.
hemlock, eastern = *Tsuga candensis* Carr
hemlock, western = *Tsuga heterophylla* Sarg.
hickory = *Carya* spp. (syn. *Hicoria* spp.)
holly, American = *Ilex opaca* Ait.
holly, European = *Ilex aquifolium* L.
hornbeam = *Carpinus betulus* L.

idigbo = *Terminalia ivorensis* A. Chev.
iroko = *Chlorophora excelsa* Benth. et Hook. f.

jarrah = *Eucalyptus marginata* Sm.
jelutong = *Dyera costulata* Hook. f.
juniper = *Juniperus* spp.

kapur = *Dryobalanops aromatica* Gaertn. f. (Malaysia), *Dryobalanops* spp. (Sagah, Sarawak)
karri = *Eucalyptus diversicolor* F. Muell

keledang = *Artocarpus lanceifolius* Roxb.
kempas = *Koompassia malaccensis* Maing. ex Benth.
keranji = *Dialium* spp.
keruing = *Dipterocarpus* spp. (Malaysia, Sabah)
kokrodua, *see* afrormosia

lancewood = *Oxandra lanceolata* Baill
larch = *Larix* spp.
larch, European = *Larix decidua* Mill
lauan = *Shorea* spp., *Pentacme* spp., *Parashorea* spp. (Philippines)
lauan, dark red, *see* lauan
lauan, light red, *see* lauan
lignum vitae = *Guaiacum officinale* L., *G. sanctum* L.
lime = *Tilia vulgaris* Hayne

mahogany = *Swietenia* spp.
mahogany, African = *Khaya ivorensis* A. Chev., *K. grandifoliola* C. DC., *K. anthotheca* C. DC.
mahogany, Brazilian = *Swietenia macrophylla* King
mahogany, Central American = *Swietenia macrophylla* King
mahogany, cherry, *see* makoré
mahogany, Cuban = *Swietenia mahogani* Jacq.
mahogany, Gaboon, *see* gaboon
mahogany, Honduras, *see* mahogany, Central American
mahogany, Philippine, *see* lauan
mahogany, sapele, *see* sapele
mahogany, Spanish, *see* mahogany, Cuban
mahogany, 'true' = *Swietenia* spp.
makoré = *Tieghemella heckelii* Pierre ex A. Chev. (syn. *Mimusops heckelii* (A. Chev. (Hutch. et Dalz.))
mansonia = *Mansonia altissima* A. Chev.
maple = *Acer* spp.
maple, Pacific = *Acer macrophyllum* Pursh
maple, Queensland = *Flindersia brayleyana* F. v. M., *F. pimenteliana* F. v. M.
maple, rock = *Acer saccharum* Marsh (principally)
melawis = *Gonystylus macrophylla* (Miq.) Airy Shaw (syn. *G. bancanus* (Miq.) Kurz)
meranti = *Shorea* spp. (Malaysia, Sarawak)
meranti, red = *Shorea* spp. (Malaysia, Sarawak)
meranti, white = *Shorea* spp. section *Anthoshorea* Brandis (Malaysia)
meranti, yellow = *Shorea* spp. section *Richetia*

merbau = *Intsia palembanica* Baker
mersawa = *Anisoptera* spp. (Malaysia)
mountain ash, *see* oak, Tasmanian
muninga = *Pterocarpus angloensis* DC.

oak = *Quercus* spp.
oak, African silky = *Grevillea robusta* A. Cunn.
oak, American red = *Quercus rubra* var. *pagodae-folia* Ashe, *Q. borealis* Michx. f., *Q. borealis* var. *maxima* Sarg., *Q. falcata* Michx., *Q. sumardii* Buckl.
oak, American white = *Quercus alba* L., *Q. montan* Willd., *Q. lyrata* Walt., *Q. prinus* L.
oak, Australian, *see* oak, Tasmanian
oak, Australian silky = *Cardwellia sublimis* F. v. M.
oak, English, *see* oak, European
oak, European = *Quercus robus* L., *Q. petraea* Liebl.
oak, Japanese = *Quercus mongolica* var. *grosseser-rata* Rehd. and Wils.
oak, Tasmanian = *Eucalyptus gigantea* Hook. f., *E. obliqua* L'Hérit., *E. regnans* F. v. M.
oak, Turkey = *Quercus cerris* L.
obeche = *Triplochiton schleroxylon* K. Schum.
olive, East African = *Olea hochstetteri* Bak.
omu = *Entandrophragma candollei* Harms
opepe = *Nauclea diderrichii* (De Wild et Th. Dur.) Merrill (syn. *Sarcocephalus diderrichii* De Wild.)

padauk = *Pterocarpus* spp.
pear = *Pyrus communis* L.
pencil cedar, *see* cedar, pencil
peroba = *Aspidosperma* spp.
persimmon = *Diospyros virginiana* L.
pine, Columbian, *see* Douglas fir
pine, contorta = *Pinus contorta* Douglas ex Loud
pine, Corsican = *Pinus nigra* Arnold subsp. *laricio* (Poir). Maire
pine, long-leaf pitch = *Pinus echinata* Mill., *P. palustris* Mill., and *P. taeda* L.
pine, maritime = *Pinus pinaster* Ait.
pine, Oregon, *see* Douglas fir
pine, Parana = *Araucaria angustifolia* (Bert.) O.Ktze.
pine, pitch, *see* pine, long-leaf pitch
pine, radiata = *Pinus radiata* D. Don
pine, Scots = *Pinus sylvestris* L.
pine, white, *see* pine, yellow
pine, yellow = *Pinus strobus* L. (United Kingdom)
podo (podocarpus) = *Podocarpus* spp.

poplar = *Populus* spp.
poplar, black = *Populus nigra* L., *P. candensis* Moench var. *serotina* Rehd., and *P. robusta* Schneid.
poplar, black Italian = *Populus serotina* (Hybrid)
poplar, Canadian = *Populus balsamifera* L. (syn. *P. tacamahaca* Mill.), *P. grandidentata* Michx.
poplar, grey = *Populus canescens* Sm.
poplar, white = *Populus alba* L.
Port Orford cedar, *see* cedar, Port Orford
punah = *Tetramerista glabra* Miq.
purpleheart = *Peltogyne* spp.

ramin = *Gonystylus* spp.
redwood, *see* redwood, European *and* sequoia
redwood, Baltic, *see* redwood, European
redwood, Californian, *see* sequoia
redwood, European = *Pinus sylvestris* L.
rengas = *Melanorrhoea* spp.
resak = *Vatica*, spp. *Cotylelobium* spp. (Malaysia)
rimu = *Dicrydium cupressinum* Sol. ex Lamb
robinia = *Robinia pseudoacacia* L.
rosewood, Indian = *Dalbergia latifolia* Roxb.

sandalwood = *Santalum album* L.
sapele = *Entandrophragma cylindricum* Sprague, and *Entandrophragma* spp.
satin walnut, *see* gum, American red
satinwood, East Indian = *Chloroxylon swietenia* DC.
satinwood, West Indian = *Fagara flava* Krug.
sequoia = *Sequoia sempervirens* Endl.
seraya = *Shorea* spp., *Parashorea* spp. (Sabah)
seraya, Borneo white = *Parashorea* spp.
seraya, white = *Shorea* spp. *Parashorea* spp. (Sabah)
sneezewood = *Ptaeroxylon obliquum* (Thunb.) Radlk.
spotted gum, *see* gum, spotted
spruce, *Picea* spp.
spruce, Canadian = *Picea glauca* Voss (principally)
spruce, European = *Picea abies* Karst
spruce, Norway, *see* spruce, European
spruce, Sitka = *Picea sitchensis* Carr.
sterculia, yellow = *Eriboma oblanga* (Mast.) Bod (Syn. *Sterculia oblonga* Mast.)
swamp gum, *see* oak, Tasmanian
sycamore = *Acer pseudoplatanus* L.

tali = *Erythrophleum suaveolens* (Guill. and Perr.) Brenan and *E. ivorense* A. Chev.

tallowwood = *Eucalyptus microcorys* F. v. M.

Tasmanian oak, *see* oak, Tasmanian

teak = *Tectona grandis* L. f.

tembusu = *Fagraea gigantea* Ridl.

terentang = *Campnosperma* spp.

tulip poplar, *see* whitewood, American

tulip tree, *see* whitewood, American

turpentine = *Syncarpia laurifolia* Ten.

utile = *Entandrophragma utile* (Dawe and Sprague) Sprague

walnut = *Juglans* spp.

walnut, African = *Lovoa trichilioides* Harms (syn. *L. klaineana* Pierre ex Sprague)

walnut, Australian, *see* walnut, Queensland

walnut, Nigerian, *see* walnut, African

walnut, Queensland = *Endiandra palmerstonii* C.T. White

western red cedar = *Thuja plicata* D. Don

whitewood = *Picea abies* Karst. and *Abies alba* Mill.

whitewood, American = *Liriodendron tulipifera* L.

whitewood, canary, *see* whitewood, American

willow = *Salix* spp.

willow, crack = *Salix fragilis* L.

willow, cricket bat = *Salix alba* var. *coerulea* Sm.

willow, white = *Salix alba* L., *S. viridis* Fr.

yang = *Dipterocarpus* spp. (Thailand)

yellow poplar, *see* whitewood, American

Appendix II:
Table of Conversion Factors

Imperial measure	Metric units	Multiply by
inches	to millimetres	25.4
inches	to centimetres	2.54
feet	to millimetres	304.8
feet	to centimetres	30.48
feet	to metres	0.3047
square inches	to square centimetres	6.45
square feet	to square metres	0.093
cubic inches	to cubic centimetres	16.39
cubic feet	to litres	28.3
pounds (avoir.)	to kilograms	0.4536
gallons	to litres	4.546
lbf per sq in	to N/mm^2	0.006895
lbf per sq ft	to N/mm^2	47.88
ft lbf	to J	1.356
Btu	to J	1055

Appendix III:
Trade Organisations in the UK with Responsibility for Certain Products

APA	The Engineered Wood Association 65 London Wall London EC2M 5TU
BWPDA	The British Wood Preserving and Damp-proofing Association Building 6 The Office Village 4 Romford Road Stratford London E15 4EA
COFI	Council of Forest Industries of British Columbia Tileman House 131–133 Upper Richmond Road London SW15 2TR
MBBPF	Mineral Bonded Board Products Federation 265 Barrowby Road Grantham Lincolnshire NG31 8NR
NPPA	National Panel Products Association Clareville House 26–27 Oxendon Street London SW1Y 4EL
TTF	Timber Trades Federation Clareville House 26–27 Oxendon Street London SW1Y 4EL
WPPF	Wood Panel Products Federation (an amalgamation of FIDOR and UKIPA) 1 Hanworth Road Feltham Middlesex England TW13 5AF

Index